Structural Prognostics and Health Management in Power & Energy Systems

Structural Prognostics and Health Management in Power & Energy Systems

Special Issue Editors

Dong Wang
Shun-Peng Zhu
Xiancheng Zhang
Gang Chen
José A.F.O. Correia
Guian Qian

MDPI • Basel • Beijing • Wuhan • Barcelona • Belgrade

MDPI

Special Issue Editors

Dong Wang
Shanghai Jiao Tong University
China

Shun-Peng Zhu
University of Electronic Science and
Technology of China
China

Xiancheng Zhang
East China University of Science and Technology
China

Gang Chen
Tianjin University
China

José A.F.O. Correia
University of Porto
Portugal

Guian Qian
Chinese Academy of Sciences
China

Editorial Office
MDPI
St. Alban-Anlage 66
4052 Basel, Switzerland

This is a reprint of articles from the Special Issue published online in the open access journal *Energies* (ISSN 1996-1073) from 2018 to 2019 (available at: https://www.mdpi.com/journal/energies/special_issues/sphm)

For citation purposes, cite each article independently as indicated on the article page online and as indicated below:

LastName, A.A.; LastName, B.B.; LastName, C.C. Article Title. *Journal Name* **Year**, *Article Number*, Page Range.

ISBN 978-3-03921-766-3 (Pbk)
ISBN 978-3-03921-767-0 (PDF)

Contents

About the Special Issue Editors

Dong Wang is Associate Professor at the Department of Industrial Engineering and Management at Shanghai Jiao Tong University. His research interests include prognostics and health management, statistical modeling, condition monitoring and fault diagnosis, signal processing, data mining, machine learning, and nondestructive testing. He has been awarded State Specially Recruited Experts (Young Talents), Hong Kong Ph.D. Fellowship, and IEEE and Elsevier Outstanding Reviewer Status. He is Associate Editor/Guest Editor of numerous international journals. He serves as Deputy Director of the Center for Systems Health Monitoring and Management, a reviewer of FONDECYT, a member of the Chinese Institute of Quality Research and the State Key Laboratory of Mechanical Systems and Vibration.

Shun-Peng Zhu is Professor of Mechanical Engineering at University of Electronic Science and Technology of China. He was an international fellow at Politecnico di Milano, Italy, during 2016–2018 and Research Associate at University of Maryland, United States, in 2010. His research, which has been published in scholarly journals and edited volumes, includes over 100 peer-reviewed book chapters, journals, and proceedings papers which explore the following aspects: fatigue design, probabilistic physics of failure modeling, structural reliability analysis, multiphysics damage modeling and life prediction under uncertainty, and probability-based life prediction/design for engineering components. He received the Award of Merit of European Structural Integrity Society (ESIS)—TC12 in 2019, Most Cited Chinese Researchers (Elsevier) in the field of Safety, Risk, Reliability, and Quality in 2018, 2nd prize for the National Defense Science and Technology Progress Award of Ministry of Industry and Information Technology of China in 2014, Polimi International Fellowship in 2015, Hiwin Doctoral Dissertation Award in 2012, Best Paper Award at several international conferences, and Elsevier Outstanding Reviewer Status. He serves as Guest Editor and Editorial Board Member of several international journals and Springer book series, Organizing Committee Co-Chair of the International Conference on Quality, Reliability, Risk, Maintenance, and Safety Engineering (QR2MSE 2013), TPC Member of QR2MSE 2014–2019, ICMR 2015, ICMFM XIX 2018, and IRAS 2019.

Xiancheng Zhang received his Ph.D. degree from Shanghai Jiao Tong University, China in 2007. He then moved to National Institute for Materials Science (NIMS) in Japan where he was a Postdoctoral Researcher for 1 year. He has contributed considerably to life design and prediction methods of high-temperature components and to the development of advanced surface manufacturing techniques. He has published more than 100 peer reviewed papers, including more than 70 SCI-indexed papers in such journals as Acta Materialia, Journal of Applied Physics, Engineering Fracture Mechanics, and Surface and Coatings Technology. Dr. Zhang has received a number of distinguished awards, including International Institute of Welding (IIW) Granjon Prize, Shanghai Outstanding Doctoral Dissertation Award, nomination for the Chinese Outstanding Doctoral Dissertation Award, and Chinese Petroleum Chemical Industry Association Technological Award, first-class of the Shanghai Natural Science Prize, first-class of the Beijing Natural Science Prize, and second-class of the National Nature Science Prize of China. He was the recipient of the New Century Excellent Talents Program Award (2011) from the Ministry of Education of China, the Outstanding Young Talent Award (2012), Shanghai Pujiang Talent (2012), the National Science Fund for Excellent Young Scholars of China (2013), Education Award for Young Teachers by FOK YING TUNG Education Foundation from the Ministry of Education of China (2014), Shanghai Young Sci-tech Talents (2014), Changjiang Young Scholars Programme of China (2015), National Science Fund for Distinguished Young Scholars of China (2017).

Gang Chen received his Ph.D. in Chemical Process Machinery from Tianjin University, China, in 2006. He was a visiting scholar at Virginia Tech in 2009 and Oak Ridge National Lab., United States, in 2014. His main research field is in the mechanical behavior of materials, low cycle fatigue, creep, creep-fatigue, and finite element analysis and constitutive modeling for electronic and conventional structural materials. He has published over 90 papers in SCI journals and been cited more than 700 times in the SCI database. Dr. Chen is a Fellow of Fatigue Institution, Chinese Materials Research Society. He is also an IEEE member. Dr. Chen has been awarded "New Century Excellent Talents in University", Ministry of Education of China, in 2013, as well as "Young Excellent Technological Innovation Talents", Tianjin Government of China, in 2014.

José A.F.O. Correia is Researcher at INEGI and CONSTRUCT/FEUP of the University of Porto (Portugal). Since 2018, he is Guest Teacher at the Engineering Structures Department of the Civil Engineering and Geosciences Faculty of the Delft University of Technology (Netherlands). He is an invited Assistant Professor at the Structural Mechanics section in the Civil Engineering Department of the University of Coimbra (since 2016/09). He obtained is BSc (2007) and MSc (2009) degrees in Civil Engineering from the University of Trás-os-Montes e Alto Douro. He was a specialist in steel and composite (steel and concrete) construction at the University of Coimbra in 2010. He was awarded his Ph.D. in Civil Engineering from the University of Porto in 2015. He is also co-author of more 70 papers in the most relevant scientific journals devoted to engineering materials and structures, as well as 150 proceedings in international and national conferences, congresses, and workshops. He is a member of scientific and professional organizations, such as Ordem dos Engenheiros, Associação Portuguesa de Construção Metálica e Mista (CMM), and Associação para a Conservação e Manutenção de Pontes (ASCP). He is Co-Chair of TC12 of European Structural Integrity Society (ESIS), the Editor of the Springer Book Series Structural Integrity, and Guest Editor of numerous international journals. His current research interests are (a) behavior to fatigue and fracture of materials and structures (steel and aluminum, riveted and bolted connections, pressure vessels, old steel bridges, wind turbine towers, offshore structures); (b) probabilistic fatigue modeling of metallic materials (including statistical evaluation, size effect, cumulative damage); (c) probabilistic design of glass structural elements; (d) mechanical behavior of materials and wooden structures (connections and characterization of ancient structures); (e) mechanical and chemical characterization of old mortars and masonry structures.

Guian Qian obtained his Ph.D. from the Institute of Mechanics, Chinese Academy of Sciences, with a major in solid mechanics, in 2009. Afterwards, he moved to Paul Scherrer Institute (PSI) and where he was Postdoctoral Fellow until 2012. Since 2013, he has been a Scientist in the Laboratory for Nuclear Materials, Nuclear Energy, and Safety Department of PSI. His current research interests lie in the fatigue and fracture analysis of nuclear components and structures. He has made significant contributions to nuclear safety assessment, especially in the pressurized thermal shock analysis of reactor pressure vessels and leak-before-break analysis of nuclear piping. He has published more than 50 peer-reviewed papers, including more than 30 SCI-indexed papers in journals such as Acta Materialia, International Journal of Solids and Structures, Engineering Fracture Mechanics, and International Journal of Fatigue. He has been invited on numerous occasions to present keynote talks at international conferences and symposiums, including the 11th International Workshop on the Integrity of Nuclear Components, 2014 International Symposium on Structural Integrity, and Nuclear Materials Symposium in Chinese Materials Conference 2017. He has been Session Organizer for the ASME Pressure Vessels and Piping Conference (2015–2017) and the 14th International Conference on Fracture Mechanics. He serves as reviewer for more than 20 international journals and several international funding committees.

Preface to "Structural Prognostics and Health Management in Power & Energy Systems"

The idea of preparing an Energies Special Issue on "Structural Prognostics and Health Management in Power & Energy Systems" is to compile information on the recent advances in structural prognostics and health management (SPHM). Continued improvements on SPHM have been made possible through advanced signature analysis, performance degradation assessment, as well as accurate modeling of failure mechanisms by introducing advanced mathematical approaches/tools. Through combining deterministic and probabilistic modeling techniques, research on SPHM can provide assurance for new structures at a design stage and ensure construction integrity at a fabrication phase. Specifically, power and energy system failures occur under multiple sources of uncertainty/variability resulting from load variations in usage, material properties, geometry variations within tolerances, and other uncontrolled variations. Thus, advanced methods and applications for theoretical, numerical, and experimental contributions that address these issues on SPHM are desired and expected, which attempt to prevent overdesign and unnecessary inspection and provide tools to enable a balance between safety and economy to be achieved. This Special Issue has attracted submissions from China, USA, Portugal, and Italy. A total of 26 submissions were received and 11 articles finally published.

The paper entitled "An Improved Signal Processing Approach Based on Analysis Mode Decomposition and Empirical Mode Decomposition" reported an improved sifting stop criterion and the combination of analysis mode decomposition and empirical mode decomposition for solving the problem of end effects and mode-mixing. Results showed that the proposed method is better than empirical mode decomposition for data preprocessing.

The paper entitled "A Lithium-Ion Battery RUL Prediction Method Considering the Capacity Regeneration Phenomenon" reported a prognostic method that solved the problem of degradation jumps caused by capacity regeneration phenomenon. Results showed that the proposed method can achieve high remaining useful life prediction accuracies in the case of battery degradation with capacity regeneration phenomenon.

The paper entitled "Weighted Regression-Based Extremum Response Surface Method for Structural Dynamic Fuzzy Reliability Analysis" reported a weighted regression-based extremum response surface method for improving structural dynamic fuzzy reliability analysis. The main contribution of this paper provided a method for structural dynamic reliability evaluation with respect to working processes.

The paper entitled "GA–BP Neural Network-Based Strain Prediction in Full-Scale Static Testing of Wind Turbine Blades" reported a method for strain prediction of wind turbine blades based on genetic algorithm back propagation neural networks. Results showed that the proposed method can predict the strain of unmeasured points of wind turbine blades accurately.

The paper entitled "Early Fault Detection of Wind Turbines Based on Operational Condition Clustering and Optimized Deep Belief Network Modeling" reported a generalized wind turbine health monitoring framework based on supervisory control and data acquisition (SCADA) data. A demonstration on structural health monitoring and fault detection was shown to support the effectiveness of the proposed idea.

The paper entitled "A Data-Driven Predictive Prognostic Model for Lithium-ion Batteries Based on a Deep Learning Algorithm" reported a battery prognostic method based on deep neural networks. Results showed that in the case of NASA battery degradation data, the deep neural network-based prognostic method can predict the battery state of health better than other traditional machine learning algorithms, including support vector machine (SVM), k-nearest neighbors (k-NN), artificial neural networks (ANN), and linear regression (LR).

The paper entitled "Initial Design Phase and Tender Designs of a Jacket Structure Converted into a Retrofitted Offshore Wind Turbine" reported an investigation of the possibility of converting actual structures for gas extraction into offshore platforms for wind turbine towers. The proposed method simplified the structural study of jacket structures that are commonly used in the Adriatic Sea for extracting natural gas.

The paper entitled "Dynamic Study of a Rooftop Vertical Axis Wind Turbine Tower Based on an Automated Vibration Data Processing Algorithm" reported an investigation of ambient dynamic responses of a rooftop vertical axis wind turbine. This paper revealed that blade rotation speed is the greatest contributing factor to vibration responses.

The paper entitled "Study on Vibration Transmission among Units in Underground Powerhouse of a Hydropower Station" reported on field structural vibration tests conducted in an underground powerhouse of a hydropower station on Yalong River. This paper provided guidance for further study on the vibration of underground powerhouse structures.

The paper entitled "A Non-Probabilistic Solution for Uncertainty and Sensitivity Analysis on Techno-Economic Assessments of Biodiesel Production with Interval Uncertainties" reported a non-probabilistic strategy for uncertainty analysis of technoeconomic assessments of biodiesel production. Results showed that the proposed nonprobabilistic reliability index in a focused biodiesel production of interest is 0.1211. Moreover, the price and cost of biodiesel, feedstock, and operating can considerably affect technoeconomic assessments of biodiesel production.

The paper entitled "Remaining Useful Life Estimation of Aircraft Engines Using a Modified Similarity and Supporting Vector Machine (SVM) Approach" reported an aircraft engine prognostic method based on the hybrid of a similarity method and SVM. Results showed that the proposed method is effective in analyzing 2008 PHM data challenge competition data and performed well in remaining useful life prediction.

The authors of this Special Issue covered very important topics connected with structural prognostics and health management in power & energy systems and contributed their knowledge to this research community. Future directions in structural prognostics and health management in power & energy systems will go towards prognostics and health management of more complicated and multiple components in power & energy systems, rather than individual components. Moreover, varying and nonstationary operating conditions must be considered in order to make prognostics and health management more practical.

Dong Wang, Shun-Peng Zhu, Xiancheng Zhang, Gang Chen, José A.F.O. Correia, Guian Qian
Special Issue Editors

energies

MDPI

Article

A Data-Driven Predictive Prognostic Model for Lithium-Ion Batteries Based on a Deep Learning Algorithm

Phattara Khumprom * and Nita Yodo

Industrial and Manufacturing Engineering, North Dakota State University, Fargo, ND 58102, USA;
nita.yodo@ndsu.edu
* Correspondence: phattara.khumprom@ndsu.edu; Tel.: +1-701-231-9818

Received: 17 January 2019; Accepted: 15 February 2019; Published: 18 February 2019

Abstract: Prognostic and health management (PHM) can ensure that a lithium-ion battery is working safely and reliably. The main approach of PHM evaluation of the battery is to determine the State of Health (SoH) and the Remaining Useful Life (RUL) of the battery. The advancements of computational tools and big data algorithms have led to a new era of data-driven predictive analysis approaches, using machine learning algorithms. This paper presents the preliminary development of the data-driven prognostic, using a Deep Neural Networks (DNN) approach to predict the SoH and the RUL of the lithium-ion battery. The effectiveness of the proposed approach was implemented in a case study with a battery dataset obtained from the NASA Ames Prognostics Center of Excellence (PCoE) database. The proposed DNN algorithm was compared against other machine learning algorithms, namely, Support Vector Machine (SVM), k-Nearest Neighbors (k-NN), Artificial Neural Networks (ANN), and Linear Regression (LR). The experimental results reveal that the performance of the DNN algorithm could either match or outweigh other machine learning algorithms. Further, the presented results could serve as a benchmark of SoH and RUL prediction using machine learning approaches specifically for lithium-ion batteries application.

Keywords: data-driven; machine learning; deep learning; DNN; prognostic and Health Management; lithium-ion battery

---

1. Introduction

In the past, nickel–cadmium batteries were generally the only electrical power source for various portable equipment, until nickel metal hybrid and lithium-ion batteries were developed in the 1990s [1]. In the present-day, lithium-ion battery technology is rapidly growing, and it is the most reliable electrical power source for numerous appliances. Lithium-ion batteries are extensively equipped in both high-power applications and low-power electronics products, such as hybrid-motor engines, electric cars, smartphones, tablet, laptops, etc. To date, lithium-ion technology is considered to be a standard power source, and its performance continues to improve. There is currently no any other technology that has proven to perform better than the lithium-ion battery. Therefore, there will be no other battery technologies that lithium-ion anytime soon, and the main focus of the ongoing technology is still aimed at improving the lithium-ion system in term of both its performance and reliability. The following are the main advantages of lithium-ion batteries: (1) high energy density (up to 23–70 Wh/kg), (2) high efficiency (close to 90%), and (3) long life cycle (provides 80% capacity at 3000 cycles) [2].

To ensure that the lithium-ion battery system performing reliably, there must be a method that helps to track and to determine the state of health (SoH) of the battery system, along with its remaining useful life (RUL). This method gives useful information for the prediction of when the battery should be removed or replaced. This type of evaluation is known as the system's prognostic and health management (PHM). There have been many advancements contributed by researchers from various disciplines to PHM of lithium-ion batteries. Downey et al. proposed a physics-based prognostic approach that considered multiple concurrent degradation mechanisms [3]. Susilo et al. studied the estimation of the lithium-ion battery SoH with the combination of Gaussian distribution data and the least square support vector machines regression approach [4]. Mejdoubi et al. employed the Rao-Blackwellization particle filter to evaluate the aging condition of lithium-ion batteries, and to estimate SoH and RUL of the battery system [5]. Bai et al. developed a generic model-free approach based on ANN and the Kalman filter, to help to improve the health management system of the lithium-ion battery [6]. Other filtering techniques, for example, particle filtering [7] or its variation of the unscented particle filtering technique [8] had been employed in the PHM aspect for lithium-ion batteries. Recently, Li et al. proposed Gauss–Hermite particle filter (GHPF) technique for battery state-of-charge estimation, which is another extension of the particle filter technique, which not only improves the estimation accuracy, but also reduces the number of sampling particles, which reduces the complexity of the algorithm [9]. Another interesting work also aims to predict the health state of the lithium-ion battery, as proposed by Wang et al. This work employed the Brownian motion technique, which is the combination of the Kalman filter and the Gaussian distribution state space technique, to determine battery prognostics based on the drift coefficient [10].

A data-driven model based on the deep learning approach for lithium-ion battery prognostics is the main focus of this paper. Although various approaches had been proposed to improve the PHM prediction of lithium-ion batteries, the deep learning approach for PHM is still limited. The advancement of computational tools and big data algorithms have largely impacted the development of this approach. The machine learning algorithms, in particular, ANN, have been proven to be able to empirically learn and recognize the more complex patterns of the system's data in many applications. This feature of machine learning algorithms also benefits prognostic analysis modeling as well. This paper presents the preliminary development of a data-driven model using Deep Neural Networks (DNN) to predict the SoH and RUL of lithium-ion batteries. DNN is a deep learning approach that was developed based on Artificial Neural Networks with multiple hidden layers, to analyze more complex data and features. Although some deep learning algorithms, such as Recurrent Neural Network (RNN) and Long Short-Term Memory Network (LSTM), are employed to model prognostic of lithium-ion battery recently, to date, there is no work that has employed a DNN model to perform similar tasks. In addition, there are limited works that have performed a deep learning approach against other data-driven algorithms. For this reason, this paper can also act as a benchmarking reference for employing a deep learning approach to prognostic data in general. The effectiveness of the proposed approach was tested in the lithium-ion battery dataset derived from the NASA Ames Prognostics Center of Excellence (PCoE). A DNN approach was employed to predict the SoH and RUL and the results were compared against other machine learning algorithms such as Linear Regression (LR), k-Nearest Neighbors (k-NN), Support Vector Machine (SVM), and ANN. This paper is constructed with the following sections: Section 2 discusses the overview of the PHM application and the characteristics of the lithium-ion battery used in this paper, Section 3 provides a concise literature review of the proposed approach for DNN analysis and modeling, Section 4 details the experimental results and the comparison of DNN and other machine learning algorithms, and Section 5 concludes the findings and investigates possible future work.

2. Prognostics and Health Management

This paper extends the application of artificial intelligence through machine learning in PHM applications, specifically for lithium-ion battery PHM applications. In this section, an overview of data-driven prognostics and the general prognostic approach for the lithium-ion battery will be discussed briefly.

2.1. Overview of Data-Driven Prognostics

The PHM of the battery has to be included as part of the condition-based maintenance (CBM) plan of the system. The CBM plan is considered as a preventive strategy, which means that maintenance tasks will be performed only when need arises. This need can be determined by continuously evaluating health status of a particular system's components, or the health state of the system as a whole [11]. CBM has included two major tasks: diagnostics and prognostics. Diagnostics is the process of the identification of faults and part of the current health status of the system, which is described as an SoH, whereas prognostics is the process of forecasting the time to failure. The time left before observing a failure is described as the remaining useful life (RUL) of such a system [12]. To avoid severe negative consequences when systems run until failure, the maintenances must be performed when the system is still up and running. These type of maintenance require early plans and preparation [13]. Thus, CBM must properly be included as part of the system's operation, especially for the critical systems. The prognostic of the system is a crucial factor in CBM.

The prognostic process additionally involves two phases. The first phase of prognostics aims to assess the current health status or state of health (SoH). Terms that are usually used to describe this phase in most of the literature are severity detection and degradation detection, which can also be considered under diagnostics. Classification or clustering techniques can be utilized to perform tasks such as pattern recognition in this phase. The second phase aims to predict the failure time by forecasting the degradation trend, and by identifying the remaining useful life (RUL). Trend projection, tracking techniques, or time series analysis are included in this phase. Most of the academic articles regarding prognostics analysis only consider the first phase [14]. This paper aims to construct and analyze both SoH and RUL, in which focus is made on both the first and second phases of prognostics for the battery system.

Generally, there are two existing major approaches for prognostics evaluation; the data-driven model, and physics-based models. Data-driven methods require adequate data or samples from systems that were run until failure, while physics-based methods evaluate the system's failures via the physics of failure progression. Both the data-driven and physics-based model also have different requirements and use cases, and both also have different advantages and drawback as well. Table 1 summarizes the information on the differences and advantages of each model.

Table 1. Difference between data-driven and physics-based models for diagnostics and prognostics.

	Data-Driven Model [15]	**Physics-Based Model** [16,17]
Based on	The empirical lifetime data and the use of previous data of the operation of the system	Physical understanding of the physical rules of the system, the exact formulas that represent the system
Advantages	The real behavior of the complex physical system is not required.	Higher accuracy because the model is based on an actual (or near-actual) physical system
	Models are less complex, easier to employ into a real application	The model represents a real system, the model can be observed and judged in a more realistic manner
Drawbacks	Needs a large amount of empirical data in order to construct a high accuracy model	Highly complex, requires extensive computational time/resources, which may not be very suitable for employment in real-world applications
	The models do not represent the actual system, it requires more effort to understand the real system behavior based on the collected data	Limitations in modeling, especially in cases of large and complex systems with non-measurable variables

One of the data-driven model approaches for prognostics and diagnostics mentioned earlier are machine learning approaches, which will be the main discussion topic of this paper.

2.2. Prognostics of the Lithium-ion Battery

The lithium-ion battery data employed in the prognostics analysis of this work was retrieved from the NASA Ames Prognostics Center of Excellence (PCoE) data repository [18]. This dataset contains the test results of commercially available lithium-ion 1850-sized rechargeable batteries, and the experiment has been performed under controlled conditions in the NASA prognostics testbed [19].

Experimental data were obtained from three different lithium-ion battery-operational test conditions: charge, discharge, and impedance. All experiments were performed at room temperature. The charge was performed at a constant current of 1.5 A until the voltage reached 4.2 V, and then it continued charging at a constant voltage until the charge current dropped to 20 μA. The discharge was also performed at a constant current of 2 A until the voltage dropped to 2.7 V, 2.5 V, 2.2 V, and 2.5 V. These same tests were performed for batteries No. 05, No. 06, No. 07, and No. 18. The impedance test was done by using EIS (Electrochemical Impedance Spectroscopy) frequency adjustment from 0.1 kHz to 5 kHz. By repeatedly performing charge and discharge tests in multiple cycles, this accelerated the aging characteristics of the batteries. This aging effects of the lithium-ion battery can be explained by using the physics-based model established in [20]. The tests were stopped when the batteries reached the end of life criteria, which was defined as a 30% fade from the rated capacity.

Figure 1 is the schematic diagram of the tested battery. The parameters of the schematic diagram included the Warburg impedance (R_W) and the electrolyte resistance (R_E), the charge transfer resistance (R_{CT}), and the double-layer capacitance (C_{DL}). The two parameters R_W and C_{DL} showed a negligible change over the aging process of the battery, and these might be excluded from further analysis [21]. Based on the schematic diagram of the tested battery, below is the characteristic profile of battery No. 05, which will be used as a training data set. Figure 2 shows some details of the current and voltage behaviors during the charging and discharging cycles of battery No. 05. Figure 1 is the schematic diagram of the tested battery. The parameters of the schematic diagram included the Warburg impedance (R_W) and the electrolyte resistance (R_E), the charge transfer resistance (R_{CT}), and the double-layer capacitance (C_{DL}). The two parameters RW and C_{DL} showed a negligible change over the aging process of the battery, and these might be excluded from further analysis [21]. Based on the schematic diagram of the tested battery, below is the characteristic profile of battery No. 05, which will be used as a training data set. Figure 2 shows some details of the current and voltage behaviors during the charging and discharging cycles of battery No. 05.

In order to evaluate the prognostics of the battery, the SoH of the battery must be defined. The prognostics of the battery data are often based on the identification of the SoH of the battery. Therefore, it is important to understand the clear definition of SoH, as the SoH will be the main prediction attribute of the proposed data-driven model, along with RUL. It is also important to note that in this work, all attributes from the test data will be used as training attributes. Some of the attributes (or parameters), and the definition of State of charge (SoC) and SoH in the battery dataset for the prognostics analysis of the battery will be discussed in the following paragraphs.

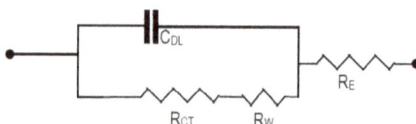

Figure 1. The schematic diagram of the tested battery.

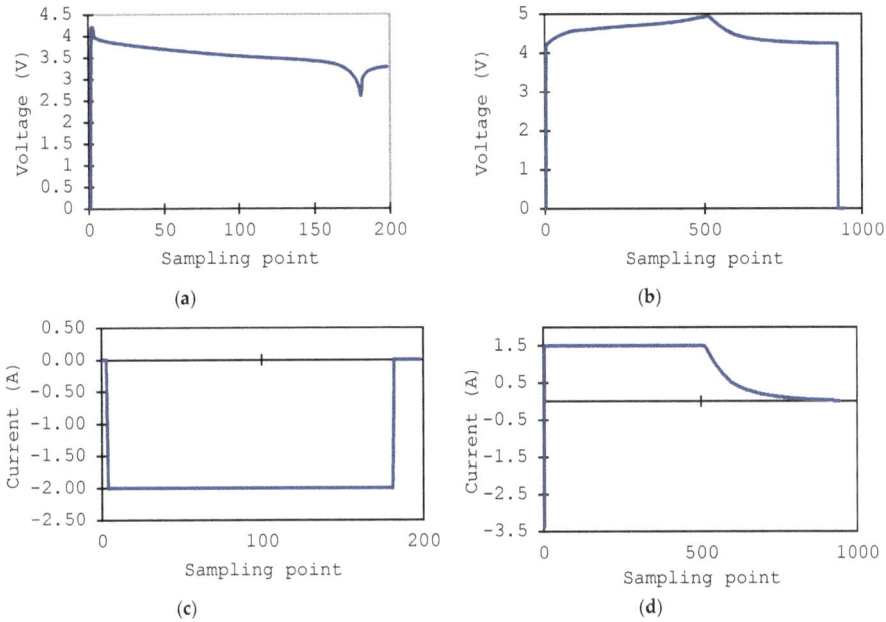

Figure 2. The current and voltage during the discharging and charging of battery No. 05: (**a**) the current of discharging, (**b**) the current of charging, (**c**) the voltage of discharging, and (**d**) the voltage of charging.

The SoC of the battery indicates the reliability of the battery system. In the literature, the ratio between the available amount of charge and the maximum amount of charge is commonly referred to as the SoC [6]. In some cases, the available amount of charge can also be replaced by the rated capacity (or nominal capacity) provided by battery manufacturers. The SoC can be mathematically expressed as:

$$Soc = \frac{Q_{available}}{C_N} \tag{1}$$

where $Q_{available}$ represents the available amount of charge and C_N represents the rated capacity from battery manufacturers.

The SoC definition from Equation (1) seemed to be a straightforward and easy to employ the formula. However, there are some problems using SoC as battery health measurement. First, the only way to derive the rated capacity of a battery is through experiments under a constant discharge rate within a controlled experimental environment. This reason explains the difficulty in using a rated capacity as a reference point in real-world applications [22]. Second, SoC is not considered to have a strong correlation with battery capacity. This is a vital point for making a long-term estimation of the battery's health, since the capacity is the main indication of the battery's health, which will fade over time.

Many alternative SoC equations are defined in several studies to address the aforementioned issues. One interesting definition is practical state-of-charge, or SoCN [23]. This definition uses the maximum practical operational capacity, instead of the manufactured rated capacity, as the maximum amount of charge. SoCN can be expressed as:

$$SoC_N = \frac{Q_{available}}{C_{max,p}} \tag{2}$$

where $C_{max,p}$ represents the maximum practical capacity as measured from the operating battery at the current time. $C_{max,p}$ may fade over time, due to the effect of battery aging.

Apart from the different ways of quantifying SoC, SoH is another important parameter for battery health management. SoH is the direct indication of the health condition of the battery system. SoH can be generally defined as:

$$SoH = \frac{C_{max,p}}{C_N} \tag{3}$$

One of the most important tasks in prognostics health management of a battery is to accurately estimate the $C_{max,p}$, as $C_{max,p}$ is required in both Equations (2) and (3) for SoC and SoH estimations, respectively. Our tested battery dataset contained all the aging information of the battery, and the battery SoH was calculated from cycle 0 to cycle 168. As shown in Figure 3, the estimated SoH of battery No. 05 exponentially degraded as the cycle number increased. The acceptable predicted results must be within the 95% confidence bound [24]. The regression model for SoC and SoH estimation, which aimed to perform similar tasks, was also proposed in [25]. This work introduced a new variable to directly indicate the voltage drop of the battery cell as the prediction variable. This work delivered very interesting results. However, it is not within the scope of our deep learning approach. Our work aimed to use only existing test variables to train and generate the deep learning model for the SoH and RUL estimation of lithium-ion batteries.

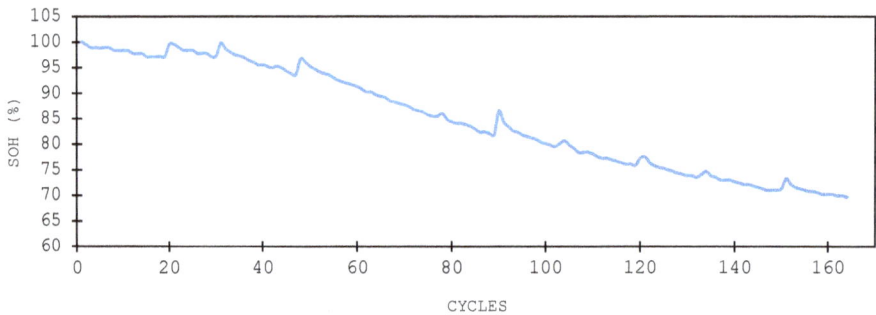

Figure 3. The state of health of battery No. 05.

As a quantification metric to evaluate the performance of the prediction model in this work, the root mean square error (RMSE) was employed for SoH, and the error of RUL cycle (E_{RUL}) was employed for RUL. The following are the formulas of RMSE and E_{RUL}:

$$RMSE = \sqrt{\frac{1}{n} \sum_{i=1}^{n} [x_i - \bar{x}_i]^2} \tag{4}$$

$$E_{RUL} = \left| RUL_{real} - RUL_{prediction} \right| \tag{5}$$

where n is the number of prediction datasets, x_i is the real value of testing and monitoring the battery capacity, and $\bar{x}_i$ is the prediction value. RMSE and E_{RUL} are used as the key performance measures of the performance of all traditional machine learning approaches and the proposed deep learning algorithm. RMSE and ERUL will be calculated within the testing phase of the modeling framework, which will be discussed in the next section.

3. Data-Driven Prognostic Analysis and Modeling

The reason that the machine learning approach works well with the prognostic data, in general, is due to the condition monitoring system, where it collects massive data from equipment. This large amount of data benefits data-driven PHM models, which requires large empirical data in order to create a high accuracy model of the systems. The traditional machine learning methods, called shallow learning models employed in this paper, include: Linear Regression (LR) [26] with the Akaike Information Criterion (AIC) [27,28], k-NN [29], SVM [30,31], and ANN [32–36]. Although there are multiple machine learning algorithms implemented in this work, the main focus is the concept of deep learning, which will be extensively discussed in this section. The rest of the algorithms are well-documented, and will not be further discussed in this paper. Interested readers are encouraged to refer to the associated references for further information. As the deep learning concept was developed based on the ANN, the first discussion in this section will be an initial description of ANN, then the deep learning algorithm and its applications to prognostics assessments will be discussed accordingly.

3.1. Artificial Neural Networks

An ANN, usually called a neural network, is a mathematical model or computational model that is inspired by the structural and functional aspects of biological neural networks [32]. A single neural network call, or a perceptron, has an interconnected group of artificial neurons, which process computational information by using a connectionist approach from node to node. An ANN is considered to be an adaptive system, which means that it is able to change its shape based on the different structure of information flow gains for the learning phase. Basically, there are two configuration modes in ANN. First are the feed-forward, and second is the back-propagation ANNs. For the feed-forward network, the connections between the units or nodes do not form a completed back-and-forth cycle. Instead, the information in the network moves only one way forward from the input units, through the hidden units, to the output units. Meanwhile, back-propagation moves the information backward, in order to update weights in the network.

Back-propagation is a supervised learning method that has two phases, the propagation phase and the weight update phase [33]. These two phases are repeated until the performance of the network is satisfied. In back-propagation algorithms, the output values from the network are compared with the actual or correct value, through the calculation of the error-function value. This error-function value is fed back through the network as a reference, to make an appropriate adjustment of the weights of each connection. The goal is to reduce the value of the error-function by selecting the proper weights. This process is repeatedly performed in the training cycle until the condition is satisfied. Usually, the network will converge to a certain state where the error of the calculations is small. This scenario can be considered as if the network has the capability to learn a certain target function.

A Multi-Layer Perceptron (MLP) is a feed-forward ANN model that maps multiple sets of input data onto a set of appropriate output. An MLP consists of multiple layers of nodes, with each layer being connected to the next one, except for the input nodes. Each node in MLP is an artificial neuron apply with a nonlinear activation function. MLP employs back-propagation to train the network, while multiple network layers consist of many computational units that are interconnected in a feed-forward fashion. Many applications apply a sigmoid function as an activation function.

There are multiple activation functions that can be implemented in ANN, for example, linear or identity functions, binary step functions, hyperbolic tangent function, and sigmoid function. In this paper, the sigmoid activation function of the hidden layer used in the implementation (Section 4) is a Gaussian spheroid function, expressed as follows:

$$y(x) = e^{-\left(\frac{||x-c||^2}{2\sigma^2}\right)} \tag{6}$$

The output of the hidden neuron gives a measure of distance between the input vector x and the centroid c of the data cluster. The parameter σ represents the radius of the hypersphere, which is generally determined by using an iterative process of selecting an optimum width. In addition to the activation function of the neural network, another condition that needs to be considered in order to construct the classification model is the learning or training algorithm of the neural network. A learning algorithm is a systematic step-by-step procedure through which the connection weights among the neurons are adjusted to minimize the difference between the predicted and the actual values of an output variable [34]. This adjustment was performed in this study by using the most popular method of training, as mentioned earlier, which is known as the back-propagation learning algorithm. In addition to its broad employments in various applications, the back-propagation learning algorithm has been shown to be more efficient than other learning algorithms for solving most regression problems [35].

There are three main reasons that are impacted by the efficiency of the back-propagation learning algorithm. First, this learning algorithm is simple to perform. Second, the back-propagation learning algorithm is able to provide reasonably accurate results for complicated applications in which the input and output relationships are nonlinear [36]. Finally, and most importantly, the back-propagation learning algorithm has revealed an acceptable level of generalization ability. The performance of the MLP networks trained by the back-propagation learning algorithm is usually controlled by the learning rate and momentum. Varying in a range between 0 and 1, the learning rate is a parameter that affects how the connection weights within a network are updated. These updates also include a portion of the last weight change, to accelerate the training convergence, and to improve the training precision. This portion is defined by the momentum, which, like the learning rate, varies over a range of 0–1. Similar to the number of hidden neurons, a specific rule that determines the best values for the learning rate and momentum has not yet been proposed in the literature. As a result, this determination was performed in this study by examining different values from 0–0.9, with a constant step size of 0.1.

3.2. Overview of the Deep Learning Concept

The deep learning concept was first suggested by Geoffrey Hinton in 2006, and it has become well-known in both academia and industry [37]. Deep learning is an improvement of a Multi-Layer Perceptron with a better power of learning representation, which holds the potential to overcome the deficiencies in traditional machine learning methods [38]. The notable advantage of deep learning is that it is able to capture the representation of information from raw data through multiple complex non-linear transformations and approximations. In order to accurately evaluate the state of systems, to decide whether or not the equipment in the systems need to be maintained or not as part of CBM, fault diagnosis and prognostics may benefit from the utilization of deep learning.

The main algorithms of deep learning include the Deep Neural Network, the Convolutional Neural Network (CNN), the Recurrent Neural Network (RNN), and the expansion of CNN and RNN, such as the Long short-term memory network (LSTM). There are also hybrid networks that combine different types of stacked layers. The following are the characteristics of each deep learning algorithm.

The **Deep Neural Network** (DNN) is generally a stack of multiple hidden layers instead of only one hidden layer in the standard ANN architecture [39]. The DNN hidden layers are the multiple feed-forward layers that are trained with a back-propagation stochastic gradient descent. The hidden layers consist of neurons nodes with tanh, rectifier (ReLU), and maxout activation functions. DNN has features such as an adaptive learning rate, rate annealing, momentum training, dropout, and regularization. These features are believed to enable a higher predictive accuracy compared to the regular ANN.

The **Convolutional Neural Network** (CNN) is basically composed of layers of convolutions consisting of neurons, with tanh, ReLU being applied to the results. CNN uses convolutions over the input layer to compute the output. An individual layer of CNN applies different types of filters. The edges of the layers capture the shape of the data, and then they use these shapes to determine higher-level features. The last layer classifies the output by using these high-level features.

The **Recurrent Neural Network** (RNN) makes use of sequential information. The RNN defines the inputs and outputs as a dependent variable based on a time sequence. RNN performs the same task for every element of a sequence. The output at the last time step of RNN is dependent on the previous computations. RNN may be considered to have a "memory", as it can capture information about calculations in past sequences. However, RNN has a limitation in capturing the length of the data. This leads to the development of the LSTM network, which can capture longer sequences of information [40,41].

3.3. Employment of Deep Learning to Prognostic Data

In the diagnostics and prognostics fields, the developing trend of employing the deep learning approach has evolved from fault detection and failure diagnosis to degradation pattern recognition and time series predictive analysis. The modeling methods have grown from using only a single algorithm such as DNN, CNN, and RNN, to the Hybrid model, or a combination of multiple layer types and traditional algorithms. The application range of using deep learning has also been expanding continually over the years, from machinery, electrical, and electronics systems, to wind-power and high-end aerospace equipment [38].

In this paper, only DNN was employed to model the SoH and RUL of the battery data against other traditional machine learning algorithms. Each of these deep learning algorithms have their own advantages and disadvantages. It has apparently been discovered that ANN and DNN are more suitable for tackling one-dimensional data. CNN is better handling multidimensional data, as it has adopted types of convolutional techniques. RNN is suitable for applications that deal with time series or dependent input data, and DNN is usually employed for extracting global features from fault data, which will be suitable for the lithium-ion battery data. Additionally, as aforementioned, the layers of CNN and RNN are far more complex than those of DNN. Therefore, CNN and RNN take more time for training the model, which is their major drawback. These reasons make DNN more suitable for employment in real applications for the most case.

3.4. The Deep Neural Network Framework and Model for Prognostic Data

Currently, deep neural network (DNN) has become a well-employed approach in machine learning, due to its promising performance and advantages. DNN employs a multi-layered feed-forward neural network that is similar to the vanilla artificial neural network, but with densely stacked or fully-connected hidden layers instead of one hidden layer. In order to develop a DNN framework for prognostic data, the similar practice of developing the ANN framework has been used.

In this work, the DNN framework for prognostic data was developed based on a Cross-industry standard process for data mining (CRISP-DM) [42], and this is illustrated in Figure 4. This can be considered as the general framework of fault prognostics by using a deep learning algorithm. The framework is generally divided into five phases: the definition states phase, the pre-processing phase, the training phase, testing phase, and evaluating phase. The details of each phase are as follows:

1. *Definition states phase.* This phase specifically focuses on defining the failure of the system, identifying the prognostic problem, and evaluating system health states.
2. *Pre-processing phase.* In this phase, sensory data are collected according to the predefined health state, in order to build a raw dataset for the experiment. The raw datasets are preprocessed and normalized, and then divided into a training and a testing dataset.
3. *Training phase.* In this phase, initial parameters are developed, and the classification model is trained by the training dataset, based on deep learning theory. It is particularly important to fine-tune the classification model through misclassification errors (such as RSME).
4. *Testing phase.* In this phase, the testing dataset is put into the trained classification model to identify prognostic predictions or projection results.
5. *Evaluating phase.* This phase mainly finishes with computing the accuracy, reporting on, and evaluating the diagnosis results from the final model.

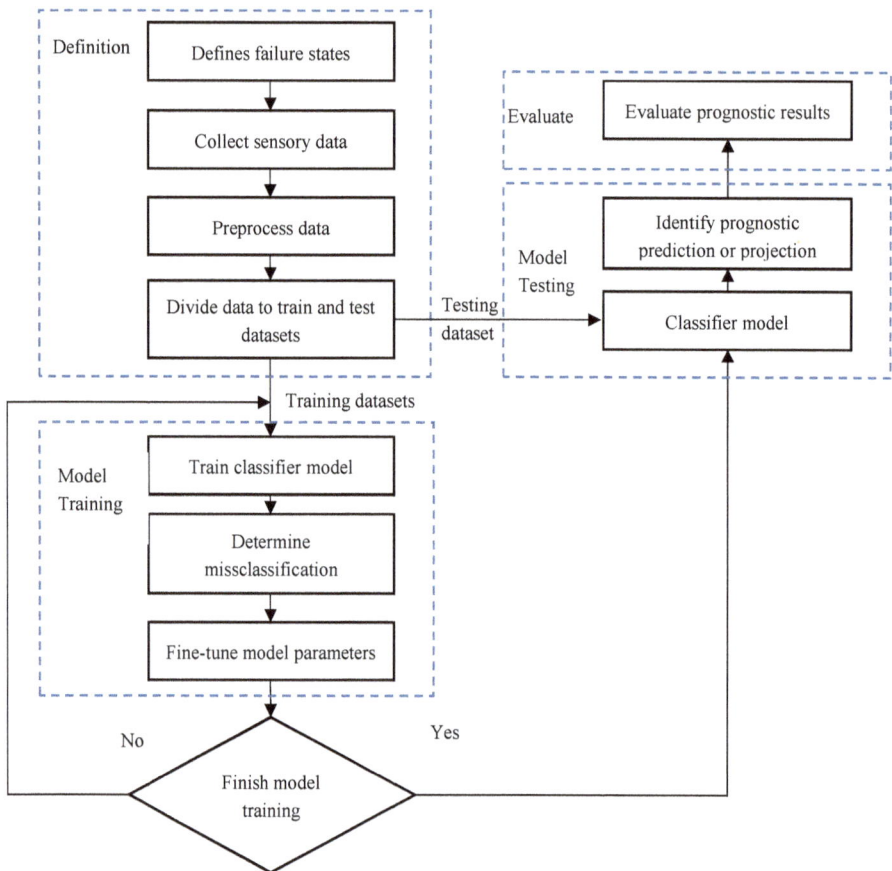

Figure 4. The process of a prognostic framework using deep learning.

The prognostic model of the battery data using DNN was developed based on the aforementioned framework. The experiment with the data was constructed by varying the number of dense layers in DNN. In this experiment, the hidden layer was varied to analyze the SoH of battery data until it delivered the best RMSE results. In addition, the dropout layer was also applied as the last layer before the output layer, to prevent the overfitting problem. The dropout layer applied to the last layer of DNN, to randomly drop neurons during the model training, as shown in Figure 5. Each neuron is retained with a fixed probability, p which is independent of other neurons. The neural network after being sampled, the so-called "thinned" network, will contain only the surviving neurons (Figure 5b).

By training a neural network with some dropouts, the whole network can be trained more often than training regular networks without dropout, because the network is thinned so that it can be trained at less frequency. The network then becomes less sensitive to some specific weights. This results in the network being better at generalization. In this work, a p-value of 0.25 is applied to the network, as suggested, to be the optimal dropout rate for the network to avoid overfitting, but to still maintain the best prediction accuracy [43].

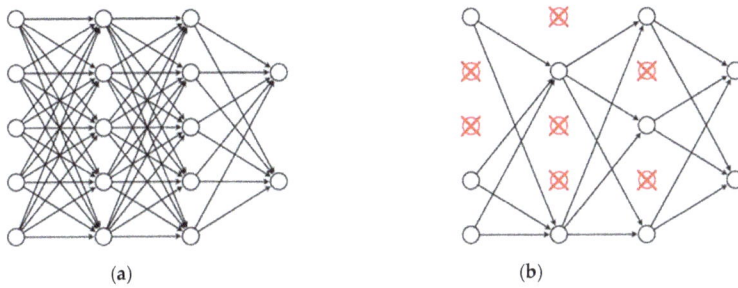

Figure 5. Dropout deep neural network model: (a) A standard network with two hidden layers; (b) the network after applying dropout.

By varying the hidden layers of DNN from two layers to four layers, the RMSE results from Table 2 show that the best formation of DNN consists of three stacked dense or fully-connected hidden layers, with the ReLU activation function (see Appendix A for the explanation of the ReLU function), as described in Figure 6. It might also be worthwhile to note that there are some predictions of fluctuation in the DNN model that are implemented by using the Keras library, which is the open source neural network library that is employed in this experiment. Therefore, each DNN experiment was performed with 10 trials. The RMSE results of the final model (three dense layers) are as shown in Table 3. The best result from the trials was chosen to be the final model.

Table 2. RMSE results of each stacked hidden layer model.

Number of Hidden Layers	RMSE
2	3.815
3	3.247
4	3.275

Table 3. RMSE results of 10 trials for a model with three stacked hidden layers.

Trials	1	2	3	4	5	6	7	8	9	10
RMSE	3.917	3.877	3.667	3.507	3.487	3.321	3.296	3.253	3.249	3.247

From this section, the preliminary model of deep learning was developed, based on the deep neural network. The objective of this paper was to prove that the deep learning algorithm outperformed other traditional machine learning algorithms, and ultimately, to provide a complete benchmark of SoH and RUL prediction for the lithium-ion battery. In the next section, the prognostic of the lithium-ion battery data mentioned in Section 2 will be tested by using the machine learning approach, and a deep learning model that has been mentioned previously.

4. Case Study

In this section, an analysis of battery No. 06, No. 07, and No. 18 degradation datasets obtained from NASA Ames Prognostics Center of Excellence (PCoE) database [18] was conducted to validate the effectiveness of the developed DNN approach. The dataset of battery No. 05 was used as a training dataset for all algorithms. A detailed description of the experimental data has been provided in Sections 2 and 3, along with the validation method for SoH and RUL. The SoH experimental results from tradition machine learning algorithms and the developed DNN will be presented in Section 4.1. The RUL results from the tradition machine learning algorithms and the developed DNN will be shown in Section 4.2, and all results for both the SoH and RUL estimations from all algorithms will be discussed in Section 4.3.

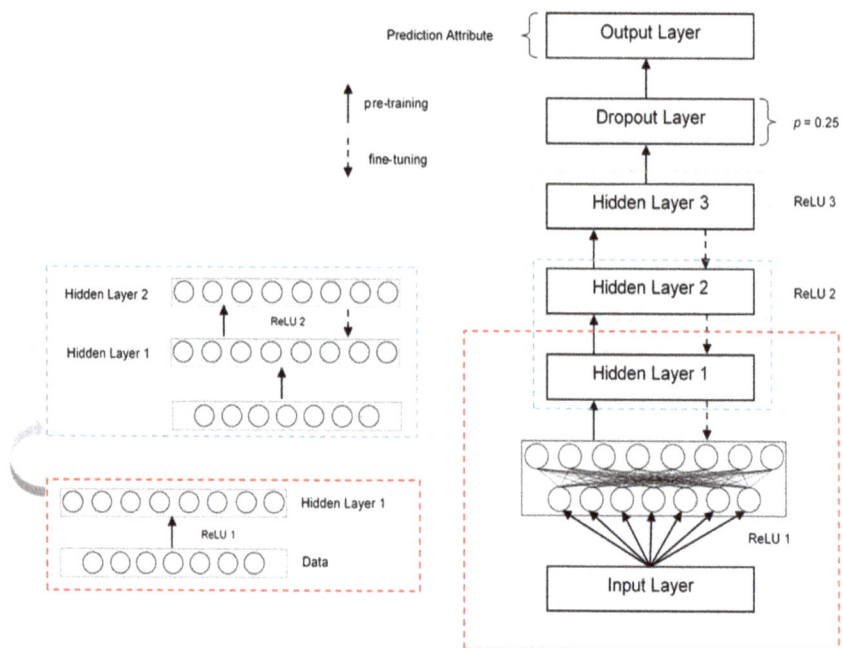

Figure 6. The proposed deep neural network model.

4.1. Results for SoH Estimation

In this experiment, the discharge data for all 164 cycles and 11,345 sample points from battery No. 05 were employed. The SoH was calculated from the initial capacity as being 1.9 AHr. The result is considered to be a long-term SoH estimation of the battery. Figure 7a–c show the SoH estimation performance for batteries No. 06, 07, and 18, using k-NN, LR, SVM, ANN, and DNN respectively. The x-axis represents the cycles, and the y-axis represents the SoH. The triangle marked with a light blue line curve shows the true SoH, and the rest of the curves show the predicted SoH by the k-NN, LR, SVM, ANN, and the developed DNN. The SoH estimators will be employed in each discharge cycle for the SoH estimation.

It might also be worthwhile to mention that the SVM formulation employed in this work was based on the radial basis kernel function, with a regularization parameter of 200, and tolerance-of-loss function of 0.1. LR employed the greedy algorithm with 0.1 minimum tolerance parameters. Additionally, k-NN employed the Euclidean distance measurement to evaluate distances among the neighbor data points.

It is clearly shown in the figures that due to the accurate curve fitting of the trained DNN model with batteries No. 06, 07, and 18, the RMSEs of the SoH estimated by the proposed model are much less than the ones estimated by k-NN, LR, SVM, and ANN. In addition, after the batteries aged from the first cycle to the 164th cycle, it is obvious that the proposed DNN approach could capture the degradation pattern far better than the other algorithms. The input capacity in the developed DNN model could provide sufficient information for the stability of the SoH estimation when the batteries were aged. On the other hand, the lack of the knowledge of other approaches resulted in an increasing error for the SoH estimation in the aged cycles. Furthermore, the performance of the capacity convergence by the proposed DNN approach was better, since the knowledge of capacity fade could be captured better by using the DNN model. Considering the result illustrated in Figure 7, it is also important to note that the results from battery No. 06 performed slightly worse when compared to battery No. 07 and 18. This could be due to the aging pattern of battery No. 06 being slightly different from the training dataset. Additionally, there was a greater distribution of the data of battery No. 06, compared to the other batteries.

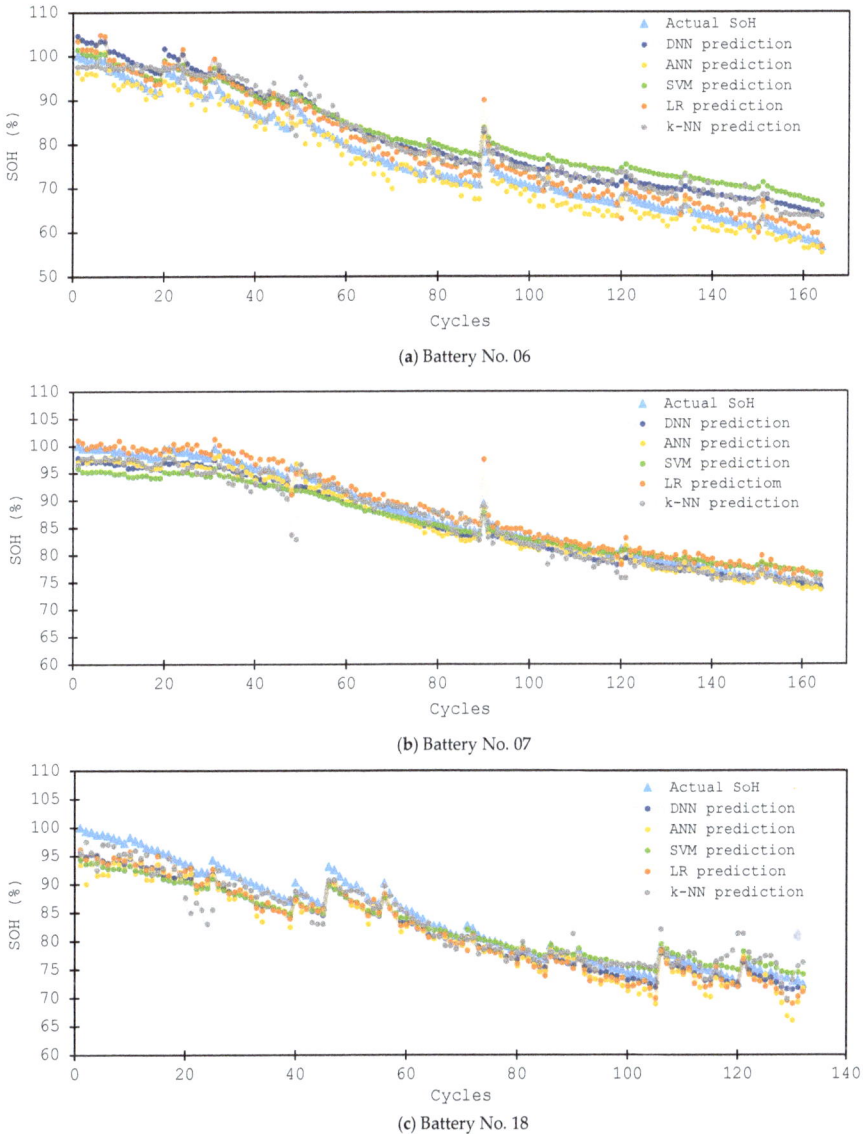

(a) Battery No. 06

(b) Battery No. 07

(c) Battery No. 18

Figure 7. The SoH estimation with all algorithms for battery No. (**a**) 06, (**b**) 07, and (**c**) 18.

The RMSE results between traditional machine learning, k-NN, LR, SVM, and ANN, along with the developed DNN, are shown in Figure 7 and Table 4. SVM has some drawbacks in terms of capturing the patterns of data; yet, it still outperformed other algorithms based on RMSE. When comparing other algorithms against DNN, DNN was shown to perform the best among the four approaches in terms of capturing both data patterns and RMSE. Additionally, the models constructed from each algorithm could also be observed in detail, from Table 5.

Table 4. RMSE of the SoH estimation by using DNN and traditional machine learning algorithms.

	k-NN	LR	SVM	ANN	DNN
RMSE	5.598	4.558	4.552	4.611	3.427

Table 5. Models created from the training dataset.

Algorithm	Model Description		
k-NN	22-Nearest Neighbor model for regression The model contains 624 examples with seven dimensions		
LR	228.765 * Voltage_measured + 237.439 × Current_measured − 1.495 * Temperature_measured − 1098.506 × Current_charge + 50.156 * Capacity − 918.727		
SVM	Total number of Support Vectors: 613 Bias (offset): −85.065 w[Voltage_measured] = 42686654.125 w[Current_measured] = −17208.396 w[Temperature_measured] = 243822393.316 w[Current_charge] = 3952.097 w[Voltage_charge] = 0.000 w[Time] = 0.000 w[Capacity] = 16430099.458 number of classes: 2 number of support vectors: 613		
ANN	Node 1 (Sigmoid) Voltage_measured: −0.172 Current_measured: −0.448 Temperature_measured: 2.894 Current_charge: −1.458 Voltage_charge: 0.005 Time: 0.042 Capacity: −0.155 Bias: −2.726	Node 2 (Sigmoid) Voltage_measured: 1.954 Current_measured: 0.328 Temperature_measured: −1.124 Current_charge: −0.397 Voltage_charge: 0.036 Time: −0.014 Capacity: 0.943 Bias: −1.930	Node 3 (Sigmoid) Voltage_measured: 0.406 Current_measured: 1.254 Temperature_measured: 1.472 Current_charge: 1.391 Voltage_charge: −0.049 Time: −0.036 Capacity: 1.107 Bias: −1.055
	Node 4 (Sigmoid) Voltage_measured: −3.468 Current_measured: −0.975 Temperature_measured: 0.080 Current_charge: −0.018 Voltage_charge: 0.044 Time: −0.020 Capacity: 2.457 Bias: −0.108	Node 5 (Sigmoid) Voltage_measured: −7.072 Current_measured: −0.455 Temperature_measured: 2.095 Current_charge: 2.091 Voltage_charge: −0.004 Time: 0.045 Capacity: −0.464 Bias: −4.078	Output Regression (Linear) Node 1: 1.278 Node 2: 1.460 Node 3: 0.865 Node 4: 1.214 Node 5: −1.134 Threshold: −0.819

Neural Network created:

Input   Hidden 1   Output

DNN	Layer (type)	No. of Hidden Nodes	No. of Parameters	
	dense_1 (Dense)	8	64	Total parameters: 217
	dense_2 (Dense)	8	72	Trainable parameters: 217
	dense_3 (Dense)	8	72	Non-trainable parameters: 0
	dropout_1 (Dropout)	8	0	
	dense_4 (Dense)	1	9	

There are some other aspects of the DNN that should also be considered further, which are the optimizer of the network, and the loss function. DNN in this work was performed by employing "Adam" or Adaptive Moment Estimation, as an optimizer (see Appendix B for more details). Additionally, based on the nature of the battery dataset in this work, the absolute error function was employed as the loss function [44]. Absolute errors measured the mean absolute value of the difference between the elementwise inputs. The absolute error formula used as the loss function can be expressed by the following equation:

$$Absolute\ error\ loss = \frac{1}{k}\sum_{i=1}^{k}|y_i - \bar{y}_i|^2 \tag{7}$$

where y_i and $\bar{y}_i$ are, respectively, the predicted data and the input data of each iteration or epoch i, and k is the number of iterations or epochs. In this work, the total number of iterations or epochs was set to be equal to 1024, as suggested in reference [45].

4.2. Results for RUL Estimation

In addition to the SoH prediction of the batteries from the previous section, another aspect of the prognostic analysis of the battery data was to predict the RUL of the batteries. RUL prediction focuses on projecting the degradation results from a certain cycle until the EoL of the batteries, which is different from that of the SoH prediction, which focuses on detecting the pattern of degradation. In this experiment, the goal was to compare the RUL prediction result by using k-NN, LR, SVM, and ANN against the proposed DNN algorithm.

The RUL predictions experiments were performed from three different starting points, which were at the 40th cycle, 80th cycle, and the 120th cycle of battery No. 05. The threshold of the EoL of the battery data was set to be at 30% remaining capacity, or at the 164th cycle. This was deemed to be the rule of thumb of the EoL threshold, for the battery to remain active. The data before the starting cycle was used as a training dataset to make the prediction from each starting cycle, and the error of RUL (Equation (5)) was calculated to compare the accuracy of each machine learning algorithm. The accumulated errors of the RUL results are as shown in Table 6, and the projection results of RUL are as shown in Figure 8. Note that the RUL results focus on making a projection, not to recognize the data's pattern.

Table 6. The error of RUL estimation by using DNN and traditional machine learning algorithms.

	Starting Points	k-NN	LR	SVM	ANN	DNN
Error of RUL	40th cycle	24	19	12	6	5
	80th cycle	17	12	10	3	2
	120th cycle	19	9	4	1	1

The results from Table 6 and Figure 8 show that, overall, the proposed DNN algorithm outperformed all other machine learning algorithms. The prediction result at the 120th cycle of DNN and ANN are the same. However, DNN still performed better than ANN in term of accuracy while having a smaller set of training data. As shown by the result, DNN provided a slightly better result when starting at the 40th cycle and the 80th cycle. Additionally, the trend of the result also showed that having more training data improves the prediction result for every algorithm in this experiment.

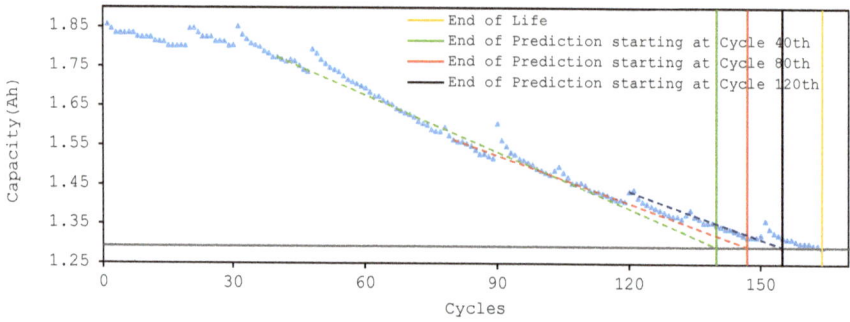

(a) The RUL estimation using k-NN

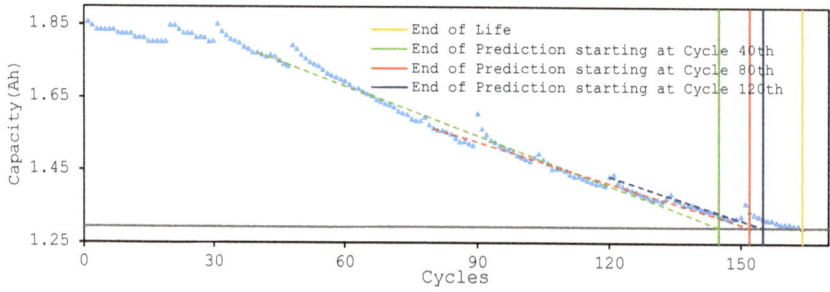

(b) The RUL estimation using LR

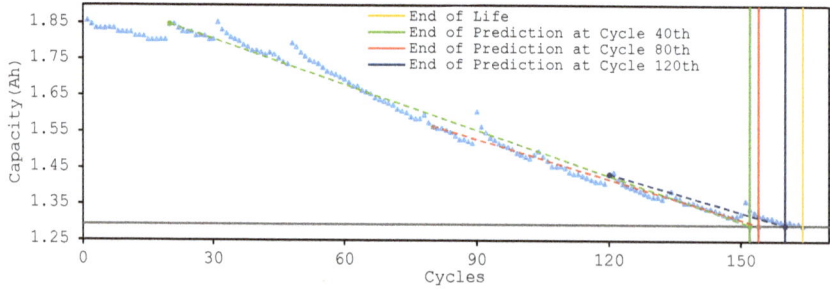

(c) The RUL estimation using SVM

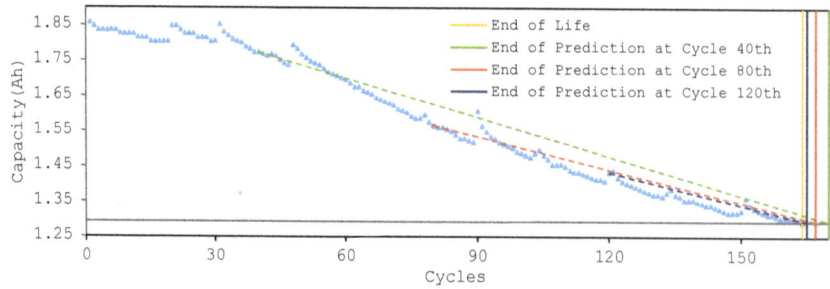

(d) The RUL estimation using ANN

Figure 8. *Cont.*

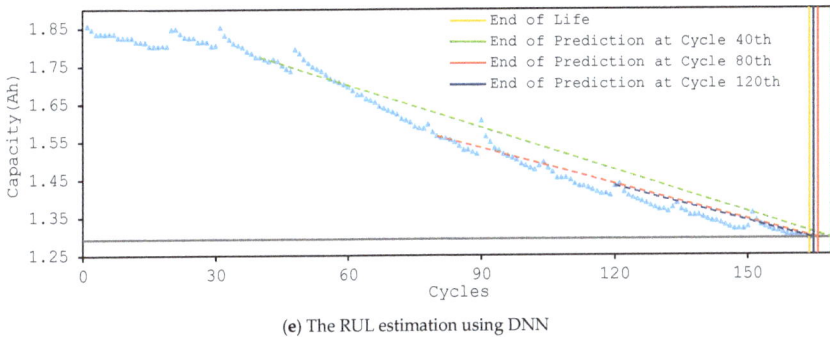

(e) The RUL estimation using DNN

Figure 8. The RUL estimation of battery No. 05 using (**a**) k-NN, (**b**) LR, (**c**) SVM, (**d**) ANN, and (**e**) DNN.

4.3. Discussion and Future Work

From the experimental results in Sections 4.1 and 4.2, it is obvious that the proposed DNN algorithm can outperform k-NN, LR, SVM, and ANN in these specific lithium-ion battery datasets. However, there are two points that need to be addressed here. First, the DNN proposed in this work can outstandingly capture the degradation pattern based on the prediction of the SoH result in Section 4.1. In contrast, in terms of predicting the projection of the RUL (Section 4.2), the performance of DNN is only comparable to ANN. This is suspected to be due to the fundamentals of DNN being based on ANN. The second point is in the case of having smaller training dataset, the DNN performed better overall. This can be observed from the RUL prediction results. DNN provided better results when started from the smaller amount of training data at the 40th and 80th cycles, compared to the typical neural network.

The results obtained from this work also prove that the deep learning algorithm is effective and suitable for employment for prognostic and diagnostic data modeling, particularly in the prognostics of the battery data set. The prognostic results will eventually aid in condition-based monitoring of maintenance activities, to obtain the best time to replace the batteries without causing a long downtime in the main systems. Based on this experiment, the downsides of using a deep learning algorithm include: (1) a higher computational time and (2) more resources are required by DNN than for the other two algorithms. These drawbacks are also true for other deep learning algorithms as well. This conclusion is that deep learning is more suitable for studies that require a higher accuracy, but which may not be the most suitable for works that need real-time processing. In the battery PHM application, real-time processing is not very crucial, since the prediction should be prior to the end-of-life of the batteries. In addition, with the advancement of the computational tools, the real-time processing concern could be minimized, and the computational time will be improved. In the future, real-time processing might no longer be an issue for implementing deep learning in most PHM cases.

The deep learning model in this paper was only developed based on the Deep Neural Network algorithm (DNN). As mentioned in Section 3.4, there are other more complex deep learning algorithms that have been developed over the years, such as the Convolutional neural network (CNN), the Recurrent neural network (RNN), and the Long short-term memory network (LSTM). These deep learning algorithms will be explored in the field of prognostics analysis in the future. It is also worth noting that some researchers have started to employ the LSTM network for similar battery prognostic data, to predict the remaining useful life (RUL) of the battery [46]. Although this has already been done, it is incomparable to the experiments in this work based on the fact that the experiment on LSTM in the literature implemented a different dataset, and more importantly, the experiment only focused on testing the LSTM network in the model, and did not provide a complete comparison to other models that use traditional machine learning algorithms. This leaves gaps to be explored in the future, particularly with benchmarking all deep learning algorithms.

In addition to benchmarking deep learning approaches with other machine learning algorithms, in the future, physical experiments will be explored, to bridge the gap between data-driven models and physics-based models for PHM applications. Data-driven models will be employed to help in easing the modeling complexity in physics-based models of lithium-ion batteries. Thus, an accurate battery models that mimics models in real-world applications could be obtained without the need for extensive outputs of time and resources. The advancements in artificial intelligence and machine learning algorithms play an important role in defining future approaches in PHM for lithium-ion batteries, as well as many other engineering applications.

5. Conclusions

This work aims to accomplish two tasks. First, a complete benchmarking of the data-driven model by using a machine learning algorithm with the battery prognostic data is made. Second, a preliminary data-driven model is developed by using a deep learning algorithm for the prognostic data. This paper has achieved its goal to aid, as a benchmark, the prognostic data-driven model for battery data using machine learning algorithms, and based on the results from the case studies, it shows that the deep learning algorithm provides a promising outcome for predicting and modeling the prognostic data, especially in the battery prognostic and health management applications. Based on the accuracy archived, we also believe that the traditional physics-based model may be replaced by data-driven models in the near future, in various fields and applications. The reliable data-driven model has many advantages over a traditional physics-based model. The first major advantage is that it overcomes the complexity of the physics-based model. This attribute of less complexity in a data-driven model helps to reduce the involvement of the domain experts in particular fields. In the future, the predictive model might be able to be generated and constructed without any opinion or knowledge from experts at all. The second advantage is that data-driven models can be employed in real-time situations, due to the shorter computational time needed, when compared to physics-based models in general. The last point is that the data-driven model is more cost-effective to construct and to employ in real applications. As an example, a data-driven model can be generated and monitored by using only regular personal computing devices, without the need for exclusive and excessive resources. This future trend of data-driven models is in line with the recent achievement of deep learning algorithms and artificial intelligence. These methodologies are believed to be the main approaches in the further development of data-driven models. However, the accuracy of prediction and the higher performance of using deep learning algorithms also comes with the drawback of higher computational time. With rapid advancements in technology, the computational time could be substantially reduced. The future direction of this work will focus on developing a hybrid-deep learning model that could be universally applicable to multiple types of prognostic data.

Author Contributions: Conceptualization, P.K. and N.Y.; methodology, P.K..; software, P.K.; validation, P.K. and N.Y.; formal analysis, P.K.; investigation, P.K.; resources, P.K.; data curation, N.Y.; writing—original draft preparation, P.K.; writing—review and editing, N.Y.; visualization, P.K.; supervision, N.Y.; project administration, N.Y.; funding acquisition, N.Y.

Funding: This research received no external funding.

Conflicts of Interest: The authors declare no conflict of interest.

Appendix A

Rectified Linear Units (ReLU) is an activation function of neural networks, defined as:

$$f(x) = x^+ = \max(0, x) \tag{A1}$$

where x is the input to a neuron, and + represents the positive part of its arguments. ReLU has been demonstrated to achieve better training for deeper networks [47–49] compared to other activation functions such as the logistic sigmoid and the hyperbolic tangent [50,51].

Appendix B

The Adaptive Moment Estimation (Adam) optimizer keeps an exponentially decaying average of past gradients $M(t)$, similar to momentum [52]. $M(t)$ and $V(t)$ are values of the first moment, which is the Mean, and the second moment, which is the Un-centered variance of the gradients, respectively. The following is the formulas for the First Moment (Mean), and the Second Moment (Variance):

$$\hat{m}_t = \frac{m_t}{1 - \beta_1^t} \tag{A2}$$

$$\hat{v}_t = \frac{v_t}{1 - \beta_2^t} \tag{A3}$$

The following is the final formula for the Parameter update:

$$\theta_{t+1} = \theta_t - \frac{\eta}{\sqrt{\hat{v}_t} - \varepsilon} \tag{A4}$$

The value for β_1 is 0.9, and 0.999 for β_2 and $10*\exp(-8)$ for ε.

References

1. Berecibar, M.; Gandiaga, I.; Villarreal, I.; Omar, N.; Van Mierlo, J.; Van den Bossche, P. Critical review of state of health estimation methods of Li-ion batteries for real applications. *Renew. Sustain. Energy Rev.* **2016**, *56*, 572–587. [CrossRef]
2. Corey, G.P. Batteries for stationary standby and for stationary cycling applications part 6: Alternative electricity storage technologies. In Proceedings of the Power Engineering Society General Meeting, Toronto, ON, Canada, 13–17 July 2003; Volume 1, pp. 164–169.
3. Downey, A.; Lui, Y.H.; Hu, C.; Laflamme, S.; Hu, S. Physics-Based Prognostics of Lithium-Ion Battery Using Non-linear Least Squares with Dynamic Bounds. *Reliab. Eng. Syst. Saf.* **2019**, *182*, 1–12. [CrossRef]
4. Susilo, D.D.; Widodo, A.; Prahasto, T.; Nizam, M. State of Health Estimation of Lithium-Ion Batteries Based on Combination of Gaussian Distribution Data and Least Squares Support Vector Machines Regression. In *Materials Science Forum*; Trans Tech Publications: Princeton, NJ, USA, 2018; Volume 929, pp. 93–102.
5. El Mejdoubi, A.; Chaoui, H.; Gualous, H.; Van Den Bossche, P.; Omar, N.; Van Mierlo, J. Lithium-ion Batteries Health Prognosis Considering Aging Conditions. *IEEE Trans. Power Electron.* **2018**. [CrossRef]
6. Bai, G.; Wang, P.; Hu, C.; Pecht, M. A generic model-free approach for lithium-ion battery health management. *Appl. Energy* **2014**, *135*, 247–260. [CrossRef]
7. Saha, B.; Goebel, K. Modeling Li-ion battery capacity depletion in a particle filtering framework. In Proceedings of the Annual Conference of the Prognostics and Health Management Society, San Diego, CA, USA, 27 September–1 October 2009; pp. 2909–2924.
8. Miao, Q.; Xie, L.; Cui, H.; Liang, W.; Pecht, M. Remaining useful life prediction of lithium-ion battery with unscented particle filter technique. *Microelectron. Reliab.* **2013**, *53*, 805–810. [CrossRef]
9. Li, B.; Peng, K.; Li, G. State-of-charge estimation for lithium-ion battery using the Gauss-Hermite particle filter technique. *J. Renew. Sustain. Energy* **2018**, *10*, 014105. [CrossRef]
10. Wang, D.; Tsui, K.L. Brownian motion with adaptive drift for remaining useful life prediction: Revisited. *Mech. Syst. Signal Process.* **2018**, *99*, 691–701. [CrossRef]
11. Camci, F.; Chinnam, R.B. Health-state estimation and prognostics in machining processes. *IEEE Trans. Autom. Sci. Eng.* **2010**, *7*, 581–597. [CrossRef]
12. Jennions, I.K. (Ed.) *Integrated Vehicle Health Management: Perspectives on an Emerging Field*; SAE International: Warrendale, PA, USA, 2011.

13. Eker, Ö.F.; Camci, F.; Jennions, I.K. Major challenges in prognostics: Study on benchmarking prognostic datasets. In Proceedings of the First European Conference of the Prognostics and Health Management Society 2012, Cranfield, UK, 25 May 2012.
14. Qiu, H.; Lee, J.; Lin, J.; Yu, G. Robust performance degradation assessment methods for enhanced rolling element bearing prognostics. *Adv. Eng. Inform.* **2003**, *17*, 127–140. [CrossRef]
15. Heng, A.; Zhang, S.; Tan, A.C.; Mathew, J. Rotating machinery prognostics: State of the art, challenges and opportunities. *Mech. Syst. Signal Process.* **2009**, *23*, 724–739. [CrossRef]
16. Jianhui, L.; Namburu, M.; Pattipati, K.; Qiao, A.L.; Kawamoto, M.A.K.M.; Chigusa, S.A.C.S. Model-based prognostic techniques [maintenance applications]. In Proceedings of the AUTOTESTCON 2003 IEEE Systems Readiness Technology Conference, Anaheim, CA, USA, 22–25 September 2003; pp. 330–340.
17. Zhang, H.; Kang, R.; Pecht, M. A hybrid prognostics and health management approach for condition-based maintenance. In Proceedings of the IEEE International Conference on Industrial Engineering and Engineering Management, Hong Kong, China, 8–11 December 2009; pp. 1165–1169.
18. Saha, B.; Goebel, K. *Battery Data Set, NASA Ames Prognostics Data Repository*; NASA Ames: Moffett Field, CA, USA, 2007. Available online: http://ti.arc.nasa.gov/project/prognostic-datarepository (accessed on March 2018).
19. Saha, B.; Goebel, K. Uncertainty management for diagnostics and prognostics of batteries using Bayesian techniques. In Proceedings of the Aerospace Conference, Big Sky, MT, USA, 1–8 March 2008; pp. 1–8.
20. He, W.; Pecht, M.; Flynn, D.; Dinmohammadi, F. A Physics-Based Electrochemical Model for Lithium-Ion Battery State-of-Charge Estimation Solved by an Optimised Projection-Based Method and Moving-Window Filtering. *Energies* **2018**, *11*, 2120. [CrossRef]
21. Saha, B.; Goebel, K.; Christophersen, J. Comparison of prognostic algorithms for estimating remaining useful life of batteries. *Trans. Inst. Meas. Control* **2009**, *31*, 293–308. [CrossRef]
22. Meissner, E.; Richter, G. Battery monitoring and electrical energy management: Precondition for future vehicle electric power systems. *J. Power Sources* **2003**, *116*, 79–98. [CrossRef]
23. Santhanagopalan, S.; White, R.E. Online estimation of the state of charge of a lithium ion cell. *J. Power Sources* **2006**, *161*, 1346–1355. [CrossRef]
24. Liu, D.; Pang, J.; Zhou, J.; Peng, Y. Data-driven prognostics for lithium-ion battery based on Gaussian Process Regression. In Proceedings of the 2012 IEEE Conference on Prognostics and System Health Management (PHM), Beijing, China, 23–25 May 2012; pp. 1–5.
25. Huang, S.C.; Tseng, K.H.; Liang, J.W.; Chang, C.L.; Pecht, M.G. An online SOC and SOH estimation model for lithium-ion batteries. *Energies* **2017**, *10*, 512. [CrossRef]
26. Bianco, A.M.; Martínez, E. Robust testing in the logistic regression model. *Comput. Stat. Data Anal.* **2009**, *53*, 4095–4105. [CrossRef]
27. Akaike, H. A new look at the statistical model identification. *IEEE Trans. Autom. Control* **1974**, *19*, 716–723. [CrossRef]
28. Burnham, K.P.; Anderson, D.R. *Model Selection and Multimodel Inference: A Practical Information-Theoretic Approach*; Springer Science & Business Media: New York, NY, USA, 2003.
29. Peterson, L.E. K-nearest neighbor. *Scholarpedia* **2009**, *4*, 1883. [CrossRef]
30. Zhang, Z.; Gu, L.; Zhu, Y. Intelligent Fault Diagnosis of Rotating Machine Based on SVMs and EMD Method. *Open Auto. Control Syst. J.* **2013**, *5*, 219–230. [CrossRef]
31. Yang, J.; Zhang, Y.; Zhu, Y. Intelligent fault diagnosis of rolling element bearing based on SVMs and fractal dimension. *Mech. Syst. Signal Process.* **2007**, *21*, 2012–2024. [CrossRef]
32. Samanta, B. Artificial neural networks and genetic algorithms for gear fault detection. *Mech. Syst. Signal Process.* **2004**, *18*, 1273–1282. [CrossRef]
33. Hegazy, T.; Fazio, P.; Moselhi, O. Developing practical neural network applications using back-propagation. *Comput. Aided Civ. Infrastruct. Eng.* **1994**, *9*, 145–159. [CrossRef]
34. Jeon, J. Fuzzy and neural network models for analyses of piles. Ph.D. Thesis, North Carolina State University, Raleigh, NC, USA, 2007.
35. Emsley, M.W.; Lowe, D.J.; Duff, A.R.; Harding, A.; Hickson, A. Data modelling and the application of a neural network approach to the prediction of total construction costs. *Constr. Manag. Econ.* **2002**, *20*, 465–472. [CrossRef]

36. Alex, D.P.; Al Hussein, M.; Bouferguene, A.; Fernando, S. Artificial neural network model for cost estimation: City of Edmonton's water and sewer installation services. *J. Constr. Eng. Manag.* **2009**, *136*, 745–756. [CrossRef]

37. Hinton, G.E.; Salakhutdinov, R.R. Reducing the dimensionality of data with neural networks. *Science* **2006**, *313*, 504–507. [CrossRef] [PubMed]

38. Zhao, G.; Zhang, G.; Ge, Q.; Liu, X. Research advances in fault diagnosis and prognostic based on deep learning. In Proceedings of the Prognostics and System Health Management Conference (PHM-Chengdu), Chengdu, China, 19–21 October 2016; pp. 1–6.

39. Schmidhuber, J. Deep learning in neural networks: An overview. *Neural Netw.* **2015**, *61*, 85–117. [CrossRef]

40. Hochreiter, S.; Schmidhuber, J. Long short-term memory. *Neural Comput.* **1997**, *9*, 1735–1780. [CrossRef]

41. Graves, A.; Mohamed, A.R.; Hinton, G. Speech recognition with deep recurrent neural networks. In Proceedings of the 2013 IEEE International Conference On Acoustics, Speech and Signal Processing (ICASSP), Vancouver, BC, Canada, 26–31 May 2013; pp. 6645–6649.

42. Wirth, R.; Hipp, J. CRISP-DM: Towards a standard process model for data mining. In Proceedings of the 4th International Conference on the Practical Applications of Knowledge Discovery and Data Mining, Manchester, UK, 11–13 April 2000; pp. 29–39.

43. Srivastava, N.; Hinton, G.; Krizhevsky, A.; Sutskever, I.; Salakhutdinov, R. Dropout: A simple way to prevent neural networks from overfitting. *J. Mach. Learn. Res.* **2014**, *15*, 1929–1958.

44. Schorfheide, F. Loss function-based evaluation of DSGE models. *J. Appl. Econom.* **2000**, *15*, 645–670. [CrossRef]

45. Gupta, S.; Agrawal, A.; Gopalakrishnan, K.; Narayanan, P. Deep learning with limited numerical precision. In Proceedings of the International Conference on Machine Learning, Lille, France, 6–11 July 2015; pp. 1737–1746.

46. Zhang, Y.; Xiong, R.; He, H.; Liu, Z. A LSTM-RNN method for the lithium-ion battery remaining useful life prediction. In Proceedings of the Prognostics and System Health Management Conference (PHM-Harbin), Harbin, China, 9–12 July 2017; pp. 1–4.

47. Hahnloser, R.H.; Sarpeshkar, R.; Mahowald, M.A.; Douglas, R.J.; Seung, H.S. Digital selection and analogue amplification coexist in a cortex-inspired silicon circuit. *Nature* **2000**, *405*, 947. [CrossRef]

48. Hahnloser, R.H.; Seung, H.S. Permitted and forbidden sets in symmetric threshold-linear networks. In Proceedings of the Advances in Neural Information Processing Systems, Vancouver, BC, Canada, 3–8 December 2001; pp. 217–223.

49. Glorot, X.; Bordes, A.; Bengio, Y. Deep sparse rectifier neural networks. In Proceedings of the Fourteenth International Conference on Artificial Intelligence and Statistics, Ft. Lauderdale, FL, USA, 11–13 April 2011; pp. 315–323.

50. Orr, G.B.; Müller, K.R. (Eds.) *Neural Networks: Tricks of the Trade*; Springer: Berlin/Heidelberg, Germany, 2003.

51. Nair, V.; Hinton, G.E. Rectified linear units improve restricted boltzmann machines. In Proceedings of the 27th International Conference on Machine Learning (ICML-10), Haifa, Israel, 21–24 June 2010; pp. 807–814.

52. Kingma, D.P.; Ba, J. Adam: A method for stochastic optimization. *arXiv* **2014**, arXiv:1412.6980.

energies

MDPI

Article

Early Fault Detection of Wind Turbines Based on Operational Condition Clustering and Optimized Deep Belief Network Modeling

Hong Wang *, Hongbin Wang, Guoqian Jiang, Jimeng Li and Yueling Wang

School of Electrical Engineering, Yanshan University, No. 438, Hebei Avenue, Qinhuangdao 066004, China;
hb_wang@ysu.edu.cn (H.W.); jgqysu@gmail.com (G.J.); xjtuljm@163.com (J.L.); yuelingw@ysu.edu.cn (Y.W.)
* Correspondence: hongw329@163.com

Received: 8 January 2019; Accepted: 7 March 2019; Published: 13 March 2019

Abstract: Health monitoring and early fault detection of wind turbines have attracted considerable attention due to the benefits of improving reliability and reducing the operation and maintenance costs of the turbine. However, dynamic and constantly changing operating conditions of wind turbines still pose great challenges to effective and reliable fault detection. Most existing health monitoring approaches mainly focus on one single operating condition, so these methods cannot assess the health status of turbines accurately, leading to unsatisfactory detection performance. To this end, this paper proposes a novel general health monitoring framework for wind turbines based on supervisory control and data acquisition (SCADA) data. A key feature of the proposed framework is that it first partitions the turbine operation into multiple sub-operation conditions by the clustering approach and then builds a normal turbine behavior model for each sub-operation condition. For normal behavior modeling, an optimized deep belief network is proposed. This optimized modeling method can capture the sophisticated nonlinear correlations among different monitoring variables, which is helpful to enhance the prediction performance. A case study of main bearing fault detection using real SCADA data is used to validate the proposed approach, which demonstrates its effectiveness and advantages.

Keywords: wind turbines; health monitoring; fault detection; optimized deep belief networks; supervisory control and data acquisition system; multioperation condition

1. Introduction

With increasing global energy demand, wind energy as a promising clean source of renewable energy has become an indispensable force in solving world energy problems. The latest annual report released by the Global Wind Energy Council (GWEC) [1] shows that the cumulative and new installed capacity in the world had reached 539,123 MW and 52,492 MW, respectively, by the end of 2017. However, wind turbines are generally situated in remote locations and have harsh operating environments, resulting in frequent failures and undesired shutdowns. High maintenance costs and downtime losses seriously affect the economic benefits of wind farms and also have a powerful impact on the healthy development of the wind power industry [2]. There is an urgent need for effective prognostics and health management (PHM) technologies to address these problems. In particular, fault detection is a premise for PHM. Therefore, it is crucial and valuable to develop advanced health monitoring and fault detection methods to detect impending wind turbine faults as early as possible in order to avoid secondary damage and even catastrophic accidents.

Vibration analysis and oil monitoring have become two commonly used techniques for wind turbine condition monitoring [3–6]. However, both techniques are sophisticated and expensive in their practical application, since additional investments, including installing extra sensors and data acquisition devices, are required. Alternatively, supervisory control and data acquisition (SCADA) systems, which have been widely installed in large-scale wind turbines, can collect and record the operational state information from wind turbines and their critical components on a regular basis [7]. Compared with the vibration and oil monitoring methods, SCADA-based monitoring has been considered to be cost-effective due to the availability of a large amount of monitoring data and no additional cost. As a result, SCADA-based wind turbine health monitoring has attracted wide attention in recent years [8], and different SCADA data analysis methods have been proposed.

Zaher et al. [9] developed normal behavior models for gearboxes and generators based on artificial neural networks by analyzing SCADA data. The case study results demonstrated that it was possible to detect faults as early as 6 months and 16 months before final replacement of the gearbox and generator, respectively. Guo et al. [10] employed a nonlinear state estimation method to construct a normal behavior model of generator temperature using 2 min and 10 min averaged SCADA data. A real case study showed that the method was able to predict generator faults about 8.5 h before the actual failure. Kusiak et al. [11] introduced a neural network to model the normal behavior of generator bearings by using 10 s SCADA data. The research showed that the method could identify anomalies about 1.5 h ahead of the eventual failure. Schlechtingen et al. [12,13] proposed an adaptive neuro-fuzzy inference system combining artificial neural network and fuzzy logic analysis and constructed 45 normal behavior models using 10 min averaged SCADA data. Case studies illustrated that the system could detect the potential failures of wind turbines months in advance and provide the root causes of these failures based on simple if–then rules. Bangalore et al. [14,15] applied artificial neural networks to establish normal behavior models of gearboxes. Case studies with 10 min averaged SCADA data showed that the proposed methods were able to detect gearbox anomalies ahead of the condition monitoring system. Bi et al. [16] presented a pitch fault detection procedure using a normal behavior model based on the performance curve and carried out six case studies. The results illustrated that the proposed method could detect pitch faults earlier than the artificial intelligence approaches investigated. Different methods have been used to model the normal behavior of wind turbines. Further, residuals between the predicted values of the models and actual measured values of the expected output variable were used to identify the anomalies of wind turbines. Practically, wind turbine operating conditions are complicated and changeable and present multiple operation regions due to varying external wind speed and a complex internal control scheme, which poses great challenges for effective and reliable fault detection. However, most existing monitoring approaches only focus on a single whole operating condition, so they cannot fully consider the dynamic operating characteristics of wind turbines, leading to unsatisfactory detection performance, such as high rates of false alarms or missed detections. On the other hand, conventional health monitoring methods, such as neural networks, naturally have classical shallow structures, which poses a difficulty in effectively capturing sophisticated nonlinear relationships among monitoring variables.

To address the above issues, a novel general health monitoring approach for wind turbines under varying operating conditions is proposed in this paper. This approach is data-driven and based on monitoring data collected from wind turbine SCADA systems. First, to consider the dynamic behavior and multiple operating characteristics of wind turbines, an operation condition partition scheme using a clustering algorithm is proposed to partition the whole operation into multiple sub-operation conditions. This is a divide-and-conquer strategy and can enable the building of local monitoring models in different sub-operation conditions, which can improve the reliability and accuracy of fault detection compared to a global monitoring model. Second, to overcome the shortcoming of traditional shallow structure–based methods, a deep learning–based modeling approach is proposed to deal with relevant SCADA data to capture the sophisticated nonlinear correlations among monitoring variables. In recent years, motivated by the powerful ability of feature learning and nonlinear modeling of deep learning methods, convolutional neural network [17], autoencoder [18], denoising autoencoder [19], and multilayered extreme learning machines [20] have been used in many classification and regression tasks. Specifically, deep belief networks (DBNs) [21], a typical class of deep learning methods, are used in this study, which are naturally probabilistic generative models with multilayered architecture. Compared with shallow neural network methods, DBNs can capture complex nonlinear features, have a powerful modeling capacity and are quite suitable for modeling complex SCADA data [22]. DBNs have received attention in the fields of wind speed prediction [23], mechanical engineering fault diagnosis [24] and complex system fault detection [25]. The performance of DBNs is largely dependent on their structural parameters. However, there is no uniform rule for parameter selection. Various optimization algorithms have shown the ability to deal with complex problems, such as particle swarm [26] and genetic algorithm [27]. In particular, chicken swarm optimization (CSO), a novel bionic heuristic optimization algorithm, is introduced for optimizing model parameters of DBNs. In summary, the main contributions of this paper are as follows:

(1) A general multioperation condition partition scheme is proposed to partition normal state data into several different clusters. Then, normal behaviors are built under different condition clusters. This divide-and-conquer strategy can help reduce false alarms caused by methods that only consider a single operating condition.

(2) An optimized DBN (ODBN) model with CSO is designed to capture the normal behavior in each cluster, which reduces the complexity of parameter selection of DBNs. To the best of our knowledge, it is the first time DBN is applied to deal with complex SCADA data from wind turbines for the purpose of fault detection.

(3) A real case from wind turbine main bearing fault was used to evaluate the performance of the proposed health monitoring approach using the SCADA data of multiple wind turbines from a real wind farm, and comparative studies were conducted.

The remainder of this paper is organized as follows. Section 2 describes the multioperation condition problem and the operation parameters studied in this paper. In Section 3, the proposed health monitoring framework is presented, the steps are explained, and the presented methodologies are described in detail. Section 4 presents the case study and discussion, and results are compared and analyzed. Conclusions are summarized in Section 5.

2. Problem Description

As critical equipment for wind power generation, a wind turbine is typically a complex electromechanical system composed of a variety of components and subsystems, including gearbox, generator, shaft, bearing, and power electronics, among many others [28]. In practical applications, wind turbines are generally located in remote areas and perennially operate under adverse weather conditions, such as storms, dust, and extreme temperature differentials. In addition, they are also affected by mechanical, electrical, and control strategies. These kinds of situations lead to operating conditions characterized by complexity and variability. As discussed in the first section, one of the primary disadvantages of existing data-driven condition monitoring approaches for wind turbines is that they only take into account a single operating condition, ignoring the characteristics that exist in the process of operating wind turbines. Due to their highly dynamic operating conditions, variations in the abnormal states of turbines are always easily masked by the condition fluctuations, making it difficult to accurately assess the health status and thereby causing frequent false alarms. In this case, it is highly desirable to develop reliable health monitoring approaches to deal with the dynamic and varying operating conditions of wind turbines.

Wind turbine SCADA data contain hundreds of monitoring parameters related to the health of the wind turbine and its critical components. Typically, these parameters include wind conditions (e.g., wind speed, wind direction), power output, blade pitch angle, generator torque and speed, temperatures (e.g., main bearing temperature, gearbox oil temperature, nacelle temperature, and ambient temperature) among others [29]. Several parameters are closely related to the wind turbine operating conditions, which can be referred to as operation parameters, describing the external and internal changes of the turbine operation due to constantly changing wind speed and complex switched control schemes. Typically, these primarily include environmental parameters such as wind speed, wind direction, and ambient temperature, and control parameters such as generator speed and torque [30]. To facilitate the understanding of the operating characteristics of the wind turbine, historical SCADA operation data were collected from a 1.5 MW turbine under normal operation from July to August 2014, shown in Figure 1. Obviously, these operation parameters segment the wind turbine operating conditions into different operation regions to varying degrees, which truly reflects the multiple operating characteristics of the wind turbine. In this case, a global health monitoring model cannot describe the turbine behavior accurately and even fails to produce reliable detection results. Therefore, it is of great practical value to investigate efficient and reliable health monitoring methods, considering varying operating conditions in order to improve the accuracy of fault prediction and reduce operation and maintenance costs.

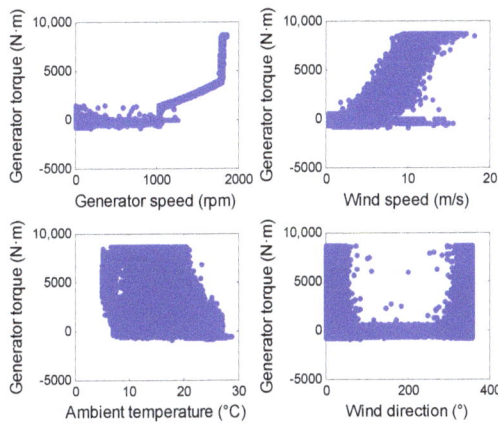

Figure 1. Operating characteristic curves of a wind turbine in normal conditions.

3. Proposed Health Monitoring Framework

In this study, a novel health monitoring framework for wind turbines under varying operating conditions is proposed, and its flowchart is shown in Figure 2. It is general and can be used for fault detection of different wind turbine subsystems and components. The main idea of the proposed framework is to build normal behavior models relying on only historical normal SCADA data from wind turbines and then perform fault detection based on the evaluation results of residuals between the predicted values and actual measured values. The changes of the residuals will give an indication of possible faults. Usually, normal test samples will produce a low residual value since they can well satisfy the learned normal model, whereas faulty test samples will produce high residual values and therefore be identified as faults. Generally, the proposed framework mainly consists of four sequential parts: operation condition partition, variable selection, model development and anomaly detection. The detailed procedures are summarized as follows:

(1) Collect normal SCADA data from multiple wind turbines on a wind farm.
(2) Choose operation parameters that characterize the complex operating conditions of wind turbines and segment the operation parameter data into K clusters using the k-means method and silhouette index. The obtained K clusters represent the corresponding K operating conditions, i.e., $[C_1, C_2, \cdots, C_K]$. Then, divide the normal state data into corresponding K parts based on the partitioned operating conditions.
(3) Select appropriate modeling variables for each operating condition by combining three variable selection techniques, and the final selected variables for different operating clusters can be represented as $[V_1, V_2, \cdots, V_K]$.

(4) Build a normal behavior model under each operating condition using ODBNs to explore the sophisticated nonlinear characteristics among modeling variables, resulting in multiple DBN models, denoted as $[DBN_1, DBN_2, \cdots, DBN_K]$ for K operating clusters.

(5) Calculate the threshold for abnormal detection under different operating conditions using the Mahalanobis distance (MD) measure to automatically identify the anomalies that occur in the operation of the wind turbines, i.e., $[MD_1, MD_2, \cdots, MD_K]$.

(6) For the new incoming SCADA data, first recognize the operating condition C_i that it belongs to, then select the corresponding modeling input variable V_i and predict the output using the constructed DBN_i. Next, compute the MD value and compare it with the threshold MD_i under condition C_i, and then output the real-time online health monitoring results.

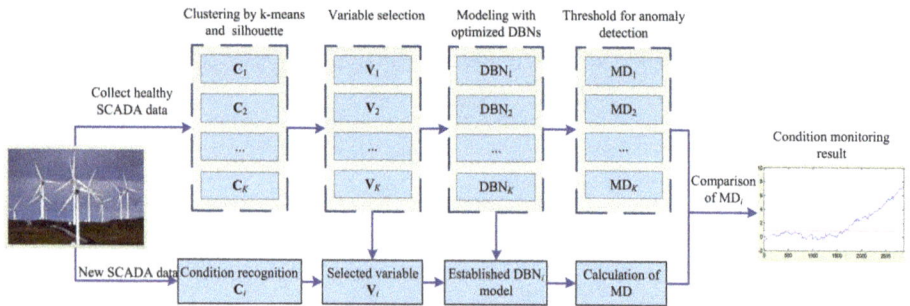

Figure 2. Flowchart of the proposed health monitoring framework. SCADA, supervisory control and data acquisition; DBN, deep belief network; MD, Mahalanobis distance.

3.1. Data Preprocessing

It should be noted that data preprocessing is a necessary step in wind turbine condition monitoring prior to modeling using SCADA data [7]. During the long-time continuous operation of a wind turbine, a large number of outliers and invalid values may be generated and included in the SCADA data because of sensor failures, communication errors, or other issues. These outliers and invalid values will directly impact the performance of the model to be trained, and they should be removed first. To reduce the effects of noise and randomness contained in SCADA data, all data are smoothed prior to selecting modeling variables. Additionally, considering that different operation parameters often have different value ranges, it is a common step to normalize the initial operation data before partitioning the conditions to ensure that each operation variable lies within the specified range between 0 and 1. Specifically, in this study, this step can be simply realized by using the following equation [31]:

$$Y_{ij} = \frac{x_{ij} - \min(x_j)}{\max(x_j) - \min(x_j)},$$ (1)

where x_{ij} is the ith value of variable j, and $\min(x_j)$ and $\max(x_j)$ are the minimum and maximum values of variable j, respectively.

3.2. Operation Condition Partition Using K-Means Clustering

The reasonable selection of wind turbine operation parameters is a prerequisite to realize the partition of operating conditions. As mentioned in Section 2, for a wind turbine, the operation parameters mainly include the following five variables: wind speed, wind direction, ambient temperature, generator speed, and generator torque, which are closely related to the operating conditions. Generally, the operating conditions can be partitioned into several typical operation regions depending on the above operation parameters. As an unsupervised learning method, k-means clustering [32] has become one of the most prevalent and widely used partitioning clustering algorithms due to its advantages of usability, efficiency, simplicity and successful experience [33]. Hence, in this study, this method is adopted for the condition partition. Certainly, other clustering methods can also be considered. The aim of k-means is to allocate all data samples into K clusters by minimizing the sum of the squared error over all K clusters, denoted as follows [33]:

$$J = \arg\min_{O} \sum_{i=1}^{K} \sum_{x_j \in O_i} \|x_j - \mu_i\|^2, \tag{2}$$

where $O = \{O_1, O_2, \ldots, O_K\}$ is the set of K clusters, μ_i is the cluster centroid of the ith cluster, $\{x_1, x_2, \ldots, x_N\}$ is the cluster samples, and N is the number of samples.

In the k-means algorithm, the number of clusters K is a key parameter. Silhouette [34] is one of the indices for evaluating the clustering number by combining the two factors of cohesion and resolution, which is employed to determine K in this paper. The silhouette value for the ith point, $S(i)$, is expressed as

$$S(i) = \frac{b(i) - a(i)}{\max\{a(i), b(i)\}}, \quad i = 1, 2, \ldots, N, \tag{3}$$

where $a(i)$ represents the average distance from the ith point to the other points in the same cluster and $b(i)$ denotes the minimum average distance from the point to points in a different cluster. The range of $S(i)$ is $[-1, 1]$. A higher value of $S(i)$ indicates that the ith point is clustered more properly. The average of all $S(i)$ is then the final silhouette value for a given cluster number.

3.3. Variable Selection

To construct the normal behavior model, it is necessary to first determine the modeling variables in each operating condition. Usually, there are multiple types of relationship among the variables and various techniques can be applied to assess each type of relationship [35]. Three typical variable selection techniques are proposed in [36–38], the Pearson, Spearman, and Kendall correlation coefficients, which are statistics for measuring the linearity, monotonicity, and dependence among variables, respectively. This paper combines the three technologies to select the input variables most relevant to the output variables. It is worth noting that the computation results of these three methods are all in the range of −1 to 1, and a higher absolute value indicates a stronger correlation between the input and output.

3.4. Proposed ODBN Method

The use of wind turbine SCADA systems becomes the primary option for most wind farms, and as a result, large amounts of monitoring data can be acquired and archived regularly. The measured SCADA data have notable features of complex nonlinearity and strong coupling due to the interdependence and interaction between the different subsystems of the wind turbines during operation. Consequently, in this section, ODBNs are proposed to capture the latent nonlinear correlations in the SCADA monitoring data, and the details are described as follows.

3.4.1. DBN Architecture

The structure of DBNs comprises probabilistic generative models composed of multiple stacked restricted Boltzmann machines (RBMs). As displayed in Figure 3, each RBM is a kind of two-layer stochastic neural network consisting of one visible layer and one hidden layer. There are connection weights between the visible layer and the hidden layer, while the units in each layer are restricted to each other.

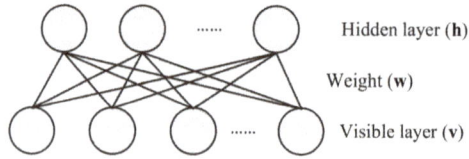

Figure 3. Topological structure of a restricted Boltzmann machine (RBM).

Assuming that the RBM is a Bernoulli–Bernoulli model (BB-RBM), for a given set of states $(\mathbf{v}, \mathbf{h})$, the energy function is defined as

$$E(\mathbf{v}, \mathbf{h}; \theta) = -\sum_{i=1}^{V}\sum_{j=1}^{H} w_{ij}v_i h_j - \sum_{i=1}^{V} b_i v_i - \sum_{j=1}^{H} a_j h_j, \tag{4}$$

where $\theta = \{\mathbf{w}, \mathbf{a}, \mathbf{b}\}$ denotes the model parameters; v_i and h_j are the visible unit i and the hidden unit j, respectively; w_{ij} is the connection weight between i and j; b_i and a_j are the biases of v_i and h_j; and V and H are the number of visible and hidden units, respectively. Given the energy function, the joint probability over the visible and hidden units can be described as follows:

$$p(\mathbf{v}, \mathbf{h}; \theta) = \frac{1}{Z} \exp(-E(\mathbf{v}, \mathbf{h}; \theta)), \tag{5}$$

where $Z = \sum_v \sum_h \exp(-E(v, h; \theta))$ is the partition function.

Since the visible–visible and hidden–hidden units are not connected, the probabilities of the visible unit v_i and the hidden unit h_j are independent. Therefore, the conditional distributions can be expressed as

$$p(h_j = 1|\mathbf{v}; \theta) = \delta(\sum_{i=1}^{V} w_{ij}v_i + a_j), \tag{6}$$

$$p(v_i = 1|\mathbf{h}; \theta) = \delta(\sum_{j=1}^{H} w_{ij}h_j + b_i), \tag{7}$$

where $\delta(x) = 1/(1+exp(x))$ represents the logistic sigmoid function. The model parameters θ of the RBM can be obtained by a contrastive divergence method [39]. The update rule for the weight $\mathbf{w}$ is written as follows:

$$\Delta w_{ij} = \varepsilon(\langle v_i h_j \rangle_{data} - \langle v_i h_j \rangle_k), \tag{8}$$

where ε refers to the learning rate, $\langle \cdot \rangle_{data}$ denotes the expectation of the training data, and $\langle \cdot \rangle_k$ represents the expectation of the sample distribution after k-step Gibbs sampling. A more detailed description of the training process of the RBM can be seen in [40].

The general architecture of a DBN model with n hidden layers is shown in Figure 4. The bottom layer of the DBN accepts input data and then passes the data to hidden layers to complete the learning process. To handle real-valued data, a Gaussian–Bernoulli RBM (GB-RBM) should be adopted in the first RBM model, and BB-RBM is applied in the rest of the RBM models. The learning process of the DBN consists of two phases, pretraining and fine-tuning. Pretraining is the process of initializing the connection weights and biases of the network in a greedy layer-wise unsupervised manner. In this phase, each RBM is trained from bottom to top individually. In the fine-tuning phase, the parameters of the DBN model are updated with the back-propagation algorithm in a supervised fashion. Thus, DBNs realize the organic combination of unsupervised and supervised learning, which can effectively improve the modeling capacity. In this study, a four-layer DBN, including the bottom input layer, the top output layer, and two hidden layers, are used for SCADA data modeling.

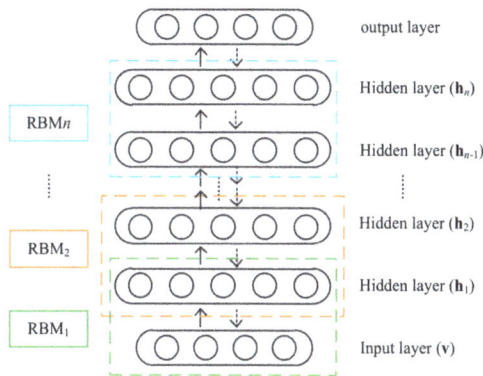

Figure 4. Structure of DBN with n hidden layers.

3.4.2. ODBN Method

It is worth noting that although DBNs can enhance the performance of prediction to some extent, the performance is highly influenced by their structural parameters. Therefore, how to determine and achieve optimal model parameters has become a primary challenge for DBNs. The main parameters of concern in this paper are the number of neurons in the two hidden layers of the DBN. In general, these parameters can be obtained experimentally, but it is time-consuming and laborious. CSO is a novel bionic heuristic optimization algorithm that mimics the hierarchy in the chicken swarm and the food search behavior [41]. The optimization algorithm obtains the optimal parameters by dividing the chicken swarm into several subgroups and competing among different subgroups. Therefore, this algorithm is used for the adaptive optimization of the model parameters. The position of each individual in the chicken swarm represents a potential solution to the optimization problem. There are three types of chickens in the CSO: roosters, hens, and chicks. To search for the optimal solution in the search space, it is necessary to update the position of each type of chicken. The position update equation for the roosters is depicted as follows:

$$x_{i,j}^{t+1} = x_{i,j}^t \cdot (1 + \text{Randn}(0, \sigma^2)), \tag{9}$$

$$\sigma^2 = \begin{cases} 1, & \text{if } f_i \leq f_k \\ \exp(\frac{(f_k - f_i)}{|f_i| + \varepsilon}), & \text{otherwise} \end{cases}, \tag{10}$$

where t is the number of iterations; $\text{Randn}(0, \sigma^2)$ is a Gaussian distribution with mean 0 and standard deviation σ^2; $i, k \in [1, \text{rsize}]$ and $i \neq k$, rsize represent the number of roosters; f_k and f_i are the fitness values of rooster particles i and k, respectively; and ε is a constant that is small enough. The position update equation for hens is as follows:

$$x_{i,j}^{t+1} = x_{i,j}^t + S1 \cdot \text{Rand} \cdot (x_{r1,j}^t - x_{i,j}^t) + S2 \cdot \text{Rand} \cdot (x_{r2,j}^t - x_{i,j}^t), \tag{11}$$

$$S1 = \exp(\frac{(f_i - f_{r1})}{|f_i| + \varepsilon}), \tag{12}$$

$$S2 = \exp(f_{r2} - f_i), \tag{13}$$

where Rand is a uniform number over [0,1]; r_1 is an index of the rooster, which is the ith hen's group mate; and r_2 is a randomly chosen index of a chicken (rooster or hen) from the swarm, and $r_1 \neq r_2$. The position update equation for chicks is as follows:

$$x_{i,j}^{t+1} = x_{i,j}^t + FL \cdot (x_{m,j}^t - x_{i,j}^t), \tag{14}$$

where FL refers to a parameter that means the chick would follow its mother to forage for food, and the range is [0, 2]; and $x_{m,j}^t$ is the position of the ith chick's mother. In this paper, the position of each individual represents the number of neurons in the two hidden layers of the DBN. Figure 5 gives the flowchart of ODBN with CSO, and the detailed procedure is explained as follows:

(1) Initialize the parameters, including number of chickens, dimensions of individual positions, maximum iteration number, updated frequency of chicken swarm, and proportions of roosters, hens, and mother hens.

(2) Randomly produce an initial population of chickens. Train the DBN and compute the fitness values, and determine the optimal individual and global fitness values and corresponding positions. Here, the root mean square error (RMSE) of the validation set is considered as the fitness function.

(3) In the next iteration, first determine the relationship between the roosters, hens, and chicks in a group, and then update their positions according to Equations (9)–(14) and calculate their fitness values. Next, update the optimal individual and global fitness values and their corresponding positions.

Repeat step 3 until the maximum iteration number is reached and output the optimal parameters of the DBN. Note that the original CSO is a method for optimizing continuous values. Since the number of neurons in hidden layers of the DBN is an integer, the CSO that optimizes the continuous values is not applicable. Hence, when initializing the population of chickens and updating the positions of roosters, hens, and chicks, discretize them to meet the requirements.

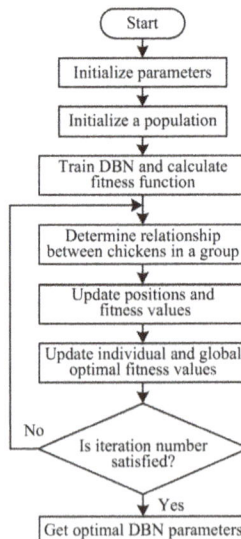

Figure 5. Flowchart of optimized DBN with chicken swarm optimization (CSO).

3.5. Anomaly Detection Approach

The purpose of establishing ODBN normal behavior models is to continuously monitor the working status of the components being modeled and identify impending faults in a timely manner, which is important to avoid major faults of components and ensure the secure and stable operation of wind turbines. In this paper, an advanced health monitoring approach called MD is utilized for operational state monitoring and abnormal behavior identification. MD has been successfully adopted in the detection of abnormalities of wind turbines [14,15].

MD is a unitless distance measurement that can capture the correlation of variables in a process or system and is defined as follows:

$$MD_i = \sqrt{(X_i - \mu)C^{-1}(X_i - \mu)^T}, \ i = 1, 2, \ldots, n, \tag{15}$$

where $X_i = [X_{i1}, X_{i2}, \ldots, X_{im}]$ is the ith observation vector, n is the number of observations, m is the total number of parameters, $\mu = [\mu_1, \mu_2, \ldots, \mu_m]$ is the vector of mean values, C is the covariance matrix, and MD_i is the MD value for the ith vector X_i.

For health monitoring, the MD values for the validation set are used to calculate the threshold for anomaly detection. During the validation stage, wind turbines are in normal operation and no abnormal behavior occurs. The MD for the validation set can be expressed as follows:

$$MD_{refi} = \sqrt{(X_{refi} - \mu_{ref})C_{ref}^{-1}(X_{refi} - \mu_{ref})^T}, \tag{16}$$

where $X_{refi} = [VE_i, TV_i]$ represents the ith vector; TV_i denotes the ith target value during the validation stage and VE_i is the corresponding validation error; μ_{ref} and C_{ref} are the mean value vector and the covariance matrix of X_{ref}, respectively. MD_{refi} refers to the MD value for the ith vector X_{refi}.

After obtaining the healthy MD values in the validation stage, the anomaly detection threshold can be determined by fitting a two-parameter Weibull probability distribution function on these MD values [42]. The two-parameter Weibull distribution is described as

$$f(t) = \beta\eta^{-\beta}(t)^{\beta-1}e^{-\left(\frac{t}{\eta}\right)^\beta}, \tag{17}$$

where β denotes the shape parameter and η stands for the scale parameter.

The MD during the condition monitoring stage is depicted as follows:

$$MD_{newi} = \sqrt{(X_{newi} - \mu_{ref})C_{ref}^{-1}(X_{newi} - \mu_{ref})^T}, \tag{18}$$

where $X_{newi} = [PE_i, MV_i]$, MV_i is the ith actual measured value from the SCADA system during the condition monitoring stage, and PE_i is the model prediction error.

In this study, in order to reduce the false alarm rate, the MD value from the condition monitoring stage is identified as an anomaly if $f(MD_{newi})$ is less than 0.1%. At this point, an alarm signal is triggered to alert the operators about the operational states of the turbine so they can take appropriate action to avoid major faults.

4. Case Study and Discussion

In this section, a real case for main bearings is investigated to demonstrate the feasibility of the proposed approach in practical applications of wind turbine health monitoring, and the results obtained in each part are presented in detail.

4.1. Data Description

The SCADA data used in this paper are from a wind farm located in Inner Mongolia, China. All wind turbines in the wind farm are variable speed constant frequency with a rated power of 1.5 MW. The sampling interval of the SCADA data is 30 s. Each record includes a total of 25 discrete pieces of information, such as turbine state, time stamp, yaw state, etc. At the same time, 49 continuous parameters are also recorded, listed in Table 1. The SCADA data for the majority of the turbines were available during the period from 1 July to 23 September 2014. In this paper, the SCADA data from 13 available turbines during the period 1 July to 31 August 2014 are investigated. Detailed descriptions of the datasets are listed in Table 2.

Table 1. Continuous parameters in SCADA data.

Continuous Parameter			
Gearbox oil temperature	Wind direction	Current phase C	Absolute wind direction
Gearbox front bearing temperature	Generator speed	Converter side speed	Blade 1 motor current
Gearbox inlet oil temperature	Gearbox speed 1	Converter side torque	Blade 2 motor current
Generator front bearing temperature	Wind speed 1	Wind speed 1 s average	Blade 3 motor current
Generator rear bearing temperature	Wind speed 2	Wind speed 1 min average	Blade 1 motor temperature
Generator stator winding temperature	Active power	Wind speed 10 min average	Blade 2 motor temperature
Converter ambient temperature	Reactive power	Ambient temperature	Blade 3 motor temperature
Gearbox rear bearing temperature	Wind speed	Main bearing temperature	Hub temperature
Wind direction 1 s average	Voltage phase A	Nacelle temperature	Cable winding angle
Wind direction 1 min average	Voltage phase B	Active power 1 s average	Generator torque
Wind direction 10 min average	Voltage phase C	Active power 1 min average	
Gearbox oil pump pressure	Current phase A	Active power 10 min average	
Gearbox inlet oil pressure	Current phase B	Hydraulic system pressure	

Table 2. Description of SCADA datasets.

Dataset	Time Stamps	Turbines Considered
Modeling	1/7/2014–31/8/2014	6, 17, 24, 33–34, 37, 49, 53, 88
Testing normal behavior	30/7/2014–2/8/2014	20
	14/8/2014–17/8/2014	46
Testing abnormal behavior	10/9/2014–14/9/2014	42
	2/7/2014–4/7/2014	13

To obtain a reliable health monitoring model of the main bearing, it is necessary to include as much data as possible to cover all normal operation regions of the turbine. Therefore, Turbines 6, 17, 24, 33–34, 37, 49, 53, 88 were randomly selected for modeling. During the period from 1 July to 31 August 2014, there were no main bearing faults in these 9 turbines, which are suitable for establishing the normal behavior model of the main bearing. The number of samples of normal SCADA data from the nine turbines in normal operation during this period is 353,131.

Similarly, Turbines 20 and 46 did not experience main bearing faults, so are used to test the performance of the normal behavior of wind turbines. Whereas Turbines 42 and 13 experienced main bearing over temperature faults, so are employed to detect the abnormal behavior of the main bearing.

4.2. Model Development

In this section, the proposed health monitoring model for wind turbine main bearings is developed in detail based on the above methods. To investigate the prediction performance of the proposed modeling method in different operating conditions, the traditional algorithms are used for comparison. Several methods are also compared without considering the operating characteristics of wind turbines.

4.2.1. Operation Condition Partition

For the SCADA data from nine turbines in normal operation from 1 July to 31 August 2014, described in Section 4.1, the operation parameters representing the wind turbine operating conditions should first be extracted. Normally, the normalization of operation data is a basic step before partitioning the operating condition depending on the operation parameters to ensure the reliability of the clustering results. In this paper, in order to be more consistent with the operating characteristics of wind turbines, the number of clusters is set from two to eight, and the calculation results of the silhouette values are shown in Figure 6. From Figure 6, it can be seen that when the number of clusters is two, the silhouette value reaches the maximum value of 0.75. This result indicates that it is optimal to segment the wind turbine operating conditions into two sub-conditions, condition 1 (C_1) and condition 2 (C_2).

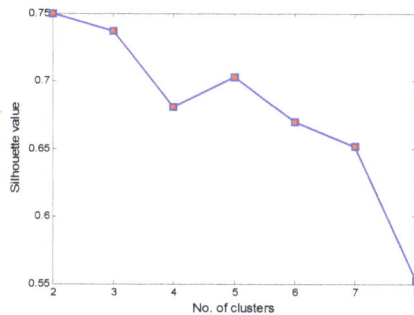

Figure 6. Calculation results of silhouette values.

To explicitly show the optimal clustering results obtained, principal component analysis is used to reduce high-dimensional operation data to a low-dimensional space for data visualization. Figure 7 shows the three-dimensional visualization of the operation parameters based on the first three principal components. It can be clearly observed that two separate operation condition spaces are presented, which proves the multiple condition characteristics of wind turbines. Table 3 summarizes the clustering distribution of operation parameters to further quantitatively understand the clustering results.

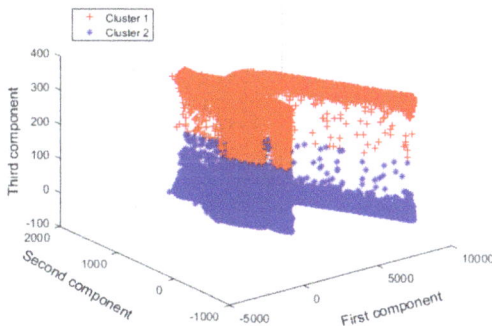

Figure 7. Optimal clustering results by k-means and silhouette index.

Table 3. Summary of clustering distribution.

Distribution	C_1	C_2
Wind speed (m/s)	0.3–29.28	0.3–25.33
Wind direction (°)	186.73–359	0.38–186.73
Ambient temperature (°C)	4.97–37.33	4.96–37.58
Generator speed (rpm)	0.17–1852.8	0.17–1859.2
Generator torque (N · m)	−970–8600	−970–8600

As can be seen in Table 3, as far as the wind direction is concerned, the ranges under the two conditions are obviously different. In the case of C_1, the range is 186.73 to 359, whereas in C_2, the range is 0.38 to 186.73. In terms of wind speed, ambient temperature, generator speed, and generator torque, there is little difference in the ranges under the two conditions. One can see from the comparison results that the wind turbine operating conditions are clearly partitioned according to this operation parameter, i.e., wind direction, and thus this parameter can be used for subsequent real-time condition recognition purposes. However, it is well known that wind direction ranges from 0° to 360°, so here, if the wind direction is between 359°–360° and 0°–0.38°, they will be automatically categorized as C_1 and C_2 separately. It should be noted that there is no theoretical (i.e., no mechanical or electrical basis) reasoning for the choice of wind direction as the partitioning parameter here and that this based purely on the analysis of the clustering data.

In the following study, the original normal SCADA data from nine turbines are divided into two portions based on the above condition partition results, and the sample numbers under C_1 and C_2 are 154,089 and 199,042, respectively.

4.2.2. Parameter Selection for Each Condition Cluster

Before dealing with forecasting problems, the integration of Pearson, Spearman, and Kendall is used for variable selection. In this study, the main bearing temperature, closely related to the health of the main bearing, is taken as the target modeling variable for the output of each model. Meanwhile, state variables that are highly correlated with the main bearing temperature should be carefully considered. The correlation coefficients between the main bearing temperature listed in Table 1 and 48 other variables are calculated. There is no doubt that data preprocessing is required before calculating the correlation coefficients, including smoothing and normalization. Figure 8 shows the 10 most relevant variables for each method under the two conditions. For these three methods, it is essential to set a threshold according to the computation results, as shown in Figure 8, and in this paper, variables whose absolute value of the correlation coefficient is greater than 0.5 are selected as the final modeling input.

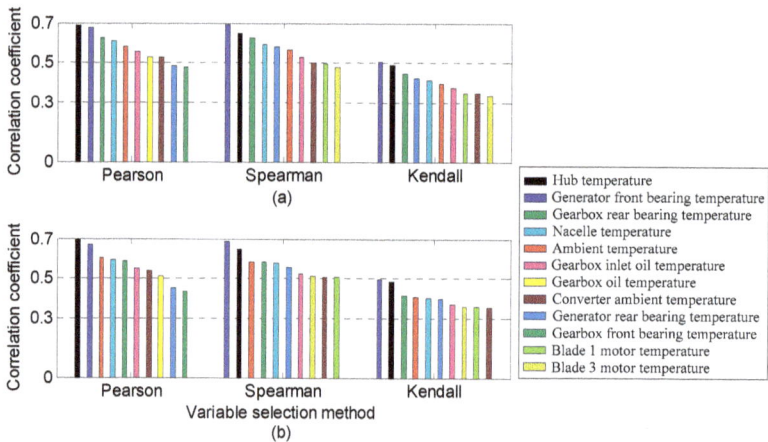

Figure 8. Correlation coefficients with different variable selection methods for (a) C_1 and (b) C_2.

As shown in Figure 8a, in C_1, nine state variables are regarded as V_1 to construct the prediction model: hub temperature, generator front bearing temperature, gearbox rear bearing temperature, nacelle temperature, ambient temperature, gearbox inlet oil temperature, gearbox oil temperature, converter ambient temperature, and generator rear bearing temperature. As can be seen in Figure 8b, in addition to the nine input variables under C_1, blade 1 and 3 motor temperatures also meet the set threshold requirements, so 11 variables are considered to be V_2 to develop the regression model under C_2. Moreover, it is not difficult to see in Figure 8 that although the Kendall technique produces relatively small values under these two conditions, a similar ordering is generated compared to the first two approaches.

4.2.3. Performance Evaluation and Comparison

In this subsection, ODBNs are employed to develop a normal behavior model of each operating condition. In this study, four-layer DBNs are used for model development. To obtain higher prediction accuracy, the CSO algorithm is used to optimize the number of neurons in two hidden layers adaptively.

Practically, a series of training parameters needs to be designed before establishing each normal behavior model. The detailed parameter settings are listed in Table 4. In each condition, the dataset is respectively divided into a training set, a validation set, and a testing set at a ratio of 80%, 10%, and 10%, respectively. The training set is utilized to train the DBN model, the validation set is applied to evaluate the performance of the model and optimize the fitness function, and the testing set is used for the final performance evaluation. Figure 9 shows the CSO optimization results under C_1 and C_2. The architecture of DBN_1 and DBN_2 is determined as 9-39-82-1 and 11-56-21-1, respectively.

Table 4. Description of parameter settings for modeling.

Description	Parameter Setting
DBN pretraining phase	size of batch training 100, training iterations 10, learning rate 1, momentum 0
DBN fine-tuning phase	size of batch training 10, training iterations 20
CSO for optimization	max iterations 20, dimension 2, population size 20, range of each dimension [1, 100], updated frequency of chicken swarm 10, proportions of roosters, hens, and mother hens 0.15, 0.7, 0.5

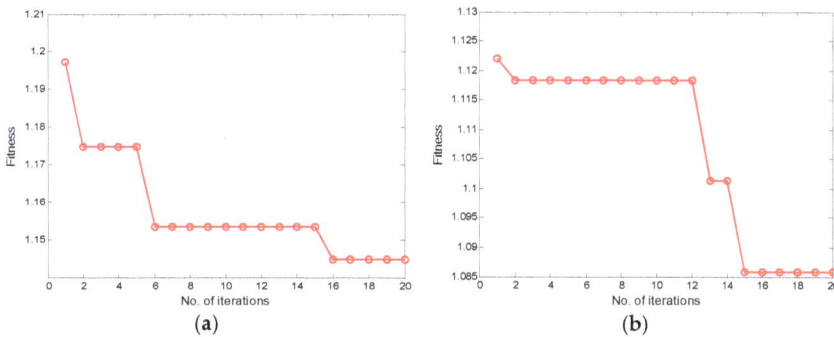

Figure 9. CSO optimization results under C_1 and C_2: (a) CSO-DBN_1 and (b) CSO-DBN_2.

In this study, to evaluate the prediction performance of the built model, four commonly used metrics, RMSE, mean absolute error (MAE), mean absolute percentage error (MAPE), and determination coefficient (R^2), are adopted, which are defined as follows [43]:

$$\text{RMSE} = \sqrt{\frac{1}{n}\sum_{i=1}^{n}(y_i - \hat{y}_i)^2}, \tag{19}$$

$$\text{MAE} = \frac{1}{n}\sum_{i=1}^{n}|y_i - \hat{y}_i|, \tag{20}$$

$$\text{MAPE} = \frac{1}{n}\sum_{i=1}^{n}\left|\frac{y_i - \hat{y}_i}{y_i}\right| \times 100, \tag{21}$$

$$R^2 = 1 - \frac{\sum\limits_{i=1}^{n}(y_i - \hat{y}_i)^2}{\sum\limits_{i=1}^{n}(y_i - \bar{y})^2}, \tag{22}$$

where y_i represents the ith measured value, $\hat{y}_i$ refers to the ith predicted value, and $\bar{y}$ is the mean value of the measurements.

Moreover, a back-propagation network with a single hidden layer (SHL-BP), a back-propagation network with double hidden layers (DHL-BP), and a support vector machine (SVM) are used for comparison. For SHL-BP and DHL-BP networks, the number of neurons in hidden layers is also optimized with CSO. For SVM, the radial basis kernel function is used to train the SVM model, and the kernel parameters and penalty factors are obtained by the cross-validation method. The quantitative evaluation results of the four models with three datasets under C_1 and C_2 are displayed in Table 5.

Table 5. Comparison of prediction results. SHL-BP, back-propagation network with single hidden layer; DHL-BP, back-propagation network with double hidden layers; SVM, support vector machine; RMSE, root mean square error; MAE, mean absolute error; MAPE, mean absolute percentage error.

Dataset	Criteria	C_1				C_2			
		$SHL-BP_1$	$DHL-BP_1$	SVM_1	DBN_1	$SHL-BP_2$	$DHL-BP_2$	SVM_2	DBN_2
Training	RMSE	1.9274	1.8044	1.8846	1.1463	1.8365	1.8325	1.9066	1.0858
	MAE	1.5034	1.4665	1.5361	0.8815	1.4444	1.4880	1.5696	0.8392
	MAPE (%)	0.6370	1.4872	0.7092	0.4393	0.3451	0.3097	0.9101	1.0432
	R^2	0.7095	0.7454	0.7223	0.8973	0.7345	0.7356	0.7138	0.9072
Validation	RMSE	1.8908	1.8014	1.8898	1.1448	1.8333	1.8191	1.9108	1.0859
	MAE	1.4865	1.4627	1.5467	0.8822	1.4428	1.4731	1.5731	0.8329
	MAPE (%)	0.6054	1.4638	0.7382	0.3806	0.2916	0.3384	0.9469	0.9889
	R^2	0.7219	0.7476	0.7222	0.8980	0.7348	0.7388	0.7119	0.9069
Testing	RMSE	1.9255	1.7982	1.8869	1.1607	1.8188	1.8198	1.9095	1.0790
	MAE	1.4941	1.4625	1.5390	0.8836	1.4319	1.4787	1.5694	0.8335
	MAPE (%)	0.6141	1.5531	0.7079	0.4485	0.3051	0.3315	0.9274	0.9726
	R^2	0.7118	0.7486	0.7232	0.8953	0.7394	0.7391	0.7127	0.9083

It can be found from Table 5 that in the forecasting performance of C_1 for the training set, validation set, and testing set, DBN_1 is better than the $SHL-BP_1$, $DHL-BP_1$, and SVM_1 models, as it offers the lowest RMSE, MAE, and MAPE and highest R^2 values. In terms of the prediction results in C_2, DBN_2 produces the lowest RMSE and MAE values and the highest R^2 values in the three datasets, whereas MAPE is slightly higher than the other three models. As indicated from quantitative evaluation results, the ODBNs generally get a higher modeling accuracy than the three traditional methods. The main reason is that the SHL-BP network is based on the principle of empirical minimization, which is prone to fall into local minima during the training process and thus produces poor results. At the same time, because it is difficult to train the depth structure effectively with the BP algorithm, the prediction accuracy of the DHL-BP model is not much different from that of SHL-BP model. The SVM algorithm also obtains poor prediction results because it is not suitable for large-scale training samples, whereas ODBNs can deeply learn and uncover the sophisticated nonlinear relationships among modeling variables by establishing a depth model, which results in better prediction accuracy. Hence, the proposed ODBN approach is used for real-time health monitoring of main bearings under varying operating conditions.

Additionally, to evaluate the monitoring performance of the proposed multioperation condition framework, the same 353,131 samples are analyzed without considering the wind turbine operating characteristics. Similarly, the three variable selection methods mentioned in Section 3.3 are employed to select the modeling input variables. A total of 10 variables (gearbox oil temperature, gearbox inlet oil temperature, generator front bearing temperature, generator rear bearing temperature, converter ambient temperature, gearbox rear bearing temperature, ambient temperature, nacelle temperature, blade 1 motor temperature, and hub temperature) are selected. After that, the samples are split into the training set, validation set, and testing set, and four models are deployed to capture the normal behavior of the main bearings. Note that the division of the dataset and the optimization of the model parameters adopt the same way of considering the multi-condition operating characteristics. The prediction performances of the four models are summarized in Table 6.

As shown in Table 6, in terms of the training set, validation set, and testing set, the MAPE generated by the DBNs is slightly higher than the SHL-BP but lower than the DHL-BP and SVM models. Furthermore, the DBNs perform better with the lowest RMSE and MAE and highest R^2 values. As the results indicate, the ODBNs achieve the best prediction performance compared to the other three conventional models, illustrating the predominance of DBN method in modeling. Thus, it is deemed to be the more appropriate model for monitoring the main bearing temperature.

Table 6. Comparison results of four models without clustering.

Dataset	Evaluation Criteria	Model			
		SHL-BP	DHL-BP	SVM	DBNs
Training	RMSE	1.9784	1.8538	1.8929	0.9615
	MAE	1.5959	1.4438	1.5574	0.7310
	MAPE (%)	0.0717	0.3721	0.7373	0.2933
	R^2	0.6930	0.7305	0.7190	0.9275
Validation	RMSE	1.9737	1.8477	1.8879	0.9509
	MAE	1.5910	1.4397	1.5518	0.7242
	MAPE (%)	0.0907	0.3959	0.7543	0.2984
	R^2	0.6941	0.7319	0.7201	0.9290
Testing	RMSE	1.9680	1.8470	1.8857	0.9536
	MAE	1.5852	1.4366	1.5518	0.7251
	MAPE (%)	0.0917	0.4039	0.7468	0.3066
	R^2	0.6955	0.7318	0.7205	0.9285

4.3. Health Monitoring Results

The examples given in this section are real wind turbine events from a wind farm recorded by SCADA systems. The MD is constructed to monitor the operating states of each wind turbine. The best performance algorithm for each condition is chosen to demonstrate the advantages of the proposed framework, and the monitoring performances are also compared with the best model considering only the single operating condition.

4.3.1. Testing Normal Wind Turbine Behavior

In this subsection, the proposed approach is used to analyze the normal behavior of wind turbines. Turbines 20 and 46 were in normal operation during 30 July to 2 August 2014 and 14–17 August 2014, respectively. The available historical SCADA data were collected and preprocessed for testing. The ODBN model was conducted for comparison. The prediction results with the two models are presented in Table 7. The computational cost is also recorded in Table 7, with the computation environment Intel Core i7 CPU @1.73 GHz, and 8.00 GB memory.

Table 7. Comparison of prediction results between two models.

Turbine	Model	Evaluation Criteria				Time (s)
		RMSE	MAE	MAPE (%)	R^2	
20	K-means based ODBNs	0.8125	0.6621	0.3725	0.9067	126.541
	ODBNs	0.8566	0.7021	0.7172	0.8963	4.308
46	K-means based ODBNs	1.5225	1.3038	4.1577	0.6272	124.126
	ODBNs	1.8409	1.5830	5.3136	0.4550	4.622

From Table 7, one can see that the models considering the operating condition characteristic produce lower RMSE, MAE, and MAPE and higher R^2 values for the two turbines than the model without this characteristic, which illustrates the superiority of the operating condition partition. In view of the better prediction performance of the proposed method, the loss of computational cost is acceptable. The condition monitoring results are displayed in Figures 10 and 11.

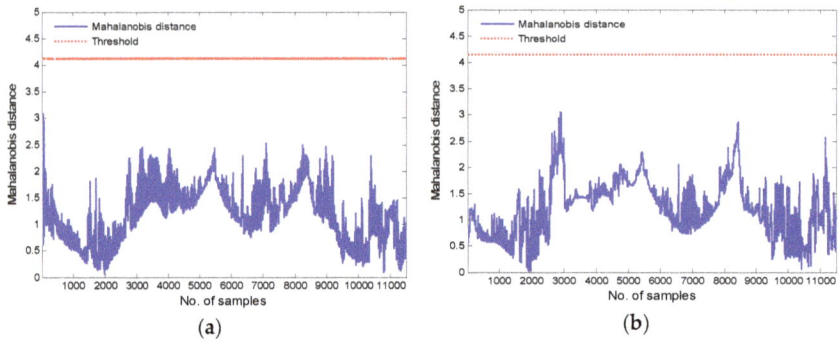

Figure 10. Condition monitoring results for Turbine 20: (**a**) K-means–based ODBNs and (**b**) ODBNs.

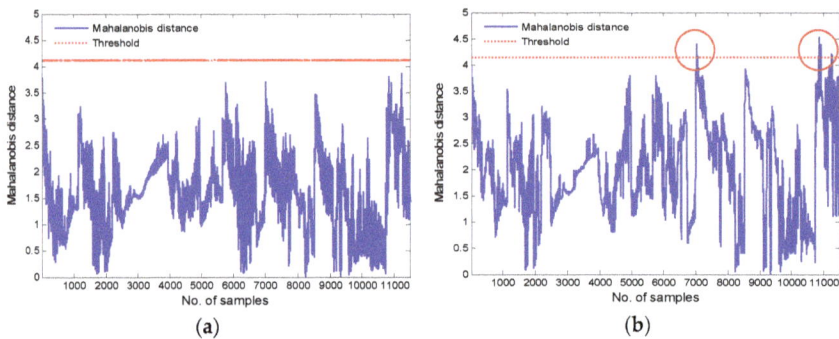

Figure 11. Condition monitoring results for Turbine 46: (**a**) K-means based ODBNs and (**b**) ODBNs.

As can be seen from Figure 10a,b, all MD values of Turbine 20 fall within the detection thresholds, indicating that both methods can precisely capture normal wind turbine behavior. All MD values in Figure 11a are within the detection thresholds, whereas outliers are detected near points 7000 and 11,000 in Figure 11b. According to the monitoring results in Figure 11, the k-means based ODBN approach successfully monitors the operating state of Turbine 46 without false alarms, whereas the ODBN model produces false alarms. Hence, it can be concluded that the condition monitoring capability of the ODBN considering the operating condition feature is generally superior to that of the ODBN without that feature, and the fault thresholds under the two conditions, that is, MD_1 and MD_2 are 4.122 and 4.127, respectively. Since the two values in this study are not much different, they are approximated as straight in Figures 10a and 11a.

4.3.2. Detecting Abnormal Main Bearing Behavior

To further verify the effectiveness of the proposed approach in detecting the abnormal behavior of wind turbine main bearings, Turbines 42 and 13 are utilized for investigation. According to the SCADA records of Turbine 42, the main bearing over temperature fault occurred at 11:25 on 14 September 2014. The 11,324 SCADA samples from 12:50 on 10 September to 11:25 on 14 September before the event occurred are applied for anomaly detection. For Turbine 13, the 5855 SCADA samples from 00:00 on 2 July to 07:48 before the main bearing over temperature fault happened at 07:48 on 4 July are used for analysis. The ODBN model is used for comparison. The forecasting results with the two models are displayed in Figures 12 and 13.

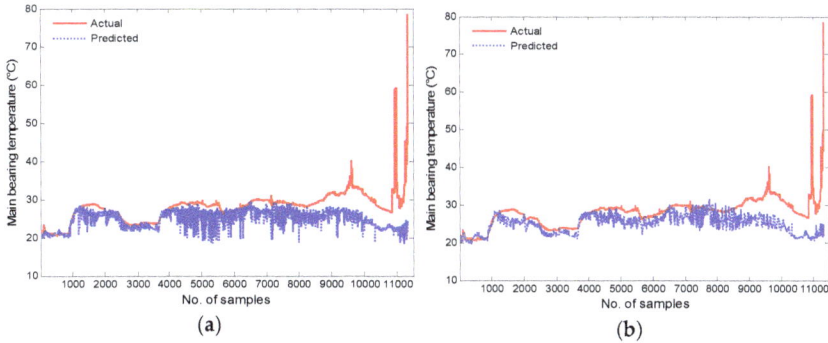

Figure 12. Forecasting results for Turbine 42: (**a**) K-means based ODBNs and (**b**) ODBNs.

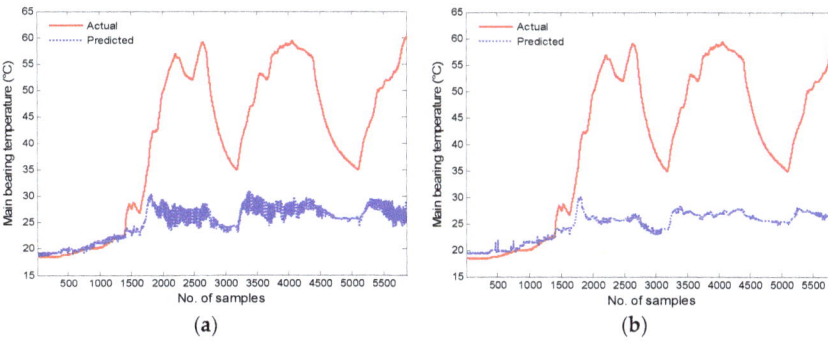

Figure 13. Forecasting results for Turbine 13: (**a**) K-means based ODBNs and (**b**) ODBNs.

Figures 12 and 13 demonstrate the trends of actual and predicted main bearing temperatures of Turbines 42 and 13 separately. It can be observed that the prediction errors between actual measured values and predicted values of the two models distinctly increased before the main bearing over temperature faults occurred, which indicates that the health conditions of the main bearings varied. The condition monitoring results are shown in Figures 14 and 15.

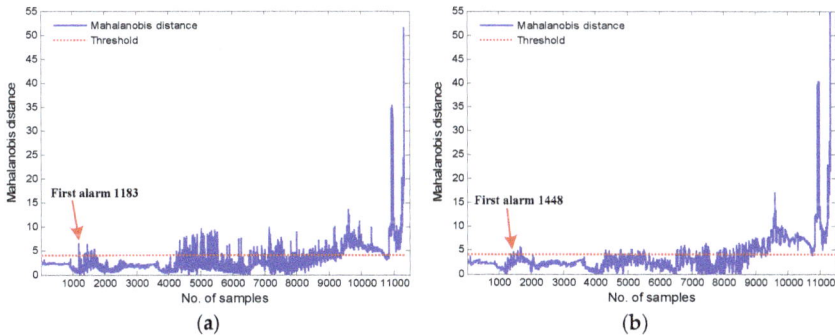

Figure 14. Condition monitoring results for Turbine 42: (**a**) K-means based ODBN and (**b**) ODBN.

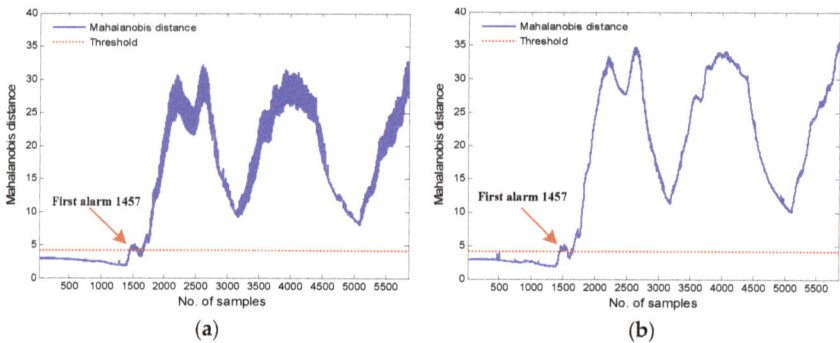

Figure 15. Condition monitoring results for Turbine 13: (**a**) K-means based ODBN and (**b**) ODBN.

Figure 14a displays the monitoring results of the proposed framework for Turbine 42, while Figure 14b illustrates the monitoring results of the ODBN model in the same turbine. As shown in Figure 14a, the MD value first crosses the fault threshold at sample point 1183, and an incipient main bearing fault is detected. However, the actual fault occurred at 11:25 on 14 September 2014, which is the 11,325th sample point. Each sample point is separated by an interval of 30 s, so the k-means based ODBN approach can detect the main bearing fault approximately 84.5 h in advance. As can be seen from Figure 14b, the MD value first exceeds the threshold at sample point 1448, and an early alarm is signaled, thus predicting the over temperature event almost 82.3 h ahead of the actual fault. From the computation results in Figure 14, both models can identify the upcoming main bearing fault successfully. Nevertheless, the proposed k-means based ODBN model is able to detect the anomaly of the main bearing nearly 2.2 h earlier than the ODBN approach. At the same time, it also means that there is only a 2.2 h improvement in using the k-means based ODBN model over the ODBN model (82.3 vs. 84.5 h), which represents only a 2.6% improvement. Still, this can also provide more time for wind farm operators to take appropriate measures. If the abnormal main bearing can be repaired and replaced in time, possible major accidents and unnecessary maintenance costs and downtime can be avoided. The computational time of k-means based ODBN and ODBN methods during the condition monitoring is 129.596 s and 4.748 s, respectively.

Figure 15a,b plot the monitoring results of the k-means–based ODBN and ODBN models, respectively, for Turbine 13. From Figure 15a, one can see that based on the fault thresholds, the alarm is first issued at sample point 1457, while the actual fault happened at 07:48 on 4 July 2014, which is the 5856th sample point. Therefore, the proposed approach can detect the fault almost 36.6 h in advance. As shown in Figure 15b, the MD value also first crosses the fault threshold at point 1457, and an early alarm is triggered. According to the computation results in Figure 15, one can conclude that the two models are able to simultaneously identify the anomaly of the main bearing nearly 36.6 h in advance, and the computational time of k-means-based ODBN and ODBN methods during the condition monitoring is 2.567 s and 68.046 s, respectively.

Based on the application analysis of the above real case, the proposed framework has certain advantages in the real-time health monitoring of wind turbines, which can mainly be attributed to the integration of multioperation condition monitoring and deep feature characterization. Meanwhile, the computational time is relatively high due to the complex procedures of the proposed divide-and-conquer strategy, but within acceptable limits. It is worth noting that the study is limited by the use of only three months of SCADA data and that the results are not valid for longer time periods.

5. Conclusions

A health monitoring method for wind turbine operational states has been proposed to consider the dynamic operating conditions of wind turbines and address the difficulty in accurately building normal behavior models. In the proposed approach, on the one hand, considering the multiple operating characteristics of wind turbines, a general multioperation condition partition scheme based on the k-means clustering method was utilized to segment the whole operation into multiple sub-operation conditions. One the other hand, ODBNs were applied to construct a healthy prediction model in each condition cluster where model parameters are optimally selected by the CSO algorithm. Compared with the conventional back-propagation and support vector machine models, the optimized modeling method can achieve higher prediction accuracy due to its deep feature representation capability.

A case for wind turbine main bearings was used to verify the effectiveness of the proposed health monitoring framework by real SCADA data. Compared with the ODBN model without considering the operating characteristics, the proposed framework does not generate false alarms under the normal conditions of wind turbines. In addition, both models are capable of detecting the anomalies of wind turbine main bearings in advance. Specifically, the proposed method could detect the faults either sooner, although only a 2.6% improvement, or at the same time. The results of normal and abnormal behavior testing demonstrate that the proposed approach generally achieves more effective and reliable detection accuracy.

This study also brings a loss of computational cost while having good detection performance. Also, due to data constraints, the study is limited by the use of only three months of SCADA data and the results are not valid for longer time periods. In future work, our developed framework will also be used for long-term monitoring of the main bearing operational states when more available SCADA data are collected. Other optimization methods will be used for the main bearing health monitoring. In addition, the approach will be extended to the application of other components in wind turbines.

Author Contributions: H.W. (Hong Wang) proposed the main idea for the paper and prepared the manuscript. H.W. (Hongbin Wang) supervised the paper and reviewed the manuscript. G.J. made suggestions for the manuscript and revised and reviewed it thoroughly. J.L. and Y.W. helped with reviewing the manuscript.

Acknowledgments: This work was supported by the National Natural Science Foundation of China (grant No. 61473248 and 61803329), the Natural Science Foundation of Hebei Province of China (grant No. F2016203496), and the Key Project of Natural Science Foundation of Hebei Province of China (grant No. F2018203413).

Conflicts of Interest: The authors declare no conflict of interest.

References

1. *Global Wind Statistics 2017*; Global Wind Energy Council (GWEC): Brussels, Belgium, 2018.
2. Takoutsing, P.; Wamkeue, R.; Ouhrouche, M.; Slaoui-Hasnaoui, F.; Tameghe, T.A.; Ekemb, G. Wind turbine condition monitoring: State-of-the-art review, new trends, and future challenges. *Energies* **2014**, *7*, 2595–2630.
3. Jiang, G.Q.; He, H.B.; Xie, P.; Tang, Y.F. Stacked multilevel-denoising autoencoders: A new representation learning approach for wind turbine gearbox fault diagnosis. *IEEE. Trans. Instrum. Meas.* **2017**, *66*, 2391–2402. [CrossRef]
4. Antoniadou, I.; Manson, G.; Staszewski, W.J.; Barszcz, T.; Worden, K. A time–frequency analysis approach for condition monitoring of a wind turbine gearbox under varying load conditions. *Mech. Syst. Signal Process.* **2015**, *64*, 188–216. [CrossRef]
5. Zhu, J.D.; Yoon, J.M.; He, D.; Bechhoefer, E. Online particle-contaminated lubrication oil condition monitoring and remaining useful life prediction for wind turbines. *Wind Energy* **2015**, *18*, 1131–1149. [CrossRef]
6. Sanchez, P.; Mendizabal, D.; Gonzalez, K.; Zamarreno, C.R.; Hernaez, M.; Matias, I.R.; Arregui, F.J. Wind turbines lubricant gearbox degradation detection by means of a lossy mode resonance based optical fiber refractometer. *Microsyst. Technol.* **2016**, *22*, 1619–1625. [CrossRef]
7. Yang, W.X.; Court, R.; Jiang, J.S. Wind turbine condition monitoring by the approach of SCADA data analysis. *Renew. Energy* **2013**, *53*, 365–376. [CrossRef]
8. Kusiak, A. Break through with big data. *Ind. Eng.* **2015**, *47*, 38–42.
9. Zaher, A.; McArthur, S.D.J.; Infield, D.J.; Patel, Y. Online wind turbine fault detection through automated SCADA data analysis. *Wind Energy* **2009**, *12*, 574–593. [CrossRef]
10. Guo, P.; Infield, D.; Yang, X.Y. Wind turbine generator condition-monitoring using temperature trend analysis. *IEEE Trans. Sustain. Energy* **2012**, *3*, 124–133. [CrossRef]

11. Kusiak, A.; Verma, A. Analyzing bearing faults in wind turbines: A data-mining approach. *Renew. Energy* **2012**, *48*, 110–116. [CrossRef]
12. Schlechtingen, M.; Santos, I.F.; Achiche, S. Wind turbine condition monitoring based on SCADA data using normal behavior models. Part 1: System description. *Appl. Soft Comput.* **2013**, *13*, 259–270. [CrossRef]
13. Schlechtingen, M.; Santos, I.F. Wind turbine condition monitoring based on SCADA data using normal behavior models. Part 2: Application examples. *Appl. Soft Comput.* **2014**, *14*, 447–460. [CrossRef]
14. Bangalore, P.; Tjernberg, L.B. An artificial neural network approach for early fault detection of gearbox bearings. *IEEE Trans. Smart Grid* **2015**, *6*, 980–987. [CrossRef]
15. Bangalore, P.; Letzgus, S.; Karlsson, D.; Patriksson, M. An artificial neural network-based condition monitoring method for wind turbines, with application to the monitoring of the gearbox. *Wind Energy* **2017**, *20*, 1421–1438. [CrossRef]
16. Bi, R.; Zhou, C.K.; Hepburn, D.M. Detection and classification of faults in pitch-regulated wind turbine generators using normal behaviour models based on performance curves. *Renew. Energy* **2017**, *105*, 674–688. [CrossRef]
17. Zhan, S.; Tao, Q.Q.; Li, X.H. Face detection using representation learning. *Neurocomputing* **2016**, *187*, 19–26. [CrossRef]
18. Wang, Y.; Wang, X.G.; Liu, W.Y. Unsupervised local deep feature for image recognition. *Inf. Sci.* **2016**, *351*, 67–75. [CrossRef]
19. Jiang, G.Q.; Xie, P.; He, H.B.; Yan, J. Wind turbine fault detection using a denoising autoencoder with temporal information. *IEEE ASME Trans. Mechatron.* **2018**, *23*, 89–100. [CrossRef]
20. Yang, Z.X.; Wang, X.B.; Zhong, J.H. Representational learning for fault diagnosis of wind turbine equipment: A multi-layered extreme learning machines approach. *Energies* **2016**, *9*, 379. [CrossRef]
21. Hinton, G.E.; Osindero, S.; Teh, Y.W. A fast learning algorithm for deep belief nets. *Neural Comput.* **2006**, *18*, 1527–1554. [CrossRef]
22. Chen, H.Z.; Wang, J.X.; Tang, B.P.; Xiao, K.; Li, J.Y. An integrated approach to planetary gearbox fault diagnosis using deep belief networks. *Meas. Sci. Technol.* **2017**, *28*, 025010. [CrossRef]
23. Wan, J.; Liu, J.F.; Ren, G.R.; Guo, Y.F.; Yu, D.R.; Hu, Q.H. Day-ahead prediction of wind speed with deep feature learning. *Int. J. Pattern Recognit. Artif. Intell.* **2016**, *30*, 1650011. [CrossRef]
24. Chen, Z.Y.; Li, W.H. Multisensor feature fusion for bearing fault diagnosis using sparse autoencoder and deep belief network. *IEEE Trans. Instrum. Meas.* **2017**, *66*, 1693–1702. [CrossRef]
25. Ren, H.; Chai, Y.; Qu, J.F.; Ye, X.; Tang, Q. A novel adaptive fault detection methodology for complex system using deep belief networks and multiple models: A case study on cryogenic propellant loading system. *Neurocomputing* **2018**, *275*, 2111–2125. [CrossRef]
26. Shao, H.D.; Jiang, H.K.; Zhang, X.; Niu, M.G. Rolling bearing fault diagnosis using an optimization deep belief network. *Meas. Sci. Technol.* **2015**, *26*, 115002. [CrossRef]
27. Unal, M.; Onat, M.; Demetgul, M.; Kucuk, H. Fault diagnosis of rolling bearings using a genetic algorithm optimized neural network. *Measurement* **2014**, *58*, 187–196. [CrossRef]
28. Qiao, W.; Lu, D.G. A survey on wind turbine condition monitoring and fault diagnosis-Part I: Components and subsystems. *IEEE Trans. Ind. Electron.* **2015**, *62*, 6536–6545. [CrossRef]
29. Kusiak, A.; Li, W.Y. The prediction and diagnosis of wind turbine faults. *Renew. Energy* **2011**, *36*, 16–23. [CrossRef]
30. Lapira, E.; Brisset, D.; Davari Ardakani, H.; Siegel, D.; Lee, J. Wind turbine performance assessment using multi-regime modeling approach. *Renew. Energy* **2012**, *45*, 86–95. [CrossRef]
31. Yang, H.H.; Huang, M.L.; Lai, C.M.; Jin, J.R. An approach combining data mining and control charts-based model for fault detection in wind turbines. *Renew. Energy* **2018**, *115*, 808–816. [CrossRef]
32. Macqueen, J. Some methods for classification and analysis of multivariate observations. In Proceedings of the 5th Berkeley Symposium on Mathematical Statistics and Probability, Berkeley, CA, USA, 21 June–18 July 1965; pp. 281–297.
33. Jain, A.K. Data clustering: 50 years beyond k-means. *Pattern Recognit. Lett.* **2010**, *31*, 651–666. [CrossRef]
34. Rousseeuw, P. Silhouettes: A graphical aid to the interpretation and validation of cluster analysis. *J. Comput. Appl. Math.* **1986**, *20*, 53–65. [CrossRef]

35. Mazidi, P.; Bertling Tjernberg, L.; Sanz Bobi, M.A. Wind turbine prognostics and maintenance management based on a hybrid approach of neural networks and a proportional hazards model. *Proc. Inst. Mech. Eng. Part O J. Risk Reliab.* **2017**, *231*, 121–129. [CrossRef]

36. Fisher, R.A. Statistical methods for research workers. *Int. J. Plant Sci.* **1954**, *21*, 340–341.

37. Best, D.J.; Roberts, D.E. Algorithm AS 89: The upper tail probabilities of Spearman's rho. *J. R. Stat. Soc.* **1975**, *24*, 377–379. [CrossRef]

38. Taylor, J.M.G. Kendall and Spearman correlation-coefficients in the presence of a blocking variable. *Biometrics* **1987**, *43*, 409–416. [CrossRef] [PubMed]

39. Hinton, G.E. Training products of experts by minimizing contrastive divergence. *Neural Comput.* **2002**, *14*, 1771–1800. [CrossRef] [PubMed]

40. Hinton, G.E. A practical guide to training restricted boltzmann machines. In *Neural Networks: Tricks of the Trade*; Springer: Berlin, Germany, 2012; Volume 7700, pp. 599–619.

41. Meng, X.B.; Liu, Y.; Gao, X.Z.; Zhang, H.Z. A new bio-inspired algorithm: Chicken swarm optimization. In *Swarm Intelligence*; Springer International Publishing: Berlin, Germany, 2014; Volume 8794, pp. 86–94.

42. Niu, G.; Singh, S.; Holland, S.W.; Pecht, M. Health monitoring of electronic products based on Mahalanobis distance and Weibull decision metrics. *Microelectron. Reliab.* **2011**, *51*, 279–284. [CrossRef]

43. Bai, Y.; Sun, Z.Z.; Zeng, B.; Deng, J.; Li, C. A multi-pattern deep fusion model for short-term bus passenger flow forecasting. *Appl. Soft Comput.* **2017**, *58*, 669–680. [CrossRef]

energies

MDPI

Article

GA-BP Neural Network-Based Strain Prediction in Full-Scale Static Testing of Wind Turbine Blades

Zheng Liu [1], Xin Liu [1,*], Kan Wang [2], Zhongwei Liang [1], José A.F.O. Correia [3,*] and Abílio M.P. De Jesus [3]

[1] School of Mechanical and Electrical Engineering, Guangzhou University, Guangzhou 510006, China; liu_best@yeah.net (Z.L.); lzwstalin@126.com (Z.L.)

[2] China General Certification Center, Beijing 100020, China; wkbjro@163.com

[3] INEGI, Faculty of Engineering, University of Porto, Porto 4200-465, Portugal; ajesus@fe.up.pt

* Correspondence: designer_liuxin@163.com (X.L.); jacorreia@inegi.up.pt (J.A.F.O.C.); Tel.: +351-966-559-442 (J.A.F.O.C.)

Received: 11 January 2019; Accepted: 4 March 2019; Published: 15 March 2019

Abstract: This paper proposes a strain prediction method for wind turbine blades using genetic algorithm back propagation neural networks (GA-BPNNs) with applied loads, loading positions, and displacement as inputs, and the study can be used to provide more data for the wind turbine blades' health assessment and life prediction. Among all parameters to be tested in full-scale static testing of wind turbine blades, strain is very important. The correlation between the blade strain and the applied loads, loading position, displacement, etc., is non-linear, and the number of input variables is too much, thus the calculation and prediction of the blade strain are very complex and difficult. Moreover, the number of measuring points on the blade is limited, so the full-scale blade static test cannot usually provide enough data and information for the improvement of the blade design. As a result of these concerns, this paper studies strain prediction methods for full-scale blade static testing by introducing GA-BPNN. The accuracy and usability of the GA-BPNN prediction model was verified by the comparison with BPNN model and the FEA results. The results show that BPNN can be effectively used to predict the strain of unmeasured points of wind turbine blades.

Keywords: wind turbine blade; full-scale static test; neural networks; strain prediction

1. Introduction

Wind turbine blades are one of the core force-bearing components of the wind turbine, and their stability and reliability directly affect the safety of the whole machine. Structural testing is a main way to check the rationality of the design and to verify the safety of manufacturing for turbine blades, and it is also a necessary means to ensure the operational reliability and safety of wind turbines [1]. The purposes of full-scale static testing of wind turbine blades are mainly to obtain two kinds of information from the blade by applying static loads to the blade. One is to verify the blade's ability under complex design loads, and another is to obtain structural characteristics, such as strain and deformation of the blade. Many studies have been conducted regarding blade structural testing. For example, Fagan et al. [2] presented an experimental testing on a 13-m long wind turbine blade and used the test results to calibrate finite element models, and then the materials used in the blade construction and manufacturing costs were reduced by optimization design using a genetic algorithm. Yang et al. [3] tested the limit loads in full-scale static testing of a wind turbine blade and the deformation situation of the blade under the limit loads. The test results can provide important technical parameters for the blade design. Pan [4] studied the effects of structural non-linearity on the full-scale static testing of wind turbine blades, and analyzed the relationship between bending moment, strain, stiffness, and deflection, and then provided more accurate stiffness data for a numerical model of load calculations for wind turbines. Thus, many achievements have been obtained in the blade load bearing capacity and parameter measurement methods, and some studies have also focused on structural characteristics and damage analysis of the blades [5–12]. Besides, some surrogate models for wind turbine blade stress/strain prediction due to the significant computational burden of physics-based simulation were constructed [13,14]. However, studies on the effects of the applied load, loading positions, and blade displacement on the test results are rare, and in the full-scale static testing of the wind turbine blade, the numbers of measuring points and strain gauges are also limited. Therefore, full-scale static testing only plays a role in the blade certification and has little significance to the blade design [4]. Blade strain is correlated to the applied load, loading position, displacement, etc. If the correlation is neglected, the test environment will have a huge deviation with the actual working conditions and the test result will be unreliable. The relation between blade strain and the applied load, loading position, displacement, etc. is non-linear, and the number of input variables is many, thus, the calculation and prediction of blade train are very complex. Back propagation neural networks (BPNNs) use an error back propagation algorithm to learn and adapt unknown information and has significant advantages in dealing with non-linear fitting and multi-input parameters. Yang et al. [15] showed that BPNNs have a good performance in solving non-linear problems by comparing with other methods in terms of absolute distance as a similarity measure. Moghaddam et al. [16] showed that BPNNs were good at solving non-linear and multi-input parameter problems. The methods of BPNNs were also introduced in the field of wind turbines. Huang et al. [17] applied the BPNN method to vibration fault diagnosis of wind turbine gearboxes and the accurate diagnostic results proved to be effective for analyzing the standard fault samples (training samples) and simulation samples (testing samples). Chen et al. [18] concluded that BPNNs can be effectively utilized to detect the incipient wind turbine faulty condition based on the data collected from wind farm supervisory control and data acquisition. Zhang et al. [19] presents an anomaly identification model for wind turbine state parameters by genetic algorithm back propagation neural network (GA-BPNN). But there are fewer studies about wind turbine blade structures. Liu [20] investigated BPNN control methods for divergent instability based on classical flutter of five degrees of freedom (DOF) wind turbine blade sections driven by pitch adjustment, and the obvious effects of fuzzy control and BPNN control are illustrated by numerical comparisons of vibration suppression from non-linear time response, amplitude of LCO, and frequency spectrum analysis.

Compared with traditional BPNN methods, BPNNs improved by genetic algorithm (GA-BPNNs) can find better initial weight and threshold and avoid falling into local optimal situations, which traditional BPNNs often meet. Moreover, the systems constructed by GA-BPNNs have better robustness and applicability in dealing with complex problems [21]. Therefore, taking the advantages of the neural network in dealing with non-linear fitting and multi-input parameters, a strain-predictive GA-BPNN model for the full-scale wind turbine blades static testing was established, and the strain value prediction of the unmeasured points was realized. The accuracy of the GA-BPNN prediction model was verified by comparison with BPNN model and FE analysis results. The applicability and usability of a neural network prediction model was verified by comparing the prediction results with the ANSYS simulation data. The study can provide the basis for the design and calibration of wind turbine blades.

The remainder of this article is designed as follows. In Section 2, the basic concepts of neural networks and the basic framework and algorithms of GA-BPNNs are introduced. The conditions and test procedures for the full-scale wind turbine blades static testing are introduced in Section 3. In Section 4, firstly, a strain-predictive GA-BPNN model was established for the center and trailing edge of the suction side based on a full-scale static testing. Secondly, the accuracy of the strain–predictive GA-BPNN model was verified by comparison with that of BPNN and FE analysis results. Then, the strain on the positions of the maximum chord length, the gravity center, and the root of the wind turbine blade are predicted. Finally, the reliability and applicability of the proposed model are proved by comparing with the ANSYS simulation result. Ultimately, conclusions are dawn in Section 5.

2. The Method of GA-BPNN

2.1. The Principles of BPNN

A relational model between the input set $\{X_j \mid j = 1,2, \ldots ,M\}$ and the dependent variable Y was established using the improved neural network algorithm. Sample $X_1, \ldots ,X_M$ was used as the input value and $Y_1, \ldots ,Y_L$ was the output value for training the dependent variable prediction model. The BPNN builds the network structure of the strain–prediction model for the full-scale wind turbine blade static testing, as shown in Figure 1.

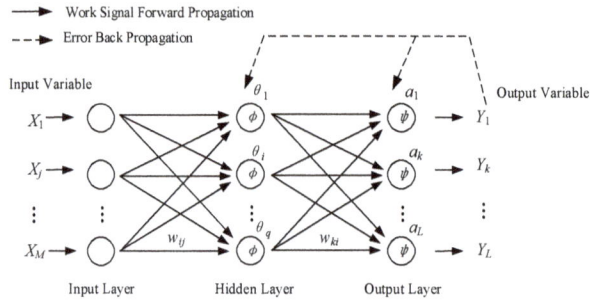

Figure 1. Back propagation neural network (BPNN) structure.

In Figure 1, X_i represents the input of the j-th node of the input layer, $j = 1, 2, \cdots , m$; w_{ij} represents the weight value between the i-th node of the hidden layer and the j-th node of the input layer; and θ_i represents the threshold value of the i-th node of the hidden layer. $\phi(x)$ represents the excitation function of the hidden layer; w_{ki} represents the weight between the k-th node of the output layer and the i-th node of the hidden layer, $i = 1, 2, \cdots , q$; a_k represents the threshold value of the k-th node of the output layer, $k = 1, 2, \cdots , l$; $\psi(x)$ represents the excitation function of the output layer; Y_k represents the output of the k-th node of the output layer. The data about the location of strain gauge, the loads with different percentage, and the displacement of loading positions were used as input data, and the data about strains and stresses were used as output data.

2.2. The Principles of GA-BPNN

Since the gradient descent method is used by BPNN algorithms, it is easy to fall into a situation of local optimization. Using a genetic algorithm to optimize the weight and threshold of BPNNs, which is improved by the Levenberg–Marquardt formula, can minimize the training error of the neural network, which can effectively avoid the training falling into a local optimization situation [22,23]. The weight and threshold of the BPNN are the chromosomes of the genetic algorithm. Each element of a chromosome is called a gene. The chromosomes with poor fitness values are eliminated, and the best genomes are selected to obtain the optimal solution by calculating the fitness values of each chromosome continuously. The method of BPNN improved on the basis of the genetic algorithm is shown in Figure 2 [24,25].

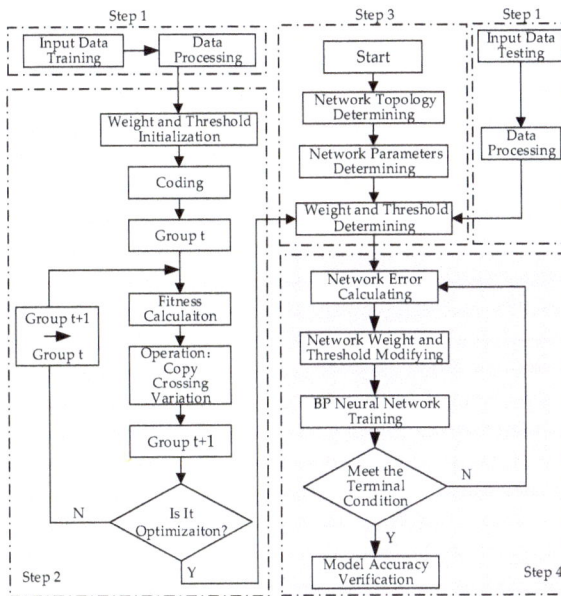

Figure 2. Flow chart of genetic algorithm (GA)-BPNN.

Step 1: Data processing. The input and output variables are determined. The input data are trained to speed-up network calculating.

Step 2: The weight and threshold are optimized. (1) The evolution numbers, population size, crossover probability, and mutation probability are initialized. (2) The network weight and threshold are encoded, and the fitness function, which is the reciprocal of the sum of errors squared is determined. (3) The selection operation: the chromosome with the fitness value "good" from the current population is selected as the parent. The higher the individual fitness value is, the greater the probability of the chromosome selected. The roulette method is used to select chromosomes. That is, a uniformly distributed random number is generated in [0, 1], and if $r \leq q_1$, the chromosome x_1 is selected. If $q_{k-1} < r \leq q_k$ ($2 \leq k \leq N$), the chromosome x_k is selected, and q_i is called the accumulation probability of chromosome x_i ($i = 1,2, \ldots ,n$), and its calculation formula is as shown in Equation (1). (4) Cross: two chromosomes are selected according to a certain probability, one or more points in the two chromosomes are exchanged with each other randomly to obtain two new chromosomes. (5) Variation: according to a certain mutation probability, in the binary coding of chromosomes, 1 becomes 0, and 0 becomes 1. This operation can effectively avoid premature convergence in the evolution process and thus falling into a local optimum. (6) Repeat steps (3), (4), and (5) until the number of evolutions is reached, then the optimal weights as well as the thresholds will be obtained.

Step 3, the BPNN model is built. The optimal initial weights and thresholds are obtained to construct the BPNN [25]. Any non-linear mapping can be realized by the three-layer BPNN in theory. The hidden-layer number, number of times, step size, and target of the BPNN are constructed. A tangent S-type transfer function as Equation (2) is used between the input layer and the hidden layer, while a linear transfer function as Equation (3) is used between the hidden layer and the output layer.

Step 4: the results are obtained by BPNN. The sample data are inputted into the BPNN model to predict the output data and then the output data are obtained if the results meet the terminal condition.

$$q_i = \sum_{j=1}^{i} P(x_j) \tag{1}$$

$$f(x) = \frac{1}{1 + e^{-x}} \qquad (2)$$

$$f(x) = kx \qquad (3)$$

3. Full-Scale Static Test of Wind Turbine Blades

3.1. The Wind Turbine Blade Specification

The full-scale static testing was conducted in cooperation with a certain blade company, and the testing result was used to verify the safety of the blade prototype, and was also used for further improvement. The testing process followed GB/T 25384-2010 [26].

The blade prototype was mainly made of fiber reinforced polymer. The blade had a mass of 15,968 kg and a natural frequency of 1.41 Hz. The maximum chord length was 3.8 m. The main elements of a wind turbine blade are shown in Figure 3.

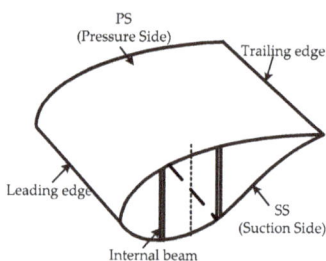

Figure 3. The structure of a wind turbine blade.

In the full-scale static testing, 56 strain gauges were attached to the surface of the wind turbine blade on the center of the pressure side (PS), the center of suction side (SS), the leading edge, and the trailing edge before testing. The locations and positions of strain gauges attached on the blade are shown in Figure 4.

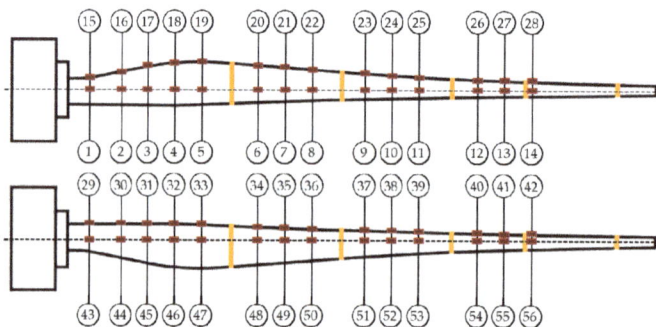

Figure 4. The locations and positions of strain gauges on the wind turbine blade.

3.2. Testing Procedure

The blade is fixed to the test platform by 64 bolts on the blade root, the limit loading was performed by pulling from one side, and the applied loading diagram is shown in Figure 5, where P1-P5 are the positions of the tensile machine, S1-S5 are the load application points. From Figure 5, we can see that the loading points were respectively arranged at a distance of 18.00 m, 30.00 m, 42.00 m, 50.00 m, and 60.00 m from the root of the blade, and the loading direction was perpendicular to the normal direction of the loading section.

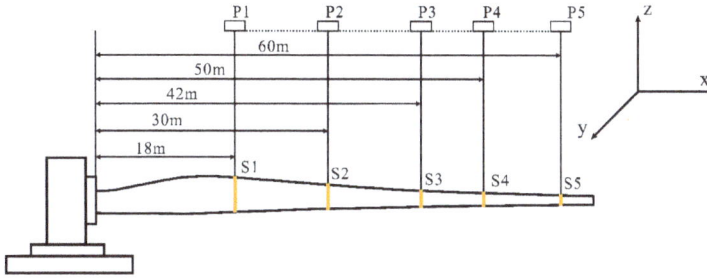

Figure 5. Applied loading diagram of the full-scale static testing.

As this paper aims at proposing a strain–predictive method, it only considers the situation of static testing in the flap+ direction as an example. The structure of every cross-section is different, thus their load bearing capacity is different. The target load of each loading point in the static testing was the design load which was obtained by finite element analysis (FEA) during the initial phase of design, and the applied target load had a certain deviation from the design value for existing equipment loading errors. In the direction of the flap+, the target load of each loading point is shown in Table 1. Following the test process of GB/T 25384-2010, the applied load, displacement, and strain of the blade were cleared before starting the test. Then, using the lateral loading device, the blade was loaded step by step according to 0%, 40%, 60%, 80%, and 100% of the target load, and the data was recorded. The load of each stage is shown in Table 2. The duration of each stage of load was not less than 10 s. After the loading was completed, the unloading was performed step by step, the blade load was unloaded to the zero state, and the displacement data and the strain gauge data were recorded during the loading process.

Table 1. The target load of each loading point.

Items	The Distance of Loading Positions from the Blade Root (m)				
	18.0	30.0	42.0	50.0	60.0
target load (kN)	94.6	143.0	59.8	104.4	68.0

Table 2. The applied load of each stage in flap+.

The Distance of Loading Positions from the Blade Root (m)	The Applied Load (kN)				
	0	40%	60%	80%	100%
18.00	0	37.85	57.48	75.76	94.74
30.00	0	57.42	86.17	114.60	143.15
42.00	0	24.09	35.85	48.16	60.15
50.00	0	42.34	62.75	83.04	104.67
60.00	0	27.25	40.98	54.51	68.07

4. GA-BPNN-Based Strain Prediction in Full-Scale Static Testing

4.1. GA-BPNN-Based Strain Prediction for the Center of Suction Side

During the loading process of the full-scale static testing, there was a non-linear mapping relationship between the strain and the applied load, loading positions, and displacements. The neural network, with its good learning method, can approximately express the non-linear mapping relationship between the above parameters through the establishment of the network model. Thereby, the strain of the blade is predicted. In the GA-BPNN model for strain prediction, the applied load, the loading positions, and the displacements were used as training inputs, and the strain of wind turbine blade was output. In the full-scale static test, there were 56 sets of data which all come from actual strain gauges, 50 set of data were trained to construct the NN models, while the remaining six sets of data were used to test the model accuracy, then a GA-BPNN model for wind turbine blade strain prediction was established. The training samples and test samples of the GA-BPNN are shown in Tables 3 and 4, respectively. The measurement method was up to standard. According to the test program, 14 strain gauges were arranged in the center of the suction side and the target load was imposed gradually by four steps with the duration of every step more than 10 s, thus 56 sets of data were obtained in the four different cases. Since the BPNN model needed enough training samples to ensure effectiveness, and the data used to test could not be selected as training data, the sample size of test data should try to be minimized without too much manual interference, so six samples were randomly selected as the test samples.

Table 3. The training samples of the GA-BPNN.

Items	The Location of Strain Gauge	Load (kN)					The Distance to the Blade Root (m)
		F1	F2	F3	F4	F5	l1
1	2000	37.85	57.42	24.09	42.34	27.25	0
2	6000	57.48	86.17	35.85	62.75	40.98	1.16
3	9000	75.76	114.6	48.16	83.04	54.51	5.69
...	...	...	...	...	...	...	...
...	...	...	...	...	...	...	...
50	48,000	75.76	114.6	48.16	83.04	54.51	5.69

Items	The Displacement of Loading Positions (m)						
	s1	s2	s3	s4	s5	s6	s7
1	159	755	2159	3717	6384	7900	−1186
2	264	1158	3257	5596	9605	11,876	−1898
3	368	1557	4350	7469	12,773	15,770	−2383
...	...	...	...	...	...	...	...
...	...	...	...	...	...	...	...
50	368	1557	4350	7469	12,773	15,770	−2971

Table 4. The test samples of the GA-BPNN.

Items	The Location of Strain Gauge	Load (kN)					The Distance to the Blade Root (m)
		F1	F2	F3	F4	F5	l1
1	2000	57.48	86.17	35.85	62.75	40.98	1.16
2	15,000	75.76	114.6	48.16	83.04	54.51	5.69
3	24,000	37.85	57.42	24.09	42.34	27.25	0
4	36,000	94.74	143.15	60.15	104.67	68.07	8.51
5	51,000	75.76	114.6	48.16	83.04	54.51	5.69
6	33,000	57.48	86.17	35.85	62.75	40.98	1.16

Items	The Displacement of Loading Positions (m)						
	s1	s2	s3	s4	s5	s6	s7
1	264	1158	3257	5596	9605	11,876	−1790
2	368	1557	4350	7469	12,773	15,770	−2748
3	159	755	2159	3717	6384	7900	−1964
4	477	1976	5490	9405	16,019	19,741	−4849
5	368	1557	4350	7469	12,773	15,770	−2490
6	264	1158	3257	5596	9605	11,876	−3047

The specific procedure of the GA-BPNN is set as follows: The input dimension is 13, and the output dimension is 1. Seven neurons are set in the hidden layer, and a tangent S-type transfer function such as Equation (2) was used between the input layer and the hidden layer. A linear transfer function such as Equation (3) between the hidden layer and the output layer was used. The network maximum number of training steps was 2000 steps, the network learning rate was six, the momentum factor is one, the training target was allowed to have a minimum convergence error of 1e-3, and the training result were displayed at intervals of 50 steps. The learning process of the training samples was simulated. Set the genetic algorithm population size to 1800 and the genetic iteration to 200. Call the GAOT which is the genetic algorithm toolbox in MATLAB and get the predicted correlation values. In order to verify the accuracy and validity of the GA-BPNN, traditional BPNN was also used to predict the strain. The specific procedure of the BPNN was set as follows: The input dimension was 16, and the output dimension was 1. There were seven neurons in the hidden layer, with a tangent S-type transfer function between the input layer and the hidden layer. In addition, a linear transfer function between the hidden layer and the output layer was used. The network maximum number of training steps was 2000 steps, the network learning rate was two, the training target was allowed to have a minimum convergence error of 1e-3, and the training result was displayed at intervals of 50 steps. The learning process of the training samples was simulated. The comparison results are shown as Figure 6.

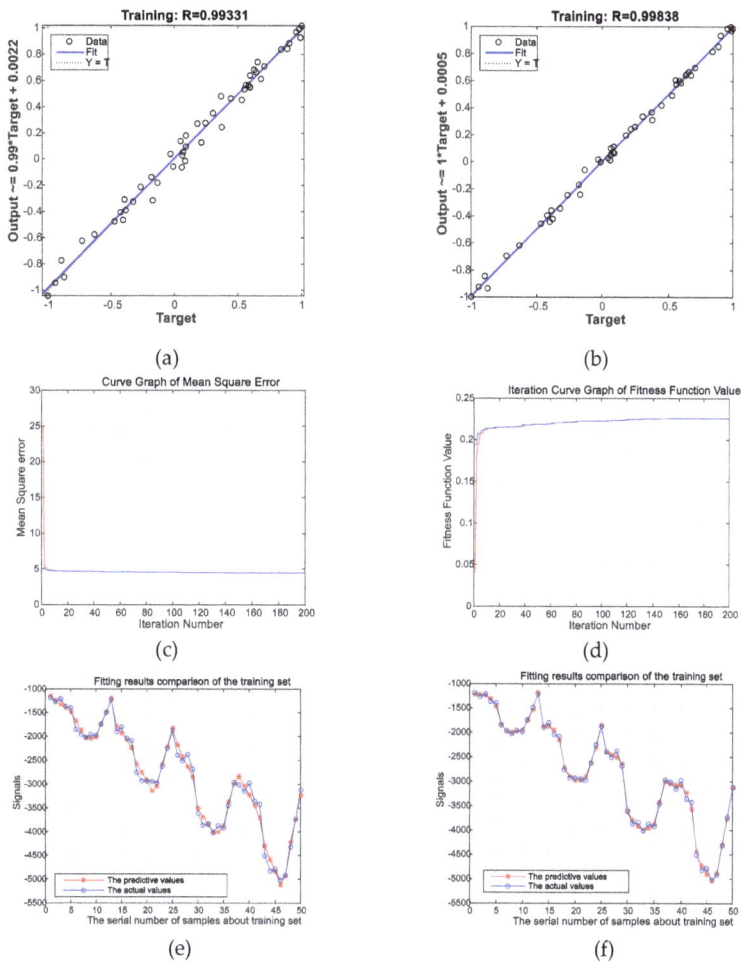

Figure 6. (**a**) Regression curve of the BPNN model error; (**b**) regression curve of the GA-BPNN model error; (**c**) GA iteration curve graph of mean square error; (**d**) GA iteration curve graph of fitness function value. (**e**) The fitting results comparison of the training set about BPNN; (**f**) The fitting results comparison of the training set about GA-BPNN.

Figure 6a,b shows regression curve of BPNN model error and GA-BPNN model error, respectively. In Figure 6a, it shows regression analysis of the training samples by BPNN model, the relevant regression coefficient was 0.99331, The relevant regression coefficient was good, which means strain prediction in full-scale static testing of wind turbine blades has good performance based on BPNN. However, Figure 6b shows that the relevant regression coefficient of GA-BPNN was 0.99838, the relevant regression coefficient was closer to 1. Figure 6c is the curve graph of mean square error, where the blue line represents the minimum sum of squared errors and the red line represents the sum of squared errors; Figure 6d is the iteration curve graph of the fitness function value, the best fitness function value is shown by a blue line, whereas the average fitness function value is shown by a red line. The genetic algorithm runs 200 times during an iteration step. In addition, from Figure 6e,f, the curve of the prediction values fitted by GA-BPNN was more similar to that of the actual values than BPNN, so we can conclude that the fitting results of GA-BPNN were better than BPNN, which means GA-BPNN has a better performance than BPNN. From Tables 5 and 6, the input weighting values of the traditional BPNN method and the GA-BPNN method have been presented, respectively. The input weighting values were a 13 × 7 matrix because of the input data with 13 variables and the 7 hidden-layer nodes and all of the BPNN and GA-BPNN were set this way.

Table 5. The input weight values of the BPNN.

−1.624	−0.512	−0.417	−0.084	−0.001	0.320	0.260	0.712	0.008	0.111	−0.382	−0.780	0.660
−0.444	−0.136	0.609	−0.492	−0.509	−0.182	−0.598	−0.761	−0.186	−0.243	−0.786	−0.296	−0.198
0.752	0.050	−0.181	0.496	0.566	0.578	−0.792	−0.285	−0.212	−0.684	−0.236	−0.142	−0.203
0.958	−0.057	0.577	−0.194	0.442	0.591	−0.815	−0.558	−0.676	0.162	0.675	0.206	0.091
−0.635	−0.264	0.651	−0.449	0.632	0.333	0.226	0.340	−0.109	−0.219	0.298	−0.844	0.505
0.217	0.182	0.478	−0.769	0.072	−0.530	0.283	−0.203	−0.106	0.436	0.570	−0.286	−0.373
1.450	−0.685	0.204	0.173	0.117	−0.745	−0.620	0.052	0.535	−0.263	0.598	−0.487	−0.182

Table 6. The input weight values of the GA-BPNN.

−3.288	0.581	−0.291	−0.754	−0.506	0.911	0.032	−0.245	0.185	−0.532	−0.308	0.648	0.230
−2.997	0.292	−0.991	−0.479	0.125	0.2931	−0.114	−0.357	−0.606	1.1067	−0.465	0.768	0.473
0.188	−0.658	−0.602	−0.798	0.145	−1.308	−0.993	−0.486	−0.466	−1.517	0.0098	0.063	0.233
0.115	0.649	−0.239	−0.401	0.574	−0.624	−0.898	−0.949	−0.045	0.681	−0.834	−0.593	−0.559
−0.840	0.154	−0.366	1.023	0.784	−0.651	−0.378	0.3140	0.922	0.233	−0.807	−0.184	−0.164
−2.537	0.575	−0.386	0.625	0.151	−0.056	−0.062	−0.721	−0.067	−0.654	0.579	−0.715	0.440
0.415	1.046	0.047	−0.508	1.002	−0.638	−0.023	−0.274	−0.669	−0.590	0.483	0.112	0.636

The test sample was used to verify the recognition ability of the GA-BPNN compared with the traditional BPNN, the comparison results are shown as Figure 7. The errors are calculated by difference between the true values which were used as testing samples and the predictive values which were trained by GA-BPNN and BPNN, respectively. It can be seen from Figure 7 that the GA-BPNN corresponding to the variable forecasting results were much more accurate, and the relative error rate of the test sample output was within 6.5%. Moreover, the relative error of every test sample analyzed by GA-BPNN was less than those analyzed by BPNN. So, GA-BPNN was more accurate than BPNN.

In order to verify the reliability and availability of the BPNN and GA-BPNN, the prediction results were used to compare it with the simulation data made by ANSYS. The unmeasured points on the center of the suction side at 33.00 m, 42.00 m, 48.00 m, 52.00 m, 54.00 m, 56.00 m, 58.00 m, 63.00 m, and 65.00 m from the root of the blade were chosen to predict their strain by using BPNN. Comparing the prediction data with the simulation data, the comparison results are shown in Figure 8. It can be seen from Figure 8 that both BPNN and GA-BPNN have a high accuracy to predict the strain, and that the GA-BPNN had a smaller error.

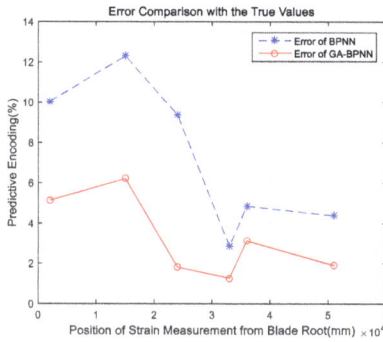

Figure 7. The comparison results of prediction errors for the center of suction side.

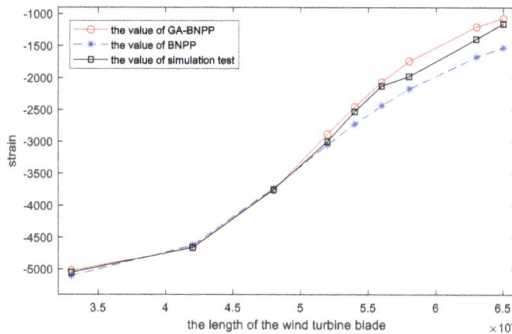

Figure 8. The comparison of GA-BPNN, BPNN, and the simulation test.

4.2. GA-BPNN-Based Strain Prediction for the Trailing Edge

As the same with the previous approaches in Section 4.1, the strain forecasting model based on GA-BPNN for the trailing edge was established by GA-BPNN. The training samples and test samples of the GA-BPNN are shown in Tables 7 and 8, respectively.

The comparison results of the GA-BPNN and traditional BPNN for the strain prediction of the full-scale static testing of the wind turbine blade are shown as Figure 9.

Figure 9a,b show the training state of BPNN and GA-BPNN, respectively; Figure 9c,d show the regression of BPNN and GA-BPNN, respectively; Figure 9e is the curve graph of mean square error, where the blue line represents the minimum sum of squared errors and the red line represents the average sum of squared errors; Figure 9f is the iteration curve graph of fitness function value, where the best fitness function value is shown by a blue line, whereas the average fitness function value is shown by a red line.

Figure 9a,b show the regression curve of BPNN model error and GA-BPNN model error, respectively. In Figure 9a, it is regression analysis of training samples by the BPNN model, the relevant regression coefficient was 0.91038, the training result is modest according to the relevant regression coefficient. The regression results trained by the BPNN are more different than the theoretical values compared with the regression results from the center of suction side. The reason for this result is that the trailing edge is the joint of two different materials. In Figure 9b, the relevant regression coefficient of GA-BPNN was 0.96706, which is closer to 1, and the number of relevant regression coefficients of GA-BPNN training was bigger than that of the BPNN, which means the regression results of the GA-BPNN was better than the BPNN. In addition, from Figure 9e,f, the fitting results of the GA-BPNN were better than the BPNN, so the GA-BPNN training had the better performance than the BPNN training. From Tables 9 and 10, the input weighting values of the traditional BPNN method and the GA-BPNN method are presented, respectively. The input weighting values are a 13 × 5 matrix because of the input data with 13 variables and the five hidden-layer nodes, and all of the BPNN and GA-BPNN models were set by this way.

Table 7. The training samples of the GA-BPNN.

Items	The Location of Strain Gauge	Load (kN)					The Distance to the Blade Root (m)
		F1	F2	F3	F4	F5	l1
1	2000	37.85	57.42	24.09	42.34	27.25	0
2	6000	57.48	86.17	35.85	62.75	40.98	1.16
3	9000	75.76	114.6	48.16	83.04	54.51	5.69
...	...	...	...	...	...	...	...
...	...	...	...	...	...	...	...
50	48,000	75.76	114.6	48.16	83.04	54.51	5.69

Items	The Displacement of Loading Positions (m)						
	s1	s2	s3	s4	s5	s6	s7
1	159	755	2159	3717	6384	7900	−146
2	264	1158	3257	5596	9605	11,876	−418
3	368	1557	4350	7469	12,773	15,770	−497
...	...	...	...	...	...	...	...
...	...	...	...	...	...	...	...
50	477	1976	5490	9405	16,019	19,741	−396

Table 8. The test samples of the GA-BPNN.

Items	The Location of Strain Gauge	Load (kN)					The Distance to the Blade Root (m)
		F1	F2	F3	F4	F5	l1
1	24,000	37.85	57.42	24.09	42.34	27.25	0
2	21,000	75.76	114.6	48.16	83.04	54.51	5.69
3	33,000	37.85	57.42	24.09	42.34	27.25	0
4	36,000	94.74	143.15	60.15	104.67	68.07	8.51
5	21,000	94.74	143.15	60.15	104.67	68.07	8.51
6	24,000	75.76	114.6	48.16	83.04	54.51	5.69

Items	The Displacement of Loading Positions (m)						
	s1	s2	s3	s4	s5	s6	s7
1	159	755	2159	3717	6384	7900	−189
2	368	1557	4350	7469	12,773	15,770	−532
3	159	755	2159	3717	6384	7900	−147
4	477	1976	5490	9405	16,019	19,741	−652
5	477	1976	5490	9405	16,019	19,741	−717
6	368	1557	4350	7469	12,773	15,770	−598

Table 9. The input weight values of the BPNN.

0.182	0.134	−0.272	−1.017	−1.216	−0.874	−1.074	−0.827	−0.489	−0.865	−0.668	0.041	0.167
−0.216	1.769	2.422	1.784	1.773	2.682	1.436	1.961	1.403	1.449	1.070	1.780	2.273
0.0524	−0.305	0.762	−0.029	−0.317	−0.078	−0.368	0.324	−0.364	0.009	−0.116	0.725	−0.416
−0.317	2.778	1.776	3.031	1.922	3.017	2.795	2.940	2.743	2.104	2.824	2.203	2.066
0.115	−1.131	−0.772	−0.829	−0.806	−0.417	−1.82	−0.654	−1.715	−1.022	−1.058	−0.487	−1.129

Table 10. The input weight values of the GA-BPNN.

41.803	−0.991	−1.52	−0.70	2.601	−1.894	0.510	1.772	1.574	0.787	0.038	−1.27	−1.056
0.763	−0.671	−0.05	0.95	−1.021	−0.596	11.960	−0.777	0.365	−0.601	−1.262	−1.40	0.277
−1.333	2.607	1.381	1.42	3.559	0.850	−15.57	3.153	3.654	2.847	2.374	3.27	2.519
−28.76	5.738	4.997	4.46	6.254	5.188	5.005	5.217	6.533	5.527	5.109	4.64	5.508
1.322	3.529	3.600	2.85	−8.248	3.232	3.252	−4.561	−4.716	−3.261	−2.641	1.17	1.212

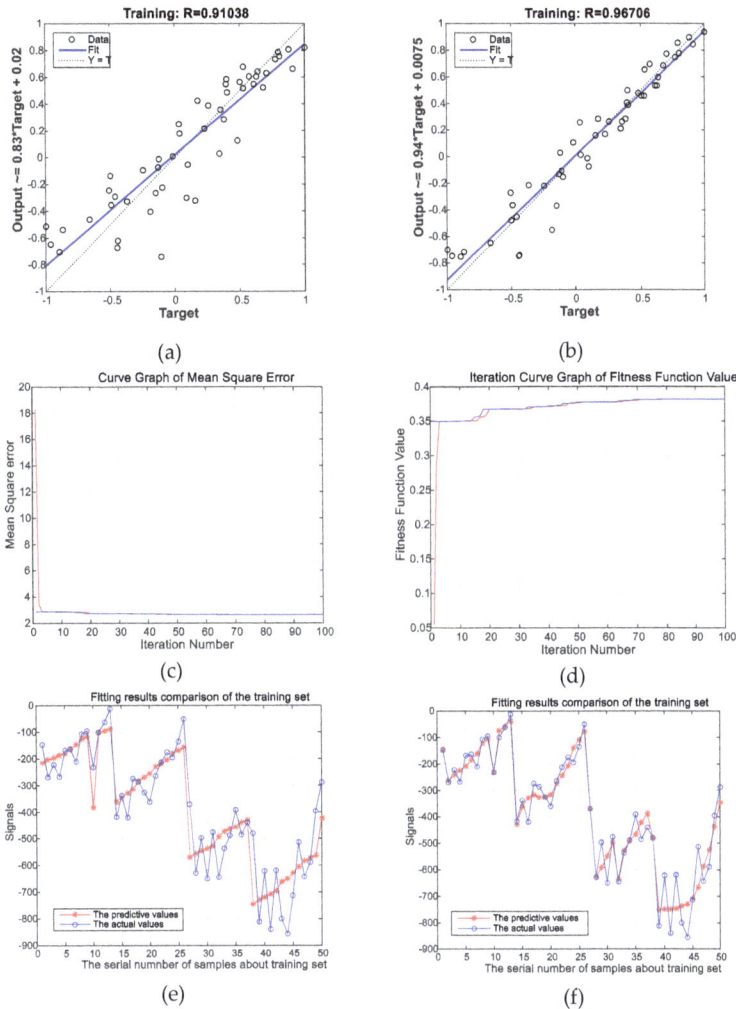

Figure 9. (**a**) Regression curve of the BPNN model error; (**b**) regression curve of the GA-BPNN model error; (**c**) GA-iteration curve graph of mean square error; (**d**) GA-iteration curve graph of fitness function value. (**e**) The fitting results comparison of the training set about BPNN; (**f**) The fitting results comparison of the training set about GA-BPNN.

The test sample was used to verify the recognition ability of the GA-BPNN in contrast to traditional BPNN, and the test results were compared as shown in Figure 10. It can be seen from Figure 10 that the average error of GA-BPNN was smaller than that of BPNN which means the GA-BPNN corresponding to the variable prediction results were more accurate, and the relative error rate of the test sample output was within 18%. Compared with the prediction results of the center of the suction side, the error was relatively larger. For the trailing edge is the faying surface of the suction side and pressure side, the strain was influenced by more factors such as binder type, binder parameters, physical dimension, etc.; thus, more inputs are needed to get a more accurate prediction.

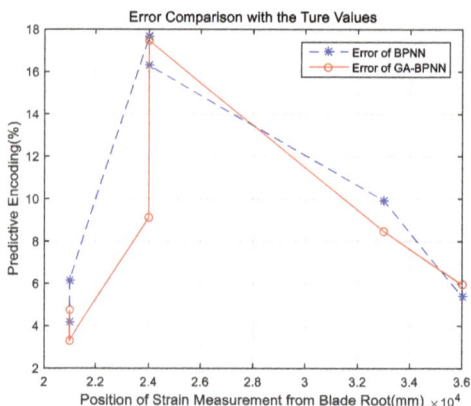

Figure 10. The comparison results of prediction errors for the trailing edge.

The unmeasured points on trailing edge at 33.00 m, 42.00 m, 48.00 m, 52.00 m, 54.00 m, 56.00 m, 58.00 m, 63.00 m, and 65.00 m from the root of the blade were chosen to predict their strain by using the BPNN. The contrast figures of BPNN, GA-BPNN, and the ANSYS simulation data are shown in Figure 11. The conclusion is the same as the analysis of the center of the suction side, both BPNN and GA-BPNN had a high accuracy to predict the strain, and the GA-BPNN had a smaller error. Thus, GA-BPNN is more suitable for the strain forecast of the full-scale static testing of wind turbine blades.

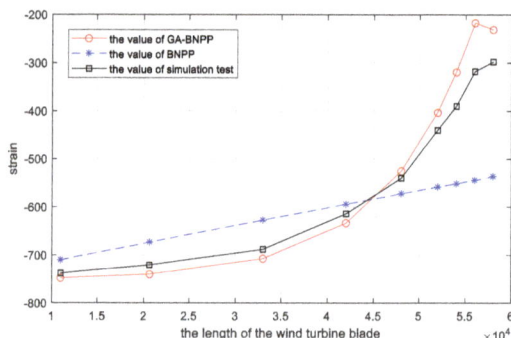

Figure 11. The comparison of BPNN, GA-BPNN, and the simulation test.

5. Conclusions

The calculation and prediction of blade strain in the full-scale static testing of wind turbine blades are very complex and difficult by traditional numerical methods, and the numbers of measuring points as well as strain gauges arranged on the blade are limited, so the test data have insufficient significance to the calibration of the blade design. As a result of these concerns, this paper proposed a strain prediction method for wind turbine blades using a GA-BPNN with applied loads, loading positions, and displacement as inputs, and tried to provide more data for the wind turbine blades' health assessment and life prediction when the measurement points in full-scale static testing of wind turbine blades are limited:

(1) Taking advantage of the neural network in dealing with complex problems, this paper established a strain–predictive GA-BPNN model for the center and the trailing edge of the suction side based on the full-scale static testing results of a certain wind turbine blade.

(2) The GA-BPNN had a better performance on strain prediction in full-scale static testing of wind turbine blades than the BPNN. In the training process, the relevant regression coefficient trained by GA-BPNN was closer to 1 than the BPNN. In the test process, all the average errors of GA-BPNN were smaller than those of the BPNN. In the prediction process, the values analyzed by the GA-BPNN were closer to the theoretical values (simulation test values) than those analyzed by the BPNN.

(3) The strain of unmeasured points at the center and the trailing edge of the suction side were predicted by strain–predictive BPNN model, respectively. For strain prediction of the points at the center of the suction side, the relative error rate of the test sample output was within 6.5%. While for strain prediction of the points at the trailing edge of the suction side, the relative error rate of the test sample output was within 18%. Compared with the prediction results of the center of suction side, the error of the trailing edge was relatively larger. For the trailing edge is the faying surface of suction side and pressure side, and the strain is influenced by more factors such as binder type, binder parameters, physical dimension, etc., thus, more inputs are needed to get a more accurate prediction.

(4) The unmeasured points at 33.00 m, 42.00 m, 48.00 m, 52.00 m, 54.00 m, 56.00 m, 58.00 m, 63.00 m, and 65.00 m from the root of the blade were chosen to predict their strain using the BPNN. Comparing the prediction results with the ANSYS simulation data, both the BPNN and GA-BPNN had a high accuracy in predicting the strain, and the GA-BPNN had a smaller error. Thus, the GA-BPNN is more suitable for strain prediction of wind turbine blade static testing.

Author Contributions: Conceptualization, Z.L. (Zheng Liu), J.A.F.O.C. and A.M.P.D.J.; methodology, Z.L. (Zheng Liu), X.L., J.A.F.O.C. and A.M.P.D.J.; software, X.L.; validation, Z.L. (Zheng Liu) and Z.L. (Zhongwei Liang); formal analysis, Z.L. (Zheng Liu) and X.L.; investigation, K.W.; resources, K.W.; data curation, X.L.; writing-original draft preparation, Z.L. (Zheng Liu), X.L. and Z.L. (Zhongwei Liang); writing—review and editing, J.A.F.O.C. and A.M.P.D.J.; supervision, X.L. and J.A.F.O.C.; project administration, J.A.F.O.C.; funding acquisition, Z.L. (Zheng Liu), X.L., Z.L. (Zhongwei Liang) and J.A.F.O.C.

Funding: This research was funded by Guangzhou University Teaching Reform Project (09-18ZX0309), the Guangzhou University Teaching Reform Project (09-18ZX0304), the Innovative Team Project of Guangdong Universities (2017KCXTD025), and the Innovative Academic Team Project of Guangzhou Education System (1201610013).

Acknowledgments: This research was partially supported by the Guangzhou University Teaching Reform Project (09-18ZX0309), the Guangzhou University Teaching Reform Project (09-18ZX0304), the Innovative Team Project of Guangdong Universities (2017KCXTD025), and the Innovative Academic Team Project of Guangzhou Education System (1201610013). Additionally, this research was also partially supported by Projects POCI-01-0145-FEDER-007457 and UID/ECI/04708/2019—CONSTRUCT—Institute of R&D in Structures and Construction funded by FEDER funds through COMPETE2020—Programa Operacional Competitividade e Internacionalização (POCI)—and by national funds through FCT—Fundação para a Ciência e a Tecnologia, and postdoctoral grant (SFRH/BPD/107825/2015) provided by FCT to the fifth author.

Conflicts of Interest: The authors declare no conflict of interest.

References

1. Malhotra, P.; Hyers, R.W.; Manwell, J.F.; McGowan, J.G. A review and design study of blade testing systems for utility-scale wind turbines. *Renew. Sustain. Energy Rev.* **2012**, *16*, 284–292. [CrossRef]
2. Fagan, E.M.; Flanagan, M.; Leen, S.B.; Flanagan, T.; Doyle, A.; Goggins, J. Physical experimental static testing and structural design optimization for a composite wind turbine blade. *Compos. Struct.* **2017**, *16*, 90–103. [CrossRef]
3. Yang, T.; Du, W.C.; Yang, H.; Ma, C. Static load strain test of wind turbine blades. *Res. Explor. Lab.* **2011**, *30*, 33–36.
4. Pan, Z.J.; Wu, J.Z. Effects of structure nonlinear on full-scale wind turbine blade static test. *J. Tongji Univ. (Nat. Sci.)* **2017**, *45*, 1491–1497.
5. Zhu, S.P.; Yue, P.; Yu, Z.Y.; Wang, Q.Y. A combined high and low cycle fatigue model for life prediction of turbine blades. *Materials* **2017**, *10*, 698. [CrossRef] [PubMed]

6.   Liao, D.; Zhu, S.P.; Correia, J.A.F.O.; De Jesus, A.M.P.; Calçada, R. Computational framework for multiaxial fatigue life prediction of compressor discs considering notch effects. *Eng. Fract. Mech.* **2018**, *202*, 423–435. [CrossRef]

7.   Meng, D.; Yang, S.; Zhang, Y.; Zhu, S.P. Structural reliability analysis and uncertainties-based collaborative design and optimization of turbine blades using surrogate model. *Fatigue Fract. Eng. Mater. Struct.* **2018**. [CrossRef]

8.   Zhu, S.P.; Liu, Q.; Peng, W.; Zhang, X.C. Computational-experimental approaches for fatigue reliability assessment of turbine bladed disks. *Int. J. Mech. Sci.* **2018**, *142–143*, 502–517. [CrossRef]

9.   Meng, D.; Liu, M.; Yang, S.; Zhang, H.; Ding, R. A fluid-structure analysis approach and its application in the uncertainty-based multidisciplinary design and optimization for blades. *Adv. Mech. Eng.* **2018**, *10*, 1–7. [CrossRef]

10. Tarfaoui, M.; Nachtane, M.; Boudounit, H. Finite element analysis of composite offshore wind turbine blades under operating conditions. *J. Therm. Sci. Eng. Appl.* **2018**. [CrossRef]

11. Tarfaoui, M.; Shah, O.R.; Nachtane, M. Design and optimization of composite offshore wind turbine blades. *J. Energy Resour. Technol.* **2019**, *141*, 051204. [CrossRef]

12. Tarfaoui, M.; Nachtane, M.; Khadimallah, H.; Saifaoui, D. Simulation of mechanical behavior and damage of a large composite wind turbine blade under critical loads. *Appl. Compos. Mater.* **2018**, *25*, 237–254. [CrossRef]

13. Hu, W.; Choi, K.K.; Cho, H. Reliability-based design optimization of wind turbine blades for fatigue life under dynamic wind load uncertainty. *Struct. Multidiscip. Optim.* **2016**, *54*, 953–970. [CrossRef]

14. Hu, W.; Choi, K.K.; Zhupanska, O.; Buchholz, J. Integrating variable wind load, aerodynamic, and structural analyses towards accurate fatigue life prediction in composite wind turbine blades. *Struct. Multidiscip. Optim.* **2016**, *53*, 375–394. [CrossRef]

15. Yang, J.C.; Park, D.S. Fingerprint Verification Based on Invariant Moment Features and Nonlinear BPNN. *Int. J. Control. Syst.* **2008**, *6*, 800–808.

16. Moghaddam, M.A.; Golmezergi, R.; Kolahan, F. Multi-variable measurements and optimization of GMAW parameters for API-X42 steel alloy using a hybrid BPNN–PSO approach. *Measurement* **2016**, *92*, 279–287. [CrossRef]

17. Huang, Q.; Jiang, D.; Hong, L.; Ding, Y. *Application of Wavelet Neural Networks on Vibration Fault Diagnosis for Wind Turbine Gearbox*; Springer: Berlin/Heidelberg, Germany, 2008; Volume 5264, pp. 313–320.

18. Chen, J.S.; Chen, W.G.; Li, J.; Sun, P. A generalized model for wind turbine faulty condition detection using combination prediction approach and information entropy. *J. Environ. Inf.* **2018**, *32*, 14–24. [CrossRef]

19. Zhang, Y.; Zheng, H.; Liu, J.; Zhao, J.; Sun, P. An anomaly identification model for wind turbine state parameters. *J. Clean. Prod.* **2018**, *195*, 1214–1227. [CrossRef]

20. Liu, T. The limit cycle oscillation of divergent instability control based on classical flutter of blade section. *J. Vibro Eng.* **2017**, *19*, 5114–5136. [CrossRef]

21. Ding, S.; Su, C.; Yu, J. An optimizing BP neural network algorithm based on genetic algorithm. *Artif. Intell. Rev.* **2011**, *36*, 153–162. [CrossRef]

22. Kuang, Y.; Singh, R.; Singh, S.; Singh, S.P. A novel macroeconomic forecasting model based on revised multimedia assisted BP neural network model and ant Colony algorithm. *Multimed. Tools Appl.* **2017**, *76*, 18749–18770. [CrossRef]

23. Wang, W.; Li, M.; Hassanien, R.H.E.; Ji, M.E.; Feng, Z. Optimization of thermal performance of the parabolic trough solar collector systems based on GA-BP neural network model. *Int. J. Green Energy* **2017**, *14*, 819–830. [CrossRef]

24. Xu, H.; Li, W.; Li, M.; Hu, C.; Zhang, S.; Wang, X. Multidisciplinary robust design optimization based on time-varying sensitivity analysis. *J. Mech. Sci. Technol.* **2018**, *32*, 1195–1207. [CrossRef]

25. Zhou, W.H.; Xiong, S.Q. *Optimization of BP Neural Network Classifier Using Genetic Algorithm*; Springer: Berlin/Heidelberg, Germany, 2013.

26. GB/T 25384-2010. *Turbine Blade of Wind Turbine Generator Systems-Full-Scale Structural Test of Rotor Blades*; Standards Press of China: Beijing, China, 2010.

energies

MDPI

Article

An Improved Signal Processing Approach Based on Analysis Mode Decomposition and Empirical Mode Decomposition

Zhongzhe Chen [1,*], Baqiao Liu [2], Xiaogang Yan [1] and Hongquan Yang [1]

[1] The School of Mechanical and Electrical Engineering, University of Electronic and Science Technology of China, Chengdu 611731, China
[2] Department of Computer Science, University of North Carolina, Chapel Hill, NC 27516, USA
* Correspondence: zhzhchen@163.com; Tel.: +86-139-8179-1418

Received: 25 June 2019; Accepted: 5 August 2019; Published: 9 August 2019

Abstract: Empirical mode decomposition (EMD) is a widely used adaptive signal processing method, which has shown some shortcomings in engineering practice, such as sifting stop criteria of intrinsic mode function (IMF), mode mixing and end effect. In this paper, an improved sifting stop criterion based on the valid data segment is proposed, and is compared with the traditional one. Results show that the new sifting stop criterion avoids the influence of end effects and improves the correctness of the EMD. In addition, a novel AEMD method combining the analysis mode decomposition (AMD) and EMD is developed to solve the mode-mixing problem, in which EMD is firstly applied to dispose the original signal, and then AMD is used to decompose these mixed modes. Then, these decomposed modes are reconstituted according to a certain principle. These reconstituted components showed mode mixing phenomena alleviated. Model comparison was conducted between the proposed method with the ensemble empirical mode decomposition (EEMD), which is the mainstream method improved based on EMD. Results indicated that the AEMD and EEMD can effectively restrain the mode mixing, but the AEMD has a shorter execution time than that of EEMD.

Keywords: empirical mode decomposition; analysis mode decomposition; analysis-empirical mode decomposition; mode mixing; sifting stop criterion

---

1. Introduction

The analysis of time-frequency of vibration signals is one of the most effective and important methods for fault diagnosis of rotating machinery, since the vibration signal includes massive information that reflects the running state of rotating machinery [1]. The empirical mode decomposition (EMD) is one of most commonly used methods for signal processing in the time-frequency domain. EMD can decompose a complex signal into the finite intrinsic mode functions (IMF) based on the local characteristic time scale of signals, and each IMF represents one intrinsic vibration mode of the original signal. Then the characteristic information of the signal can be extracted by analyzing such stationary stable IMFs [2]. EMD has attracted increasing attention since it appeared [3–5], and it has been widely used in economics, biomedicine and engineering science fields, especially for fault diagnosis. Ali et al. [6] applied the EMD method and artificial neural network in fault diagnosis of rolling bearings automatically. Xue et al. [7] presented an adaptive, fast EMD method and applied it to rolling bearings fault diagnosis. Yu et al. [8] introduced various applications of EMD in fault diagnosis. Cheng et al. [9] combined EMD with a Hilbert transform to conduct the recognition for mode parameters. Then, they introduced how to apply the EMD to the fault diagnosis for local rub-impact of rotors [10]. Rilling [11] investigates how the EMD behaves under the case of a composite two-tone signal.

Though EMD has been commonly utilized in reality, some shortcomings are exposed, including mode mixing, end effects, stop criteria of IMF, over-envelope and under-envelope. These deficiencies normally restrict the further promotion and application of EMD [12]. Since the criterion to judge IMF is that the average value of its upper and lower envelope spectrums is zero in the EMD method, but in the real decomposition process, that average value is impossible to be zero due to the disturbance of cubic spline interpolation, the influence of end effect and adopted frequency. Thus, it is necessary to define a valid stop criterion for the decomposition process in engineering, namely the stop criterion problem of IMF for the EMD method, also named as the sifting stop criterion. Some researchers have attempted to improve the efficiency of EMD. Among them, Pustelnik [13] mainly developed an alternative to the sifting process for EMD, based on non-smooth convex optimization allowing integration flexibility in the criteria, proposed algorithm and its convergence guarantees.

Mode mixing means that one IMF contains vastly different characteristic time scales, or similar characteristic time scales that are distributed in different IMFs, which results in the waveform mixing of the two adjacent IMFs. These mixing modes influence each other so that it is difficult to identify them. Mode mixing is the fatal flaw of EMD. It makes the physical significance of IMF components uncertain finally, which influences the correctness of signal decomposition and seriously restricts its application in engineering [14].

Until now, various methods have been developed to restrain the mode-mixing problem in the EMD. Zhao [15] directly filtered abnormal information related to the IMF and fitted filtered data segment by spline interpolation, but it only proved to dispose the mode mixing problems caused by a known transient abnormity. The masking signal method [16] and the high frequency harmonic method [17] are simple and effective, but they are susceptible to distortion and need to be reprocessed for practical engineering signals.

The ensemble empirical mode decomposition (EEMD) presented by Huang is generally considered an effective one [18]. EEMD takes advantage of the statistical characteristics of white Gaussian noise while frequency is uniformly distributed, so the signal after adding white Gaussian noise shows continuity in different scales, which solved the mode mixing problem to some extent. However, it also raised some other issues. For example, the number and the amplitude of white noises which are added in the signal are greatly subjective, and the EEMD sacrifices some adaptivity. In addition, although the number and the amplitude of added white noises are chosen reasonably, the mode mixing in low frequency may be aroused artificially while high-frequency mode mixing is restricted [19]. Moreover, the algorithm of the EEMD method is complex and it takes a long time to run the program, which will restrict its application on the signal process that demands to be processed in real-time. Accordingly, some researchers presented improved methods to overcome the deficiencies of EEMD. Among them, Lei [20] proposed adaptive EEMD to improve its adaptivity. Zheng [21] developed partial EEMD and Tan [22] presented multi-resolution EMD to solve mode mixing problem. Mohammad [23] uses approximate entropy and mutual information to improve EEMD to generate statistical features in order to increase the performance of early fault appearance detection, as well as the fault type and severity estimation.

In this paper, a new sifting stop criterion was proposed based on valid data segments to solve the problem of sifting stop criteria in the EMD, and an improved method, namely AEMD, combined the analysis mode decomposition (AMD) and EMD. It was developed to solve the mode-mixing problem. The sifting stop criteria and AEMD proposed were applied to a simulation signal and an engineering case to illustrate the validity and superiority of the proposed method. The structure of the paper is as follows: Section 2 introduces the basic principles of EMD; Section 3 narrates the proposed sifting stop criteria based on valid data segment and compares it with the original one; in Section 4, the principle and steps of AEMD are expounded firstly, then applied to decompose simulation signals and a rotor vibration signal, and it is compared with the EMD and EEMD methods; finally, Section 5 draws a brief conclusion of current work.

2. Basic Principles of EMD

EMD refers to a "sifting" process, in which, the component with the smallest extreme time feature scale is sifted out firstly, then those with larger extreme time feature scales, and the component with the largest feature scales are finally sifted out. That is to say, the average frequency of IMF components obtained from the EMD is reduced gradually, and the main steps of the EMD are introduced as follows [24]:

(1) Firstly, identify all the local maximum points and minimum points of original signal $x(t)$, then match the upper and lower envelope spectrums of extreme points with cubic spline line respectively. Furthermore, ensure that the signal $x(t)$ is between the upper and lower envelope spectrums; (2) calculate the local mean of upper and lower envelope spectrums, denoted as m_1; (3) calculate the first component $h_1(t)$ according to Equation (1):

$$h_1(t) = x(t) - m_1 \tag{1}$$

(4) Judge that whether $h_1(t)$ can satisfy the conditions to be an IMF. If not, $h_1(t)$ should be treated as an original signal to repeat steps (1), (2) and (3) until $h_1(t)$ can meet the conditions, designated as $C_1(t) = h_1(t)$. $C_1(t)$ is the first IMF component after decomposition. (5) Separate $C_1(t)$ from the signal $x(t)$, i.e., $r_1(t) = x(t) - c_1(t)$. (6) Treat $r_1(t)$ as an original signal to repeat steps (1)–(5), and after n cycles, n IMF components and 1 residual value $r_n(t)$ can be derived; i.e.,

$$\begin{cases} r_2 = r_1 - c_2 \\ r_3 = r_2 - c_3 \\ \vdots \\ r_n = r_{n-1} - c_n \end{cases} \tag{2}$$

Then, the original signal $x(t)$ can be expressed as:

$$x(t) = \sum_{i=1}^{n} c_i + r_n \tag{3}$$

where c_i refers to the ith IMF component and r_n is the residual function.

This cycle comes to the end when r_n becomes a monotonic function, but in the real decomposition process, when r_n meets the conditions of monotonic function, the cycle number is usually large, and the number of IMF components will be too large. Furthermore, a mass of EMD tests have revealed that most of these final components obtained from EMD are false IMF components, without substantive physical significance. Thus, a good sifting stop criterion can not only improve the decomposition efficiency, but also increase the decomposition accuracy.

The stop criteria of decomposition process proposed by Huang was realized by limiting the standard deviation of the two adjacent IMFs and the sifting times are eventually controlled by the iteration threshold S_d [25]. In particular, S_d is defined as:

$$S_d = \sum_{t=0}^{T} \frac{[h_k(t) - h_{k-1}(t)]^2}{h_{k-1}^2(t)} \tag{4}$$

where T refers to the time span of a signal, $h_k(t)$ and $h_{k-1}(t)$ denote the two adjacent processing sequences in the process of EMD and the value of S_d is usually between 0.2 and 0.3.

It is of great significance to determine reasonable iteration threshold S_d. If the threshold is too small, the computational cost will increase greatly and the IMF components finally obtained will be of no significance; while if it is too large, it will be difficult to satisfy the condition of IMF.

3. The Sifting Stop Criteria Based on the Valid Data Segment

The sifting stop criteria of IMF proposed by Huang is valid for most cases, but there are some problems; it can be noted from Equation (3) that when $h_k(t) = 0$, the iteration threshold will become an uncertain value. It is evidently inappropriate if the iteration threshold is used to control sifting number at the time. In addition, the sifting stop criteria ignored the influence of end effect. Consequently, the decomposition result may produce errors if the distorted endpoint data are adopted. In most cases, the end effect of EMD is obvious, though some effective measures will been taken to restrain it, the end effect still exists.

In this paper, an improved sifting stop criterion of IMF based on the valid data segment is proposed, where the valid data segment means the residual data segment after kicking out the distorted endpoint data. The sifting stop criteria fully consider the influence of end effect to sifting stop criteria, and only use the valid data segment to calculate the iteration threshold. The sifting stop criterion is expressed as follows:

If $h_k(t)$ is an IMF, it satisfies the following inequality:

$$-\delta < \widetilde{m}_{k+1} < \delta \tag{5}$$

where $\widetilde{m}_{k+1}$ denotes the valid data segment of the mean curve of upper envelope and lower envelope; δ represents the error threshold, which ranges from 0.01 to 0.1.

The EMD is applied to decompose a simulation signal respectively based on the proposed sifting stop criterion and the traditional one below. The simulation signal is

$$x(t) = x_1(t) + x_2(t), t \in [0, 1] \tag{6}$$

$$x_1(t) = (1 + 0.5\sin(2\pi * 4t)) * 2\cos(2\pi * 80t),$$
$$x_2(t) = \sin(2\pi * 24t) \tag{7}$$

The simulation signal consists of $x_1(t)$ and $x_2(t)$ (as shown in Figures 1–3) are the decomposition results by using EMD based on the proposed sifting stop criterion and the traditional one, respectively. It is worth noting that the EMD can efficiently decompose the original signal by using the two sifting stop criteria, but it is clear from the two edges of these two figures that the decomposition result c_1 and c_2 in Figure 2 are more accurate than imf_1 and imf_2 in Figure 3, which show an evident end effect.

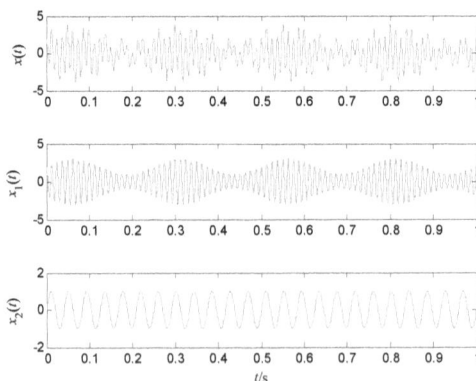

Figure 1. The simulation signal and its two components.

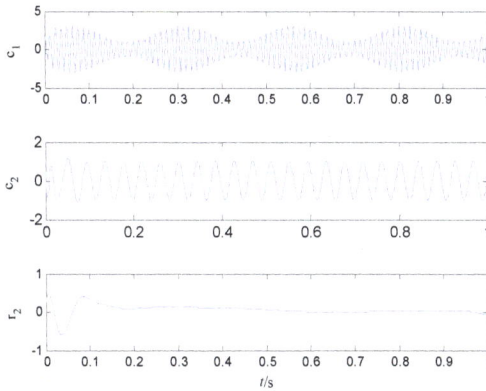

Figure 2. The result of empirical mode decomposition (EMD) method using the sifting criterion based on the valid data segment.

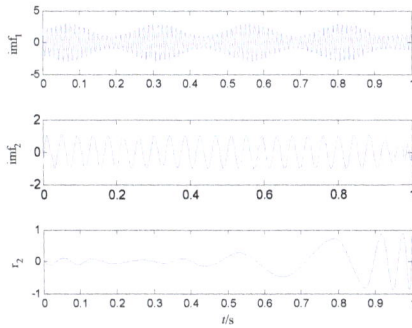

Figure 3. The result of EMD method using the sifting criterion proposed by Huang.

Figure 4 plots the vibration signal of a real gear with broken teeth. The number of gear teeth $z = 75$; module $m = 2$; the rotating frequency $f = n_2/60 \approx 13.6$. EMD based on the proposed sifting stop criteria is used to decompose the vibration signal of a gear, and obtain five IMFs. The first IMF imf_1 contains massive fault information of the gear. Note from its time domain graph (see Figure 5) that the waveform of imf_1 presents an obvious modulation feature, and its cycle of modulation wave (T, approximately 0.074 s) accordingly had a frequency of about 13.6 Hz, which is exactly the rotating frequency of the faulty gear. Thus, it can be concluded that the fault information has been exacted from the vibration signal of the practical gears by using the proposed sifting stop criterion.

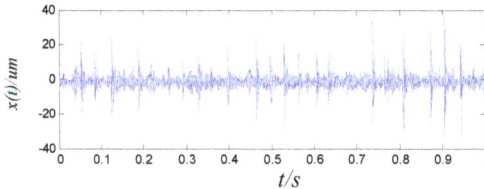

Figure 4. The vibration signal of gear with broken teeth.

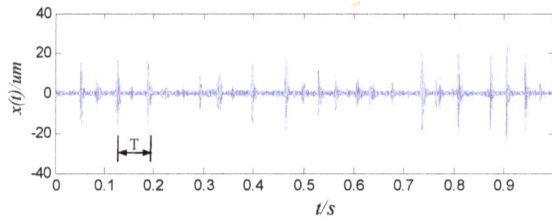

Figure 5. The first IMF of EMD for the vibration signal with broken teeth.

4. AEMD Method

4.1. Analytical Mode Decomposition (AMD) Method

AMD can separate the harmonic components of each frequency band from a multiband signal, its principle is as follows:

A signal $x_0(t)$ is separated into the two components by boundary frequency f_b, one is the fast-changing signal $x_f(t)$ and another is slowly-changing signals $x_s(t)$:

$$x_0(t) = x_f(t) + x_s(t) \tag{8}$$

Their corresponding Fourier transforms $X_f(\omega)$, $X_s(\omega)$ have no overlap in the frequency band. The Hilbert transforms of $\cos(2\pi f_b t) \cdot x_0(t)$ and $\sin(2\pi f_b t) \cdot x_0(t)$ are as follows:

$$H[\cos(2\pi f_b t) \cdot x_0(t)] = x_s(t) \cdot H[\cos(2\pi f_b t)] + \cos(2\pi f_b t) \cdot H[x_f(t)] \tag{9}$$

$$H[\sin(2\pi f_b t) \cdot x_0(t)] = x_s(t) \cdot H[\sin(2\pi f_b t)] + \sin(2\pi f_b t) \cdot H[x_f(t)] \tag{10}$$

$x_s(t)$ and $x_f(t)$ can be derived from the above formulas:

$$x_s(t) = \sin(2\pi f_b t) \cdot H[\cos(2\pi f_b t) \cdot x_0(t)] - \cos(2\pi f_b t) \cdot H[\sin(2\pi f_b t) \cdot x_0(t)] \tag{11}$$

$$x_f(t) = x_0(t) - x_s(t) \tag{12}$$

As aforementioned, the signal $x_0(t)$ is decomposed into $x_s(t)$ and $x_f(t)$ by using the analysis mode decomposition.

4.2. Steps of the AEMD Method

When the signal to be decomposed contains high-frequency intermittent signals or some similar compositions in time feature size, it is inevitable that the mode mixing appears in decomposition result by using EMD directly [11]. In this section, the AEMD that combines AMD with EMD is proposed to restrain the mode mixing in EMD. In particular, EMD is first applied to dispose the original signal, and then AMD is used to decompose these mixed modes. Then, these decomposed modes are reconstituted according to a certain principle. These reconstituted components show mode mixing phenomena alleviated. The specific steps are given as follows:

(1) EMD is applied to the original signal and obtain some IMFs and a residual component r;

(2) A fast Fourier transform (FFT) is conducted on IMFs, and the spectral diagrams of these IMFs are obtained. The first $IMF_1(t)$ with mode mixing is performed by AMD. The boundary frequency f_{b1} can be determined according to the spectrum. The average frequency of the two confused modes is used as the boundary frequency in this analysis. Then $IMF_1(t)$ is separated into the two signals $c_1(t)$ and $\hat{c}_1(t)$ by using AMD. $c_1(t)$ is the correction component of $IMF_1(t)$ and $\hat{c}_1(t)$ is the residual component of $c_1(t)$. Thus, $IMF_1(t)$ can be expressed as follows:

$$IMF_1(t) = c_1(t) + \hat{c}_1(t) \tag{13}$$

(3) Add $\hat{c}_1(t)$ to the $IMF_2(t)$ and get the renewed $IMF_2(t)$, denoted as $IMF^*_2(t)$. The boundary frequency f_{b2} can be obtained according to the amplitude spectrum of $IMF^*_2(t)$. Then $IMF^*_2(t)$ is decomposed into the two signals $c_2(t)$ and $\hat{c}_2(t)$ by using AMD. $c_2(t)$ is the correction component of $IMF^*_2(t)$ and $\hat{c}_2(t)$ is the residual component of $c_2(t)$. Similarly, $IMF^*_2(t)$ is expressed as follows:

$$IMF^*_2(t) = c_2(t) + \hat{c}_2(t) \tag{14}$$

(4) By that analogy, the similar approach is utilized to dispose all the subsequent IMF components with mode mixing. The final residual component $\hat{c}_k(t)$ is added to the residual error r and obtain the final residual error, denote as $v_k(t)$.

4.3. The Comparison of Simulation Signal Analysis by Different Methods

A simulation signal was decomposed by using EMD, EEMD and the proposed AEMD respectively, and their disposal results were compared, to demonstrate AEMD efficiency and superiority. The simulation signal $x(t)$ is:

$$x(t) = x_1(t) + x_2(t) + x_3(t); t \in [0,1] \tag{15}$$

$$x_1(t) = \begin{cases} 0.8\cos(600\pi t); & 0.12 \le t \le 0.18 \\ 0.9\cos(600\pi t); & 0.42 \le t \le 0.48 \\ 0.8\cos(600\pi t); & 0.75 \le t \le 0.81 \\ 0; & \text{else} \end{cases} \tag{16}$$

$$\begin{aligned} x_2(t) &= 2\cos(60\pi t); & t \in [0,1] \\ x_3(t) &= 2.5\cos(28\pi t); & t \in [0,1] \end{aligned} \tag{17}$$

The simulation signal is shown in Figure 6. The decomposition result of EMD, in Figure 7, shows more serious mode mixing. For the first mode, the high frequency intermittency signal $c_1(t)$ is confused with the sinusoidal signal at 30 Hz, for the second mode $c_2(t)$, the sinusoidal signal at 30 Hz with the sinusoidal signal at 14 Hz. The third and fourth modes are false IMF components. The decomposition result of EEMD is imf_1, imf_2, imf_3, imf_4, as shown in Figure 8, with ensemble average number N = 30, and the amplitude of noise is set as 0.01 standard deviation of the original signal. The decomposition result of AEMD is shown in Figure 9. A successful decomposition result should distinctly obtain the several components of the original signal. In order to compare the results of decomposition results by these different methods, the green lines in Figures 7–9 represent the several components of the original signal; The blue lines represent the composition results by these different methods. Table 1 lists the run time by these three methods using the same computer.

Note from Figures 7–9 that the goodness of fit between the blue line and green line by EEMD and AEMD methods is higher than that of EMD. The comparison of these three methods reveals that EEMD and AEMD can restrain effectively the mode confusion phenomena, but the running time of EEMD is much longer than those of AEMD and EMD, because the EEMD method adds white noise to original signals to alleviate mode mixing problem, which makes the decomposition process more complex and time consuming. The executed time of EMD is approximately equal with AEMD's, but it has inevitable mode mixing problem. Thus, AEMD has shown certain comprehensive advantages in alleviating mode mixing and making decomposition efficient.

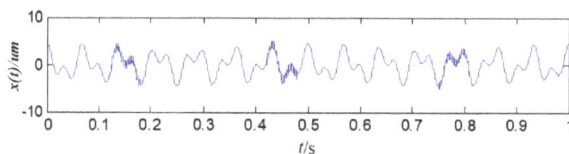

Figure 6. The time domain chart of simulation signal $x(t)$.

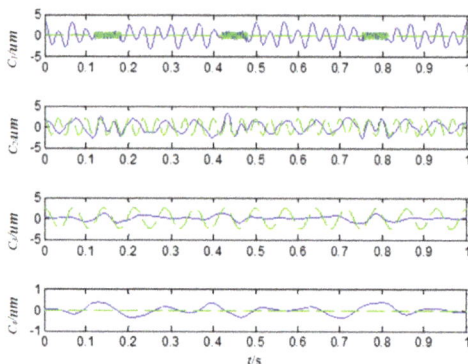

The green lines represent the components of the original signal.
The blue lines represent the composition results

Figure 7. The decomposition result of the simulation signal by EMD.

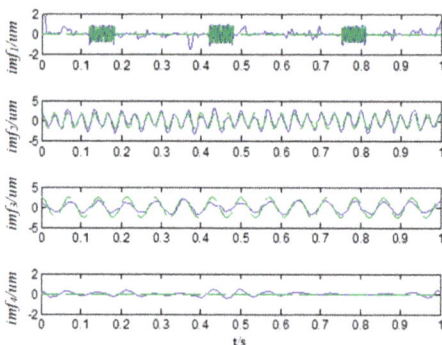

The green lines represent the components of the original signal.
The blue lines represent the composition results

Figure 8. The decomposition result of the simulation signal by ensemble empirical mode decomposition (EEMD).

Table 1. The running time of the three methods (unit: seconds).

Method	First	Second	Third	Fourth	Fifth	Sixth
EMD	1.3988	1.3860	1.3930	1.3846	1.3820	1.3940
EEMD	41.4875	44.8383	44.1800	42.8480	43.2321	44.024
AEMD	1.4027	1.3832	1.3843	1.3934	1.3880	1.3848

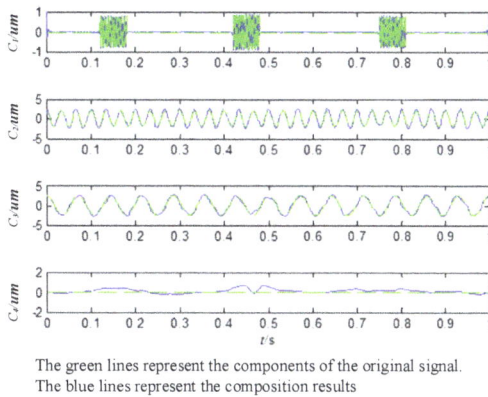

The green lines represent the components of the original signal.
The blue lines represent the composition results

Figure 9. The decomposition result of the simulation signal by AEMD (combined AMD and EMD).

4.4. Case Study

Figure 10 presents the experimental device to simulate the local rubbing fault of rotors, during which the rotate-speed was at 300 RPM, controlled by the inputted motor. The displacement sensor that was horizontally installed was employed to sample the vibration signals, with sampling frequency 4k Hz. Figure 11 plots the vibration signal of a practical rotor system with rub-fault. The decomposition result for the vibration signal by using the EMD and the AEMD are shown as Figures 12 and 13, respectively. Figure 14a,b is the spectrum diagram by fast Fourier transform (FFT) for the composition results of EMD and AEMD, respectively.

Note from Figures 12 and 14a that the mode mixing problem still distinctly exists in the decomposition results by EMD. In the imf_2 diagram, the characteristic frequency of the rotor rubbing fault is confused in multiple frequencies, caused by the slight rubbing fault of the rotor, which affects the identification for the fault. In the imf_3 diagram, the triple frequency of the characteristic frequency of the rotor rubbing fault—150 Hz is nearly invisible, which is mixed in the fundamental frequency of the rotor rubbing fault—50 Hz.

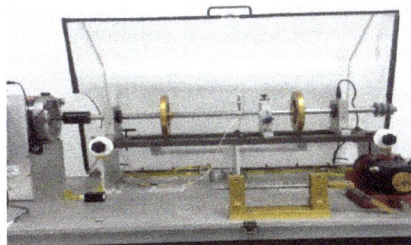

Figure 10. The experimental device for simulating local rubbing fault of rotors.

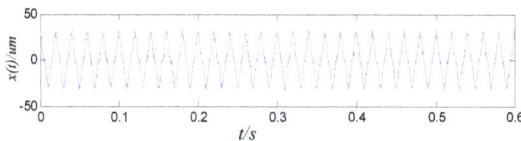

Figure 11. The vibration signal of rub-fault for rotors.

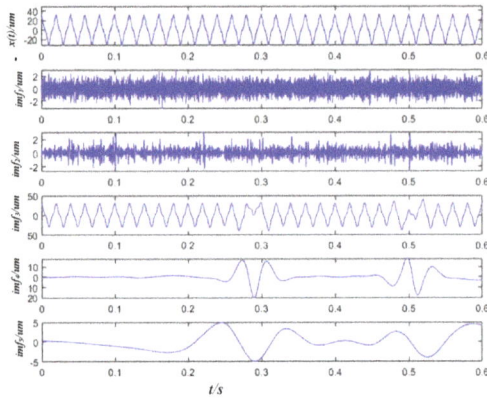

Figure 12. The decomposition result for the vibration signal of rub-fault by EMD.

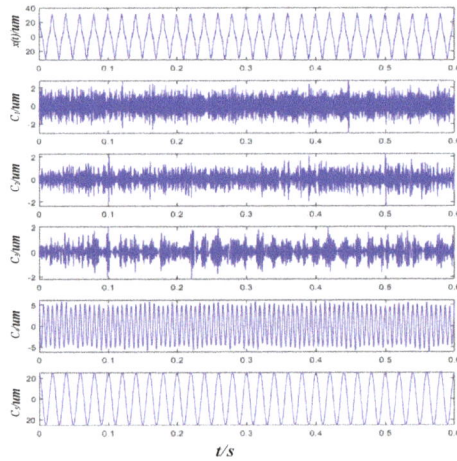

Figure 13. The decomposition result for the vibration signal of rubbing-fault by AEMD.

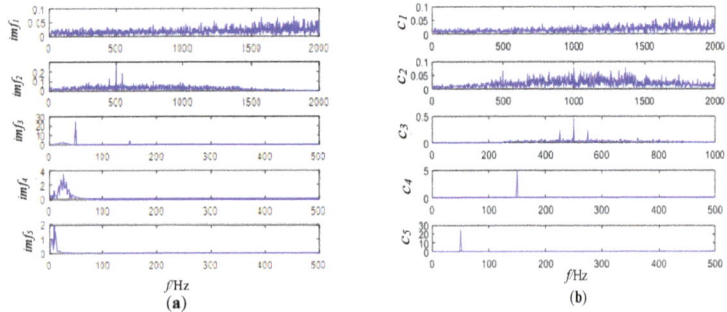

Figure 14. (a) The spectrum of intrinsic mode functions (IMFs) by EMD. (b) The spectrum of IMFs by AEMD.

In addition, note from Figures 13 and 14b that the tripling frequency 150 Hz and ten times frequency—500 Hz of the characteristic frequency of the rotor rubbing fault are distinctly decomposed. This spectrum character with multiple higher harmonics is in accordance with the slight rubbing fault of the rotors. Thus, the characteristic frequency of the rotor rubbing fault is successfully obtained by AEMD, though there is still slighter mode mixing.

5. Conclusions

In this paper, an improved sifting stop criterion based on valid data segment was proposed. Compared with the traditional one, results indicated that the newly proposed sifting stop criterion avoids the influence of end effects and improves the correctness of the EMD. In addition, a novel method, namely AEMD, which combines the AMD and EMD, was developed to solve the mode mixing problem of EMD. Model comparison was conducted between the proposed method and the EEMD. It is worth mentioning that both the AEMD and EEMD can effectively restrain the mode mixing, but the AEMD needs less execution time than that of EEMD. The proposed method overcomes some shortcomings of EMD to some extent and makes it better for use in the fault diagnosis of rotating machinery. However, the proposed method has certain imitations; the boundary frequency of AMD method requires human intervention, which means worse adaptivity than EMD. It can achieve better results for more simple signals, but its advantage is not obvious for complex engineering signals, since they take more time than EMD.

Author Contributions: Z.C. conceived the methodology, conducted the research and wrote the original draft. B.L. performed the programming to fulfil the method and edited the manuscript. X.Y. collected data and analyzed the data. H.Y. investigated the experiment and discussed the results.

Funding: This research was funded by the National Natural Science Foundation of China grant number 11672070.

Conflicts of Interest: We declare no conflict of interest.

References

1. Chen, Z.; Cao, S.; Mao, Z. Remaining useful life estimation of aircraft engines using a modified similarity and supporting vector machine (SVM) approach. *Energies* **2017**, *11*, 28. [CrossRef]
2. Han, J.; Ji, G.Y. Gear Fault Diagnosis Based on Improved EMD Method and the Energy Operator Demodulation Approach. *J. Changsha Univ. Sci. Technol.* **2015**, *12*, 66–71.
3. Yang, Y.F.; Wu, Y.F. *Application of Empirical Mode Decomposition in the Analysis of Vibration*; National Defence of Industry Press: Beijing, China, 2013; pp. 17–45.
4. Huang, D.S. The Method of False Modal Component Elimination in Empirical Mode Decomposition. Vibration. *Meas. Diagn.* **2011**, *31*, 381–384.
5. Wang, L. Fault Diagnosis of Rotor System Based on EMD-Fuzzy Entropy and SVM. *Noise Vib. Control* **2012**, *6*, 172–173.
6. Ben Ali, J.; Fnaiech, N.; Saidi, L.; Chebel-Morello, B.; Fnaiech, F. Application of empirical mode decomposition and artificial neural network for automatic bearing fault diagnosis based on vibration signals. *Appl. Acoust.* **2015**, *89*, 16–27. [CrossRef]
7. Xue, X.; Zhou, J.; Xu, Y.; Zhu, W.; Li, C. An Adaptively Fast Ensemble Empirical Mode Decomposition Method and Its Applications to Rolling Element Bearing Fault Diagnosis. *Mech. Syst. Sig. Process.* **2015**, *62–63*, 444–459. [CrossRef]
8. Yu, D.J.; Cheng, J.S. *The Hilbert-Huang Transform Method in Mechanical Fault Diagnosis*; Science Press: Beijing, China, 2006; pp. 179–190.
9. Cheng, J.; Xu, Y.L. Application of HHT Method in Structural Modal Parameter Identification. *J. Vib. Eng.* **2003**, *16*, 383–387.
10. Cheng, J.; Yang, Y. The application of EMD Method in Local Touch Friction Fault Diagnosis of Rotor. Vibration. *Meas. Diagn.* **2006**, *26*, 24–27.
11. Rilling, G.; Flandrin, P. One or Two Frequencies? The Empirical Mode Decomposition Answers. *IEEE Trans. Signal Process.* **2008**, *56*, 85–95. [CrossRef]
12. Dou, D.Y.; Zhao, Y.K. Application of Ensemble Empirical Mode Decomposition in Failure Analysis of Rotating Machinery. *Trans. Chin. Soc. Agric. Eng.* **2010**, *26*, 190–196.

13. Pustelnik, N.; Borgnat, P.; Flandrin, P. Empirical mode decomposition revisited by multicomponent non-smooth convex optimization. *Signal Process.* **2014**, *102*, 313–331. [CrossRef]

14. Lei, Y.G.; He, Z.J.; Zi, Y.Y. Application of the EEMD Method to Rotor Fault Diagnosis of Rotating Machinery. *Mech. Syst. Signal Process.* **2009**, *23*, 1327–1338. [CrossRef]

15. Zhao, J.P. Study on the Effects of Abnormal Events to Empirical Mode Decomposition Method and the Removal Method for Abnormal Signal. *J. Ocean Univ. Qingdao* **2001**, *31*, 805–814.

16. Deering, R.; Kaiser, J.F. The Use of a Masking Signal to Improve Empirical Mode Decomposition. *Acoust. Speech Signal Process.* ICASSP **2005**, *4*, 485–488.

17. Hu, A.J. *Research on the Application of Hilbert-Huang Transform in Vibration Signal Analysis of Rotating Machinery;* North China Electric Power University: Bao Ding, Hebei, China, 2008.

18. Wu, Z.; Huang, N.E. Ensemble Empirical Mode Decomposition: A Noise-Assisted Data Analysis Method. *Adv. Adapt. Data Anal.* **2009**, *1*, 1–41. [CrossRef]

19. Chen, G.D.; Wang, Z.C. A Signal Decomposition Theorem with Hilbert Transform and Its Application to Narrow Band Time Series with Closely Spaced Frequency Components. *Mech. Syst. Signal Process.* **2012**, *28*, 258–279. [CrossRef]

20. Lei, Y.G.; Kong, D.T. Adaptive Ensemble Empirical Mode Decomposition and Application to Fault Detection of Planetary Gear Boxes. *J. Mech. Eng.* **2014**, *50*, 64–70. [CrossRef]

21. Zheng, J.D.; Cheng, J.S.; Yang, Y. Partly Ensemble Empirical Mode Decomposition: An Improved Noise-Assisted Method for Eliminating Mode Mixing. *Signal Process.* **2014**, *96*, 362–374. [CrossRef]

22. Hu, J.S.; Yang, S.X. Energy-Based Stop Condition of Empirical Mode Decomposition of Vibration Signal. *J. Vib. Meas. Diagn.* **2009**, *29*, 19–22.

23. Mohammad, S.H.; Siamak, E.K.; Mohammad, S.S. Quantitative diagnosis for bearing faults by improving ensemble empirical mode decomposition. *ISA Trans.* **2018**, *83*, 261–275.

24. Yang, Y.; Yu, D.J.; Cheng, J.S. A Roller Bearing Fault Diagnosis Method Based on EMD Energy Entropy and ANN. *J. Sound Vib.* **2006**, *294*, 269–277.

25. Huang, N.E.; Shen, Z.; Long, S.R.; Wu, M.C.; Shih, H.H.; Zheng, Q.; Yen, N.C.; Tung, C.C.; Liu, H.H. The Empirical Mode Decomposition and the Hilbert Spectrum for Nonlinear and Non-stationary Time Series Analysis. *Proc. R. Soc. Lond. A* **1998**, *454*, 903–995. [CrossRef]

energies

MDPI

Article

A Lithium-ion Battery RUL Prediction Method Considering the Capacity Regeneration Phenomenon

Xiaoqiong Pang [1,*], Rui Huang [1], Jie Wen [2], Yuanhao Shi [2], Jianfang Jia [2] and Jianchao Zeng [1]

[1] School of Data Science and Technology, North University of China, No.3, XueYuan Road,
    JianCaoPing District, Taiyuan 030051, China; 18834169601@163.com (R.H.); zengjianchao@263.net (J.Z.)
[2] School of Electrical and Control Engineering, North University of China, No.3, XueYuan Road,
    JianCaoPing District, Taiyuan 030051, China; wenjie015@gmail.com (J.W.); yhshi@nuc.edu.cn (Y.S.);
    jiajianfang@nuc.edu.cn (J.J.)
*   Correspondence: xqpang@nuc.edu.cn; Tel.: +86-0351-3923623

Received: 11 May 2019; Accepted: 8 June 2019; Published: 12 June 2019

Abstract: Prediction of Remaining Useful Life (RUL) of lithium-ion batteries plays a significant role in battery health management. Battery capacity is often chosen as the Health Indicator (HI) in research on lithium-ion battery RUL prediction. In the rest time of batteries, capacity will produce a certain degree of regeneration phenomenon, which exists in the use of each battery. Therefore, considering the capacity regeneration phenomenon in RUL prediction of lithium-ion batteries is helpful to improve the prediction performance of the model. In this paper, a novel method fusing the wavelet decomposition technology (WDT) and the Nonlinear Auto Regressive neural network (NARNN) model for predicting the RUL of a lithium-ion battery is proposed. Firstly, the multi-scale WDT is used to separate the global degradation and local regeneration of a battery capacity series. Then, the RUL prediction framework based on the NARNN model is constructed for the extracted global degradation and local regeneration. Finally, the two parts of the prediction results are combined to obtain the final RUL prediction result. Experiments show that the proposed method can not only effectively capture the capacity regeneration phenomenon, but also has high prediction accuracy and is less affected by different prediction starting points.

Keywords: lithium-ion battery; remaining useful life; regeneration phenomenon; wavelet decomposition; NAR neural network

1. Introduction

Lithium-ion batteries are considered to be the best energy storage devices for many applications because of their light weight, high energy density, and long life [1–3]. From tiny Bluetooth headsets, cameras, mobile phones, and laptops to hybrid electric cars and aerospace power systems, batteries are very important and critical components. At the same time, battery failures can lead to performance degradation of the power system, or even directly lead to the failure of the task, and increase costs. Improper management of lithium-ion batteries in electric vehicles may cause fire or even explosions [4], and power system malfunction is the main cause of mission failure in the aerospace field [5–7]. Therefore, it is significant to study the prediction of RUL of lithium-ion batteries in practical applications.

In the field of lithium-ion battery RUL prediction, the gradually degraded battery capacity is often used as an effective health indicator in order to track the attenuated process of the battery. In general, a lithium-ion battery is deemed to fail when its capacity fades by 20–30% of the rated value [8]. In some online applications of lithium-ion battery health prognostics, impedance, voltage, and current are often used as HIs to reflect the battery degradation [9,10]. However, the RUL prediction using capacity as the HI is generally more accurate and effective than other HIs as the change in battery capacity directly reveals the health states of batteries [8]. Therefore, battery capacity as the HI is widely used in the RUL prediction of lithium-ion batteries. Additionally, the procedure of battery degradation is not monotonous, the working process of the battery consists of charging, discharging, and the rest stage. Overall, the capacity of a lithium-ion battery shows a degradation trend during usage because of the side reactions that occur between the electrodes and electrolyte of the battery. However, when the battery rests during charge/discharge profiles, the residual reaction products have a chance to dissipate, thus increasing the available capacity for the next cycle. This phenomenon, which emerges during the use of lithium-ion batteries, is called regeneration. The regeneration phenomenon of lithium-ion batteries can alter the trend of the capacity prediction curve, thus affecting the performance of prognostic models [11,12]. Therefore, it is important to consider the capacity regeneration phenomenon in RUL prediction of lithium-ion batteries. Furthermore, the capacity time series should be considered as a hybrid signal of multi-scale components, where global degradation and local regeneration are signals of different scales. It is necessary to decouple the correlative components from the original capacity time series in RUL prediction for extracting the most useful information [13].

In recent years, the RUL prediction methods for lithium-ion batteries can be divided into two types: model-based and data-driven [7]. The data-driven prediction method has become a current research hotspot because it does not need to analyze the complex internal mechanism and electrochemical reaction process of batteries, only relies on the existing historical monitoring data, and avoids the defects of a complex modeling process and large interference caused by environmental factors. Data-driven methods [8,13–21] for health prediction of lithium-ion batteries usually use machine learning models (support vector machines, logistic regression, and neural networks, etc.) to establish the relationship between monitoring data and system health, so as to track the battery degradation and estimate the RUL of batteries. Most of above methods [14–20] only focus on the global degradation trend of batteries and ignore the regeneration phenomenon in battery rest time. In [21], Deng et al. propose an improved empirical model based on that of Saha et al. [12], where they relax the fixed Coulombic efficiency and estimate it with measured data; the dual EKF estimation is then employed to deal with the coupled problem of parameter and state capacity. However, the experimental results show that the sudden rise in capacity during regeneration is not predicted obviously. In [13], He et al. utilize a wavelet decomposition and Gaussian regression combined method to capture the regeneration phenomenon in battery health prognostics, but the performance is not ideal for capturing the regeneration part. In [8], Yu proposes a method that combines empirical mode decomposition (EMD), logistic regression, and Gaussian regression with consideration of the regeneration phenomenon, where an adaptive moving window is added to the regression process to capture the regeneration of batteries by constantly changing the size of the window. Compared with reference [13], this method performs better in capturing the regeneration section. However, the RUL prediction accuracy of such regression methods [8,13] is greatly affected by different prediction starting points. Therefore, it is still a challenging task to design an appropriate RUL prediction method for lithium-ion batteries which can capture the capacity regeneration phenomenon well and acquire a good prediction performance.

In addition, the degradation process of batteries is a complex, dynamic, and nonlinear electrochemical process [22]; in the use of a battery, the capacity and service life of batteries show an irreversible trend of gradual decline with time. Moreover, the degradation of capacity accelerates in the later cycle life, showing a nonlinear characteristic. Therefore, battery capacity degradation data is a kind of nonlinear time series data based on monitoring. In order to predict RUL better, it is necessary to establish an appropriate model that can effectively deal with capacity degradation data with the characteristics of nonlinear time series. With the rise of neural networks, many neural network based methods have been applied in the field of health management and life prediction [18,23–31]. Among them, the NARNN [25,26,29,30] is a dynamic recurrent neural network with time series prediction capability, which can effectively simulate nonlinear processes and deal with stationary and non-stationary time series. Therefore, this paper chooses NARNN to establish the RUL prediction model of lithium-ion batteries.

Driven by the desire to capture capacity regeneration effectively and improve prediction accuracy, a novel WDT–NARNN method for lithium-ion battery RUL prediction is proposed based on a combination of WDT and NARNN. The WDT with multi-resolution characteristics is used to decompose the capacity time series in multi-scale to get both the global degradation trend and the local regeneration. The NARNN is utilized to recursively predict global degradation and local regeneration. Finally, the NARs are integrated to achieve RUL prediction of lithium-ion batteries. Therefore, the main contributions and innovations of this paper include the following: (1) The global degradation and local regeneration in battery capacity time series can be separated effectively by WDT, which will be helpful to improve the prediction performance of the prediction model; (2) A combined model based on WDT and NARNN is established to model the local and global tendency of the battery capacity changes, which enables the prediction model to capture the actual capacity decay tendency of batteries effectively.

2. Related Algorithms

2.1. Wavelet Decomposition

Wavelet decomposition technology is a powerful tool to analyze non-linear and non-stationary time series, which is widely used in various fields of engineering [13,32–35]. The global degradation and local regeneration of battery capacity can be decomposed by using wavelet decomposition technology, and the two parts can be processed respectively to achieve accurate prediction.

Wavelet analysis utilizes the wavelet function $\psi(\omega)$ and scaling function $\varphi(\omega)$ to perform the multiresolution analysis decomposition and reconstruction of the signal. Following the idea used in deriving the Meyer's wavelet, Gilles [32] defines the empirical scaling function as

$$\varphi_i(\omega) = \begin{cases} 1, & \text{if } |\omega| \leq (1-\xi)\omega_i \\ \cos(\frac{\pi\rho(\xi,\omega_i)}{2}), & \text{if } (1-\xi)\omega_i \leq |\omega| \leq (1+\xi)\omega_i \\ 0, & \text{otherwise} \end{cases} \tag{1}$$

and the empirical wavelet function as

$$\psi_i(\omega) = \begin{cases} 1, & \text{if } (1+\xi)\omega_i \leq |\omega| \leq (1-\xi)\omega_{i+1} \\ \cos(\frac{\pi\rho(\xi,\omega_{i+1})}{2}), & \text{if } (1-\xi)\omega_{i+1} \leq |\omega| \leq (1+\xi)\omega_{i+1} \\ \sin(\frac{\pi\rho(\xi,\omega_i)}{2}), & \text{if } (1-\xi)\omega_i \leq |\omega| \leq (1+\xi)\omega_i \\ 0, & \text{otherwise} \end{cases} \tag{2}$$

where $\rho(x)$ is any arbitrary function with values in the range $[0,1]$ with the following properties:

$$\rho(x) = \begin{cases} 0, & \text{if } x \leq 0 \\ \text{and } \rho(x)+\rho(1-x) & \forall x \in [0,1] \\ 1, & \text{if } x \geq 1 \end{cases} \tag{3}$$

Then, the inherent mechanism of empirical wavelet transform is based on the formation of adaptive wavelet based filters. In the wavelet analysis, the Mallat algorithm is used to decompose the signals and obtain trend information (low frequency) and regeneration information (high frequency). The decomposition equations are expressed as

$$A_{j+1,k} = \sum_m h(m-2k)A_{j,m} \tag{4}$$

$$D_{j+1,k} = \sum_m g(m-2k)A_{j,m} \tag{5}$$

where j is the decomposition scale, k and m are translation variables, h is the low-pass filter, and g is the high-pass filter.

The decomposed data is reconstructed by reconstruction algorithm to restore it to the original spatial scale. The reconstruction equation is described as

$$A_{j+1,k} = \sum_m h^*(m-2k)A_{j,m} + \sum_m g^*(m-2k)A_{j,m} \tag{6}$$

where h^* and g^* are the inverse functions of low-pass and high-pass filters respectively.

2.2. NAR Neural Network

NARNN is a kind of dynamic recurrent neural network with time series prediction ability. It forms a discrete nonlinear autoregressive system with endogenous input [26,29], and the mathematical representation of NARNN can be defined as follows:

$$y(t) = f(y(t-1), y(t-2), \ldots, y(t-d)) \tag{7}$$

where $y(t)$ is the current output, $y(t-1), y(t-2), \ldots, y(t-d)$ are the historical outputs, d is the delay of the network.

NARNN is composed of an output layer, hidden layer, and feedback layer. The function of the feedback layer is to store the previous outputs, which can be regarded as a kind of 'memory' operator. The network structure of NARNN is shown in Figure 1.

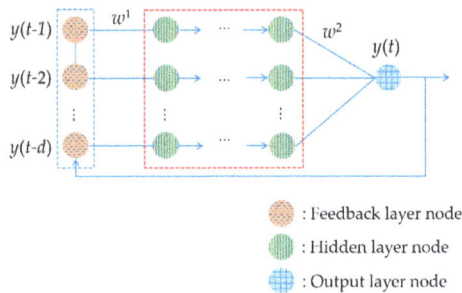

Figure 1. The structure of the Nonlinear Auto Regressive (NAR) neural network, where the w^1 is the connection matrix between input layer and hidden layer, w^2 is the connection matrix between hidden layer and output layer.

3. WDT–NARNN Prediction Method

3.1. Experiment Data Analysis

In this article, the experimental battery data used are derived from the data repository of NASA Ames Prognostics Center of Excellence (PCoE) [36]. Four typical types of 18,650 sized rechargeable batteries (#5, #6, #7 and #18) were used to illustrate the performance of our proposed approach for RUL prediction of lithium-ion batteries. Figure 2 shows the true capacity degradation curves of the four batteries. It can be seen that there is a clear descending trend, namely, the global degradation, and several capacity regenerations (black circles). Obviously, the existence of these capacity regeneration phenomena changes the normal degradation trend of the batteries.

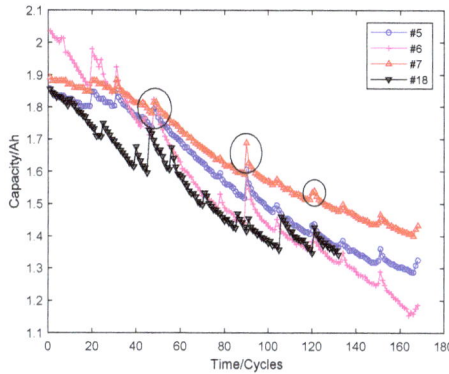

Figure 2. The capacity degradation curves of four batteries (#5, #6, #7, and #18).

A key step of the proposed model in this paper is to use WDT to decompose the capacity time series at multi-scale to obtain the global degradation trend and local regeneration. Take battery #5 as an example, the results of 6-layer decomposition by using the 'dmey' wavelet is shown in Figure 3. The curve 'a6' in Figure 3(a6) is the low frequency signal, which reveals the normal degradation tendency of the battery capacity. Curve 'C' in Figure 3(a6) is the real capacity degradation curve. d1–d6 are the high frequency signals, and the peak value of the fluctuations corresponds to the captured local regenerations. It can be seen that wavelet decomposition can get more degradation information of battery capacity at different scales, and taking these parts into account will be beneficial to improve the prediction accuracy of the RUL prediction model for lithium-ion batteries.

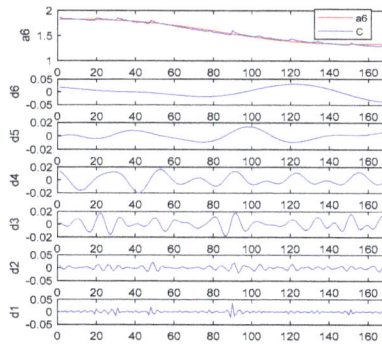

Figure 3. The wavelet decomposition of capacity for battery #5.

3.2. WDT- NARNN Modeling Process

As can be seen from the above decomposition results, the global degradation trend and local regenerations of capacity are quite different in frequency characteristics and shapes. If the prediction is performed directly, it is difficult to capture both features at the same time. Therefore, we proposed a multi-scale decomposition and fusion prediction method WDT–NARNN for the RUL prediction of lithium-ion batteries. The major procedures of the proposed WDT–NARNN modelling method is shown in Figure 4, which is mainly divided into the following steps:

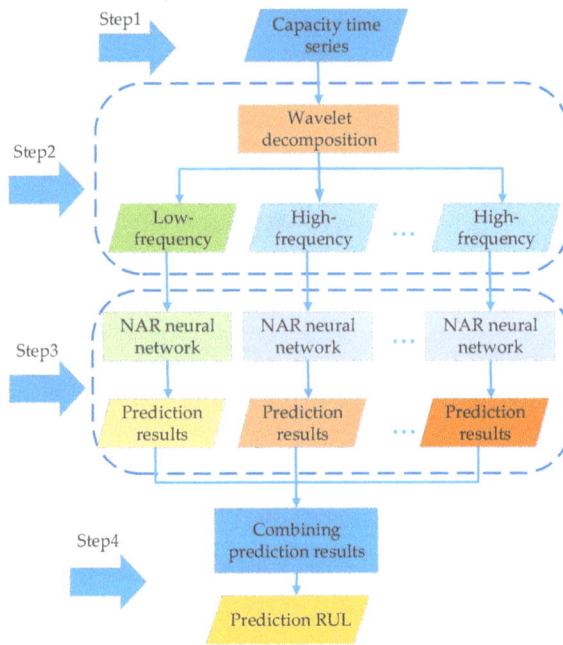

Figure 4. Procedures of lithium-ion battery Remaining Useful Life (RUL) prediction based on the proposed method.

(1) Extract the raw battery capacity series $\{Cap(i), i\}_{i=1}^{N}$, N means the total cycle number of each battery.

(2) Decomposition of capacity time series. Firstly, wavelet function $\psi(t)$ and the decomposition level l are initialized. Based on that, the raw capacity time series are decomposed into l levels. At the same time, several high frequency signals and one low frequency signal are obtained by low and high pass filters. The low frequency signal reflects the global trend of capacity degradation and high frequency signals reflect local fluctuations of capacity regeneration.

(3) Constructing prediction models for low-frequency and high-frequency parts respectively. The prediction starting point is set to T. The low- and high-frequency parts after wavelet decomposition $\{WDT\_Cap(i), i\}_{i=1}^{T-1}$ are used as the training samples respectively. Then, NAR neural networks for each decomposition component are designed and trained. Next, the trained NAR models are used to predict the later time series data of each decomposed component.

(4) Combining prediction results. The final capacity prediction result is obtained by combining and reconstructing the prediction results of all the NAR models. Then, according to the relationship between the capacity and cycle numbers, the final RUL value is calculated by the following equation:

$$N_{RUL} = N_{EOL} - N_{ECL} \tag{8}$$

where N_{RUL} is the remaining useful life. N_{EOL} is the total charge and discharge cycles when the actual battery capacity degrades to the threshold. N_{ECL} is the charge and discharge cycles at the current time of the battery.

For clarity, the integrated method with WDT and NARNN can be summarized in Algorithm 1.

Algorithm 1. The integrated method with WDT and NARNN

---

(1) Initialization:

Select the wavelet function $\psi(t)$ and the decomposition levels l;

(2) Decomposition:

Decompose the capacity series $\{Cap(i), i\}_{i=1}^{T-1}$ for the l levels to obtain the low and high-frequency signals at different scales by Equations (4) and (5);

(3) Initialize the NAR neural network:

Initialize the parameters of NAR neural network, the numbers of input layer, hidden layer, and output layer are set to N_i, N_h, and N_o respectively, the delay of the network is set to d, and the training function is set to 'trainbr';

(4) Output the prediction results:

Input the decomposed signals $\{WDT\_Cap(i), i\}_{i=1}^{T-1}$ into the NAR models to predict the following changes after time T, then prediction results $\{WDT\_Cap(i), i\}_{i=T}^{N}$ are obtained;

(5) Wavelet reconstruction:

The signals $\{WDT\_Cap(i), i\}_{i=T}^{N}$ are reconstructed from 1 to l levels by Equation (6) to obtain the fusing predicted series corresponding to capacity series, and then RUL value can be calculated by Equation (8);

(6) Evaluate the prediction results:

The evaluation is given with original testing data and prediction results through some criteria to evaluate the performance of the integrated method WDT–NARNN.

3.3. Performance Analysis

We use the following four evaluation criteria to measure and demonstrate the accuracy and stability of the proposed method:

(1) *Root Mean Square Error* (*RMSE*) to evaluate the prediction accuracy. The smaller the *RMSE* is, the better the prediction performance:

$$RMSE = \sqrt{\frac{\sum\limits_{i=1}^{n} (y_i - \hat{y}_i)^2}{n}} \tag{9}$$

(2) R^2 to evaluate the prediction performance. If the fitting degree between the prediction curve and real curve is high, R^2 will be close to 1:

$$R^2(y, \hat{y}) = 1 - \frac{\sum\limits_{i=0}^{n-1} (y_i - \hat{y}_i)^2}{\sum\limits_{i=0}^{n-1} (y_i - \bar{y}_i)^2} \tag{10}$$

(3) *Absolute Error* (*AE*) to evaluate the RUL accuracy of the prediction model:

$$AE = |R - \hat{R}| \tag{11}$$

(4) Prediction Accuracy Improvement Ratio (η_{AE}) to evaluate the RUL prediction accuracy improvement ratio of two different methods. If $\eta_{AE} > 0$, the first method is more accurate, on the contrary, the second method has higher prediction accuracy:

$$\eta_{AE} = \frac{AE_2 - AE_1}{R} \tag{12}$$

where n is the sample size, y_i is the real value of battery capacity, $\hat{y}_i$ is the predicted value of battery capacity, and $\bar{y}_i$ is the mean value of predicted battery capacity. R is the real RUL, $\hat{R}$ is the predicted RUL.

4. Results and Discussion

4.1. RUL Prediction of Lithium-Ion Battery

In order to verify the effectiveness of the proposed RUL prediction model for lithium-ion batteries based on WDT–NARNN (named M1), we design two comparative models M2 and M3 as shown in Table 1. Model M2 uses the NAR neural network to directly establish the original capacity time series prediction model without using wavelet decomposition technology to decompose the capacity series. Model M3 uses BP neural network to establish the prediction model, and the other steps are the same as model M1. In this comparison, M2 is tested to analyze the effect of WDT in the proposed model M1. M3 is considered to illustrate the ability of the NAR neural network to predict capacity time series. Experimental setting: We select the data from cycle 1 to cycle 69 as the training samples, and the predicted starting point is 70. The 'dmey' wavelet is selected to carry out 6-layer decomposition of capacity series. The Bayesian regularization algorithm (trainbr) is utilized as the training function of NARNN. The hidden layer nodes are set to 10 and the feedback delay is set to 2. The training samples of network are divided into 70% training samples, 15% verification samples, and 15% test samples.

Table 1. The proposed three models (M1–M3).

Model	Model Description
M1	WDT combine with NARNN
M2	NARNN without using WDT
M3	WDT combine with BPNN

Figure 5 shows the comparison of prediction results of different models. It can be observed that the prediction curve of model M1 is closest to the real capacity degradation curve and the capturing of capacity regeneration is most accurate. Compared with model M1, the prediction result of model M2 is less effective in capturing capacity regeneration, and is more far away from the real capacity with the increase in cycle numbers. The comparison reflects the validity of using WDT to extract capacity global degradation and local regenerations to build prediction models respectively. Meanwhile, model M3 uses the WDT to separate the global degradation and local regeneration, but the prediction curve of M3 still shows a smooth trend. This phenomenon indicates that the traditional BP neural network does not learn the regeneration phenomenon well in the training stage of network, which is mainly due to its lack of feedback connection structure, resulting in its lack of 'memory' ability. So, BP neural network is not very effective in dealing with RUL prediction of lithium-ion batteries, which is a time series prediction problem. However, the model M1 utilizes NARNN with time series prediction ability that can learn the capacity regeneration well and has a better prediction performance.

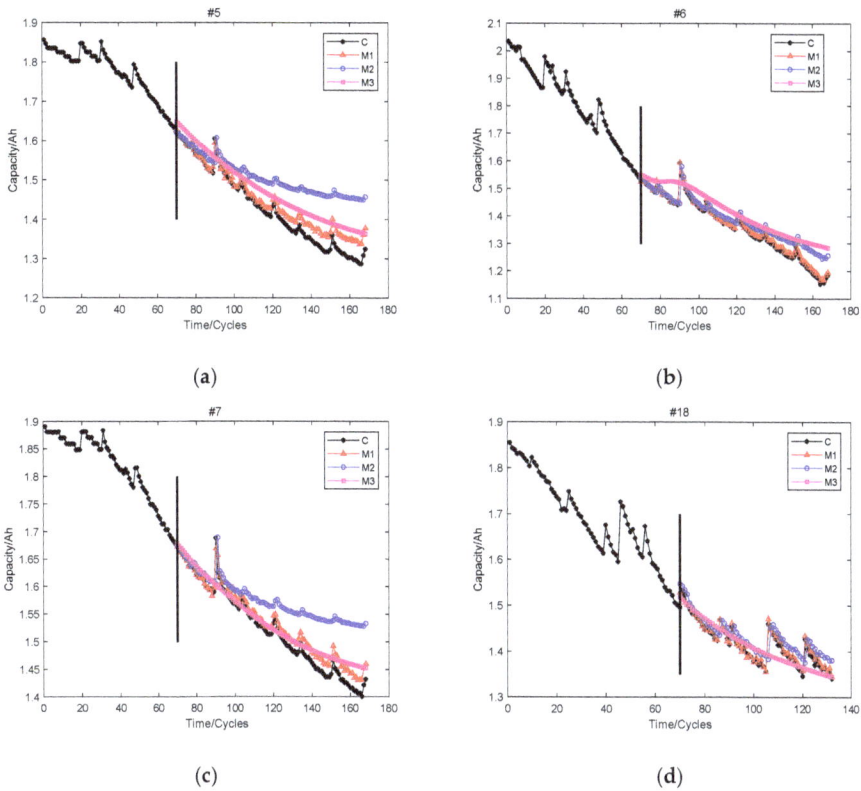

Figure 5. The comparisons for different models. (**a**) #5; (**b**) #6; (**c**) #7; (**d**) #18.

In addition, Table 2 gives the prediction performance of different models for four batteries (#5, #6, #7, and #18). It can be seen that the RMSE of four batteries predicted by model M1 are all smaller than 0.03 and R^2 are all larger than 0.9, indicating that model M1 has good prediction performance and high fitting degree to the original capacity curve. Meanwhile, as can be seen from columns 3–6 in Table 2, for RMSE, model M1 of each battery is significantly smaller than model M2 and M3. For R^2, M1 is significantly larger than model M2 and M3. The above analysis illustrates that model M1 can effectively capture the regeneration of capacity and has good prediction performance, and the prediction performance of model M1 is significantly better than that of model M2 and M3.

Table 2. Comparison of different models (M1–M3) for four batteries.

Evaluate Criteria	Model	#5	#6	#7	#18
RMSE	M1	0.0270	0.0087	0.0175	0.0064
	M2	0.0949	0.0436	0.0678	0.0260
	M3	0.0500	0.0616	0.0234	0.0253
R^2	M1	0.9226	0.9933	0.9460	0.9751
	M2	0.4151	0.8457	0.4611	0.6494
	M3	0.7745	0.7298	0.9035	0.6091

4.2. Different Starting Point Predictions and Comparison

Considering the EMD decomposition combined with logic regression and Gaussian regression model (named M-LG) proposed in reference [8] also realized the lithium-ion battery health prognostics with capturing the capacity regeneration phenomenon. In order to analyze the validity of model M1 more comprehensively, the prediction experiments of model M1 at four different prediction starting points (60, 70, 80, and 90) are designed, and the prediction results of RUL are compared with those of M2, M3, and M-LG models.

Figure 6 shows the capacity series prediction of model M1 at different prediction starting points for four batteries. It can be observed from the overall that, the later the prediction starting point is, the closer the prediction result is to the real value for the four batteries. For battery #5 and #7, the effect of prediction starting point on predicted results is relatively obvious. For battery #6, the predicted curves at the starting points 70, 80, and 90 are near to the real capacity curve. For battery #18, the prediction results at each starting point are very close to the real value, which means the prediction result is not influenced evidently by the different prediction starting point.

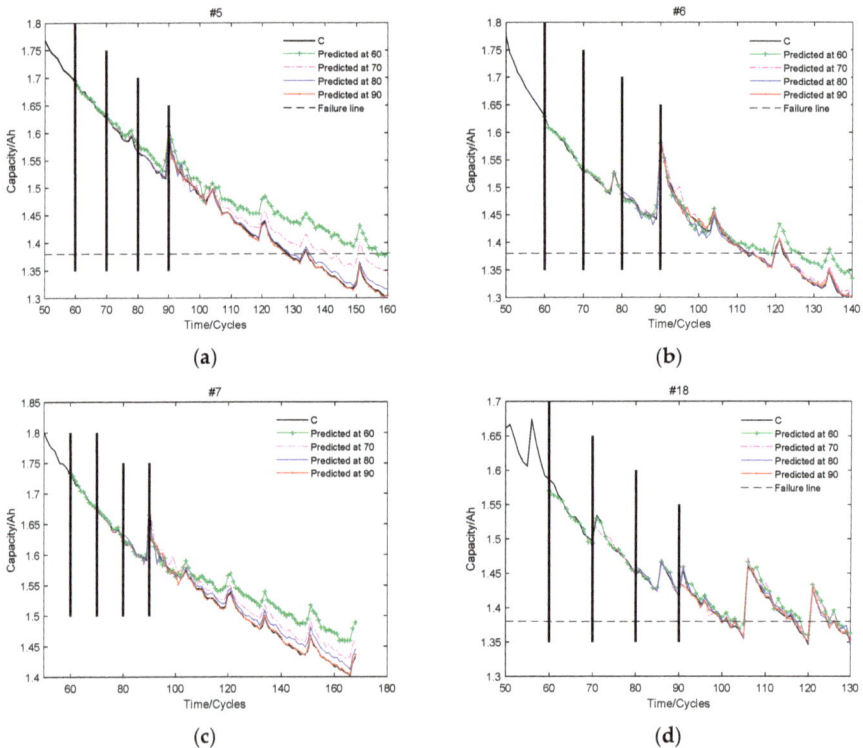

Figure 6. Capacity series prediction of model M1 at different prediction starting points: (**a**) battery #5; (**b**) battery #6; (**c**) battery #7; (**d**) battery #18.

Table 3 gives the RUL prediction results using model M1 at different starting points for battery #5, #6, and #18 (Since the degradation of battery #7 does not reach the failure threshold, no research is done for the time being). We can see that for each battery, the AE of RUL prediction results decreases with the prediction starting point moving backward. For battery #5, the RUL AE is equal to 28 and 12 at the prediction starting points 60 and 70, respectively, and the errors are relatively large. However, the RUL AE is 0 and 1 at the prediction starting point 80 and 90 respectively. For battery #6, the RUL AE at prediction starting point 60 is 5, and the RUL AE is less than or equal to 1 at the starting point 70 and later. For battery #18, the RUL AE at each prediction starting point is less than or equal to 2, which is obviously less effected by the prediction starting point. The above experimental results show that the model M1 is relatively less affected by the prediction starting point, and the prediction performance is relatively stable.

Table 3. RUL prediction results of model M1 at different prediction starting points.

Battery	Prediction starting point	Predicted RUL	RUL AE
#5	60	96	28
	70	70	12
	80	48	0
	90	37	1
#6	60	57	5
	70	42	0
	80	33	1
	90	22	0
#18	60	42	2
	70	30	0
	80	20	0
	90	10	0

Besides, we did 50 times prediction experiments for batteries #5, #6, and#18 using model M1 at different prediction starting points respectively, and gave the PDF of predicted End of Life (EOL) in Figure 7 (Color Filling Part), that is, within the PDF distribution range, the battery is likely to reach the EOL. As a whole, each distribution at different starting points is relatively concentrated, and as the predicted starting point moves backward, the center of the PDF is closer to the actual EOL, which indicates that the predicted result is more accurate.

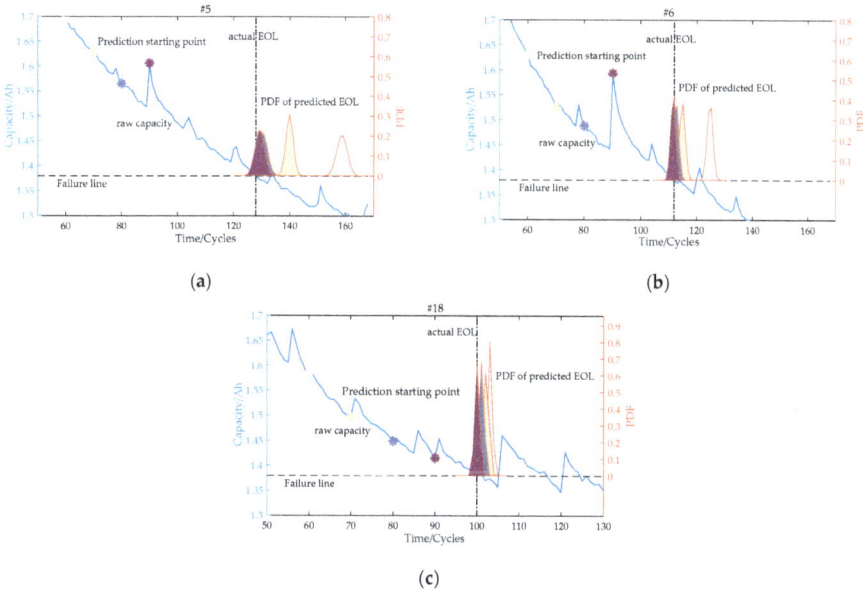

(a)

(b)

(c)

Figure 7. Uncertainty of predicted End of Life (EOL) at different prediction starting points. (**a**) battery #5; (**b**) battery #6; (**c**) battery #18.

Table 4 gives the comparison of average RUL AE and average η_{AE} under four different starting points of models M1, M2, M3, and M-LG for three batteries, where η_{AE} is the prediction accuracy improvement ratio of model M1 corresponding to other models, so the average η_{AE} of M1 is null. As can be seen from the third column, for batteries #5, #6, and #18, the average RUL AE of model M1 is obviously smaller than that of other models. From the fourth column, we can see that the average η_{AE} is between 14.3% and 54.9%, which shows the prediction performance of model M1 is obviously better than that of model M2, M3, and M-LG on the whole. All the above experiments show that the proposed WDT–NARNN method is reasonable and suitable for RUL prediction of lithium-ion batteries with capturing capacity regeneration phenomenon.

Table 4. Comparisons of the average RUL prediction accuracy at four prediction starting points under model M1 and other three methods (M2, M3, and M-LG).

Battery	Method	Average RUL AE	Average η_{AE}
#5	M1	10.3	-
	M2	14	34.2%
	M3	12	21.1%
	M-LG	16.3	14.3%
#6	M1	1.5	-
	M2	17.8	37.8%
	M3	12	34.6%
	M-LG	22.3	54.9%
#18	M1	0.5	-
	M2	6.5	35.6%
	M3	14.8	50%
	M-LG	6	25.8%

5. Conclusions

Considering that the capacity regeneration phenomenon has a great impact on the RUL prediction of lithium-ion batteries, in order to improve the RUL prediction accuracy, a WDT–NARNN method with considering capacity regeneration phenomenon is proposed. Firstly, the Wavelet decomposition technology is used to decompose the capacity time series at multi-scales to extract the global degradation part and local regeneration part. Then, the time series prediction models based on NAR neural network for the two parts are constructed respectively. Finally, the two parts of the prediction results are combined to obtain the final RUL prediction value. Experimental results show that the proposed method can effectively capture the regeneration phenomenon and has high prediction accuracy. Additionally, the prediction performance is stable and less affected by different prediction starting point.

Author Contributions: Conceptualization, X.P.; methodology, X.P.; validation, R.H.; writing—original draft preparation, R.H.; writing—review and editing, X.P.; visualization, R.H.; supervision, J.W., Y.S., J.J. and J.Z.

Funding: This work was partially supported by National Natural Science Foundation of China under Grant 61533013, the key Program of Research and Development of ShanXi Province under Grand 201703D111011, and the Natural Science Foundation of ShanXi Province under Grant 201801D121159, 201801D221208.

Conflicts of Interest: The authors declare no conflict of interest.

References

1. Liu, D.T.; Zhou, J.B.; Guo, L.M.; Peng, Y. Survey on lithium-ion battery health assessment and cycle life estimation. *Chin. J. Sci. Instrum.* **2015**, *36*, 1–16. [CrossRef]
2. Lukic, S.M.; Cao, J.; Bansal, R.C.; Rodriguez, F.; Emadi, A. Energy storage systems for automotive applications. *IEEE Trans. Ind. Electron.* **2008**, *55*, 2258–2267. [CrossRef]
3. Nishi, Y. Lithium-ion secondary batteries. *Power Sour.* **2001**, *100*, 101–106. [CrossRef]
4. Saha, B.; Goebel, K.; Poll, S.; Christophersen, J. Prognostics Methods for Battery Health Monitoring Using a Bayesian Framework. *IEEE Trans. Instrum. Meas.* **2009**, *58*, 291–296. [CrossRef]
5. Goebel, K.; Saha, B.; Saxena, A.; Celaya, J.R.; Christophersen, J.P. Prognostics in battery health management. *IEEE Instrum. Meas. Mag.* **2008**, *11*, 33–40. [CrossRef]

6. Williard, N.; He, W.; Hendricks, C.; Pecht, M. Lessons Learned from the 787 Dreamliner Issue on Lithium-Ion Battery Reliability. *Energies* **2013**, *6*, 4682–4695. [CrossRef]
7. Lu, L.G.; Han, X.B.; Li, J.Q.; Hua, J.F.; Ouyang, M.G. A review on the key issues for lithium-ion battery management in electric vehicles. *J. Power Sour.* **2013**, *226*, 272–288. [CrossRef]
8. Yu, J.B. State of health prediction of lithium-ion batteries: Multiscale logic regression and Gaussian process regression ensemble. *Reliab. Eng. Syst. Saf.* **2018**, *174*, 82–95. [CrossRef]
9. Sun, Y.H.; Jou, H.L.; Wu, J.C. Aging Estimation Method for Lead-Acid Battery. *IEEE Trans. Energy Convers.* **2011**, *26*, 264–271. [CrossRef]
10. Yu, J.B. State-of-Health Monitoring and Prediction of Lithium-Ion Battery Using Probabilistic Indication and State-Space Model. *IEEE Trans. Instrum. Meas.* **2015**, *64*, 2937–2949. [CrossRef]
11. Liu, D.T.; Pang, J.Y.; Zhou, J.B.; Peng, Y.; Pecht, M. Prognostics for state of health estimation of lithium-ion batteries based on combination Gaussian process functional regression. *Microelectron. Reliab.* **2013**, *53*, 832–839. [CrossRef]
12. Saha, B.; Goebel, K. Modeling Li-ion Battery Capacity Depletion in a Particle Filtering Framework. In Proceedings of the Annual Conference of the Prognostics and Health Management Society, San Diego, CA, USA, 27 September–1 October 2009.
13. He, Y.J.; Shen, J.N.; Shen, J.F.; Ma, Z.F. State of health estimation of lithium-ion batteries: A multiscale Gaussian process regression modeling approach. *Aiche J.* **2015**, *61*, 1589–1600. [CrossRef]
14. Song, Y.C.; Liu, D.T.; Yang, C.; Peng, Y. Data-driven hybrid remaining useful life estimation approach for spacecraft lithium-ion battery. *Microelectron. Reliab.* **2017**, *75*, 142–153. [CrossRef]
15. Gao, D.; Huang, M.H. Prediction of Remaining Useful Life of Lithium-ion Battery based on Multi-kernel Support Vector Machine with Particle Swarm Optimization. *J. Power Electron.* **2017**, *17*, 1288–1297. [CrossRef]
16. Qin, T.C.; Zeng, S.K.; Guo, J.B. Robust prognostics for state of health estimation of lithium-ion batteries based on an improved PSO-SVR model. *Microelectron. Reliab.* **2015**, *55*, 1280–1284. [CrossRef]
17. Song, Y.C.; Liu, D.T.; Hou, Y.D.; Yu, J.X.; Peng, Y. Satellite lithium-ion battery remaining useful life estimation with an iterative updated RVM fused with the KF algorithm. *Chin. J. Aeronaut.* **2018**, *31*, 31–40. [CrossRef]
18. Wu, J.; Zhang, C.B.; Chen, Z.H. An online method for lithium-ion battery remaining useful life estimation using importance sampling and neural networks. *Appl. Energy* **2016**, *173*, 134–140. [CrossRef]
19. Liu, D.T.; Yin, X.H.; Song, Y.C.; Liu, W.; Peng, Y. An On-Line State of Health Estimation of Lithium-Ion Battery Using Unscented Particle Filter. *IEEE Access* **2018**, *6*, 40990–41001. [CrossRef]
20. Peng, Y.; Hou, Y.D.; Song, Y.C.; Pang, J.Y.; Liu, D.T. Lithium-Ion Battery Prognostics with Hybrid Gaussian Process Function Regression. *Energies* **2018**, *11*, 1420. [CrossRef]
21. Deng, L.M.; Hsu, Y.C.; Li, H.X. An Improved Model for Remaining Useful Life Prediction on Capacity Degradation and Regeneration of Lithium-Ion Battery. In Proceedings of the Annual Conference of the Prognostics and Health Management Society, Saint Petersburg, FL, USA, 2–7 October 2017.
22. Liu, D.T.; Zhou, J.B.; Liao, H.T.; Peng, Y.; Peng, X.Y. A Health Indicator Extraction and Optimization Framework for Lithium-Ion Battery Degradation Modeling and Prognostics. *IEEE Trans. Syst. Man Cybern. Syst.* **2015**, *45*, 915–928. [CrossRef]
23. Zangenehmadar, Z.; Moselhi, O. Assessment of Remaining Useful Life of Pipelines Using Different Artificial Neural Networks Models. *J. Perform. Constr. Facil.* **2016**, *30*. [CrossRef]
24. Zhang, X.H.; Xiao, L.; Kang, J.S. Degradation Prediction Model Based on a Neural Network with Dynamic Windows. *Sensors* **2015**, *15*, 6996–7015. [CrossRef] [PubMed]
25. Li, M.B.; Ji, S.W.; Liu, G. Forecasting of Chinese E-Commerce Sales: An Empirical Comparison of ARIMA, Nonlinear Autoregressive Neural Network, and a Combined ARIMA-NARNN Model. *Math. Probl. Eng.* **2018**, *2018*, 1–13. [CrossRef]
26. Wang, K.W.; Deng, C.; Li, J.P.; Zhang, Y.Y.; Li, X.Y.; Wu, M.C. Hybrid methodology for tuberculosis incidence time-series forecasting based on ARIMA and a NAR neural network. *Epidemiol. Infect.* **2017**, *145*, 1118–1129. [CrossRef]
27. Zhang, Y.Z.; Xiong, R.; He, H.W.; Pecht, M.G. Long Short-Term Memory Recurrent Neural Network for Remaining Useful Life Prediction of Lithium-Ion Batteries. *IEEE Trans. Veh. Technol.* **2018**, *67*, 5695–5705. [CrossRef]
28. Du, Z.H.; Qin, M.J.; Zhang, F.; Liu, R.Y. Multistep-ahead forecasting of chlorophyll a using a wavelet nonlinear autoregressive network. *Knowl. Based Syst.* **2018**, *160*, 61–70. [CrossRef]

29. Ibrahim, M.; Jemei, S.; Wimmer, G.; Hissel, D. Nonlinear autoregressive neural network in an energy management strategy for battery/ultra-capacitor hybrid electrical vehicles. *Electr. Power Syst. Res.* **2016**, *136*, 262–269. [CrossRef]

30. Benmouiza, K.; Cheknane, A. Small-scale solar radiation forecasting using ARMA and nonlinear autoregressive neural network models. *Theor. Appl. Climatol.* **2016**, *124*, 945–958. [CrossRef]

31. Rai, A.; Upadhyay, S.H. The use of MD-CUMSUM and NARX neural network for anticipating the remaining useful life of bearings. *Measurement* **2017**, *111*, 397–410. [CrossRef]

32. Gilles, J. Empirical Wavelet Transform. *IEEE Trans. Signal Process.* **2013**, *61*, 3999–4010. [CrossRef]

33. Bhattacharyya, A.; Singh, L.; Pachori, R.B. Fourier-Bessel series expansion based empirical wavelet transform for analysis of non-stationary signals. *Digit Signal Process.* **2018**, *78*, 185–196. [CrossRef]

34. Amezquita-Sanchez, J.P.; Adeli, H. A new music-empirical wavelet transform methodology for time-frequency analysis of noisy nonlinear and non-stationary signals. *Digit. Signal Process.* **2015**, *45*, 55–68. [CrossRef]

35. Tobon-Mejia, D.A.; Medjaher, K.; Zerhouni, N.; Tripot, G. Estimation of the Remaining Useful Life by Using Wavelet Packet Decomposition and HMMs. In Proceedings of the IEEE Aerospace Conference, Big Sky, MT, USA, 5–12 March 2011.

36. Saha, B.; Goebel, K. Battery Data Set. Available online: https://ti.arc.nasa.gov/tech/dash/groups/pcoe/prognostic-data-repository/ (accessed on 20 October 2018).

energies

MDPI

Article

Weighted Regression-Based Extremum Response Surface Method for Structural Dynamic Fuzzy Reliability Analysis

Cheng Lu [1], Yun-Wen Feng [1] and Cheng-Wei Fei [2,*]

[1] School of Aeronautics, Northwestern Polytechnical University, Xi'an 710072, China; lucheng2013@mail.nwpu.edu.cn (C.L.); fengyunwen@nwpu.edu.cn (Y.-W.F.)
[2] Department of Aeronautics and Astronautics, Fudan University, Shanghai 200433, China
* Correspondence: cwfei@fudan.edu.cn

Received: 3 April 2019; Accepted: 23 April 2019; Published: 26 April 2019

Abstract: The parameters considered in structural dynamic reliability analysis have strong uncertainties during machinery operation, and affect analytical precision and efficiency. To improve structural dynamic fuzzy reliability analysis, we propose the weighted regression-based extremum response surface method (WR-ERSM) based on extremum response surface method (ERSM) and weighted regression (WR), by considering the randomness of design parameters and the fuzziness of the safety criterion. Therein, we utilize the ERSM to process the transient to improve computational efficiency, by transforming the random process of structural output response into a random variable. We employ the WR to find the efficient samples with larger weights to improve the calculative accuracy. The fuzziness of the safety criterion is regarded to improve computational precision in the WR-ERSM. The WR-ERSM is applied to perform the dynamic fuzzy reliability analysis of an aeroengine turbine blisk with the fluid-structure coupling technique, and is verified by the comparison of the Monte Carlo (MC) method, equivalent stochastic transformation method (ESTM) and ERSM, with the emphasis on model-fitting property and simulation performance. As revealed from this investigation, (1) the ERSM has the capacity of processing the transient of the structural dynamic reliability evaluation, and (2) the WR approach is able to improve modeling accuracy, and (3) regarding the fuzzy safety criterion is promising to improve the precision of structural dynamic fuzzy reliability evaluation, and (4) the change rule of turbine blisk structural stress from start to cruise for the aircraft is acquired with the maximum value of structural stress at $t = 165$ s and the reliability degree ($Pr = 0.997$) of turbine blisk. The proposed WR-ERSM can improve the efficiency and precision of structural dynamic reliability analysis. Therefore, the efforts of this study provide a promising method for structural dynamic reliability evaluation with respect to working processes.

Keywords: dynamic fuzzy reliability analysis; extremum surface response method; weighted regression; turbine blisk; fuzzy safety criterion

---

1. Introduction

In mechanical systems, the structures always endure complex loads in the extreme environment. For instance, an aeroengine turbine blisk always suffers from high temperature, high pressure and high speed under operation [1]. With the increasing complexity of a mechanical system, the requirements on structural design have become higher. A structural failure during operation could seriously threaten the safety of the entire system and could even be catastrophic. Therefore, it is worthwhile to perform reliability analysis to improve the performance of mechanical system.

In respect of a large number of investigations on the structural reliability evaluations, many methods were developed and briefly described below. Liu et al. [2] adopted a first-order reliability method (FORM) in the chatter reliability analysis of milling system. Keshtegar [3] used the hybrid conjugate search direction to improve the efficiency and robustness of FORM in structural reliability analysis. Zhang et al. [4] proposed a second-order reliability method (SORM) for mechanical reliability design. Huang et al. [5] developed a new SORM with saddlepoint approximation for reliability analysis. Hu et al. [6] explored a novel second order approximation for structural reliability analysis. Nakamura et al. [7] discussed the Monte Carlo (MC) method by the probabilistic transient thermal analysis of an atmospheric reentry vehicle structure. Martinez-Velasco et al. [8] studied the reliability of distribution systems with distributed generation using the parallel MC method. Yang et al. [9] evaluated the structural reliability of a beam pumping unit by the finite element (FE) method with the MC simulation. However, it is difficult to employ FORM and SORM in complicated calculations in structural reliability analysis for low computing accuracy. Moreover, the MC method always spends tremendous time on structural reliability analysis for the requirement of a large number of iterations and simulations.

To address the above issues, surrogate models (called response surface methods, RSM) emerged and underwent rapid development. So far, various surrogate models have been appeared, such as RSM-based polynomials, Kriging model, neural network method, support vector machine, and so forth. Yang et al. [10] used the RSM and FE model to optimize preform shapes, to improve deformation homogeneity in aerospace forgings. Allaix and Carbone [11] proposed the coupling method of the RSM and FE method for structural reliability analysis to prohibit computational cost. In the above works, the RSM has been validated to hold higher computational efficiency than the MC simulation. However, it is troublesome to apply the RSM to process the nonlinearity and transient problems of complex structural reliability analyses, because it is impossible for the RSM model to perfectly reflect the parameter features in high-dimensional space and thus ensure modeling precision. In this case, the Kriging model was developed by Danie G. Krige (after whom the method is named), and then also applied in the field of structural reliability [12,13]. As a classical implicit and intelligent algorithm, the neural network method was investigated in structural reliability analyses [14,15]. Additionally, a support vector machine is also widely focused on since the outstanding performance in overcoming high-dimensional and nonlinear features in structural reliability [16,17]. Although these methods have acceptable accuracy in modeling and reliability assessment in static reliability analysis, it is difficult for these methods to accurately evaluate structural dynamic reliability with time-varying features and the increasing limit state functions.

With respect to the solution of the above questions, extremum RSM (ERSM) was first proposed to handle the transient problem of two-link flexible rotor manipulator reliability analysis, by regarding the time-varying feature and the extremum values of output responses [18]. Later, the ERSM was extended to the dynamic probabilistic designs of aeroengine typical components such as disks and blades [19,20]. The investigations revealed that the ERSM is efficient to reduce computational burden in structural dynamic reliability design to some extent. As for the transients and nonlinearity of structural dynamic reliability analyses, however, the ERSM still face with the low and even unacceptable computational accuracy. In addition, the use of the parameters always influences the modeling precision. In the improvement of modeling accuracy, the weighted regression (WR) is an efficient way by seeking for the better values of the parameters in modeling. Broadie et al. [21] improved the risk estimation model of a financial budget via the WR. The WR technique was also applied to the surrogate modeling of structural reliability analyses. Kaymaz and McMahon [22] utilized the WR to improve the response surface model. In the related published works, it has not been found that the technique is employed in structural dynamic reliability analysis. Along with the heuristic thought, we apply the WR to structural dynamic reliability analysis to refine the modeling precision and accuracy. Meanwhile, the strong fuzziness of parameters is ubiquitous in the material property, boundary conditions, geometry sizes, safety criteria, and so forth [23–25]. Herein, the safety criterion is fuzzy when a specific failure value cannot be determined. The fuzzy safety criterion is more reasonable in structural fuzzy reliability analyses, because the analytical accuracy is improved by transforming fuzzy safety criterion to stochastic safety criterion [26–28].

To perform a structural dynamic reliability analysis with high-precision, this study proposes an efficient approach based on the ERSM and WR, called as WR-based ERSM (WR-ERSM), to improve the accuracy of surrogate modeling and reliability analysis. In the WR-ERSM, the ERSM is employed to address the transient problem of structural dynamic reliability analysis by simplifying the stochastic process of output response as a random variable, the WR is introduced to find the efficient samples for the ERSM modeling to improve modeling accuracy, and the fuzziness of the safety criterion is considered to improve the precision of dynamic reliability analysis by transforming fuzzy safety criterion into stochastic safety criterion. The proposed WR-ERSM is validated by the dynamic fuzzy reliability analysis of a turbine blisk with regard to both the randomness of input variables and the fuzziness of safety criterion.

In Section 2 WR-ERSM is developed for structural dynamic fuzzy reliability analysis. Section 3 investigates the dynamic fuzzy reliability analysis of an aeroengine turbine blisk based on the WR-ERSM by considering fluid-structure interaction and fuzzy safety criterion. The developed WR-ERSM is validated by the comparison of methods in Section 4. The conclusions on this study are summarized in Section 5.

2. Basic Theory on Dynamic Fuzzy Reliability Analysis

In this section, we discuss the basic principle of the WR-ERSM for structural dynamic fuzzy reliability analysis as drawn in Figure 1.

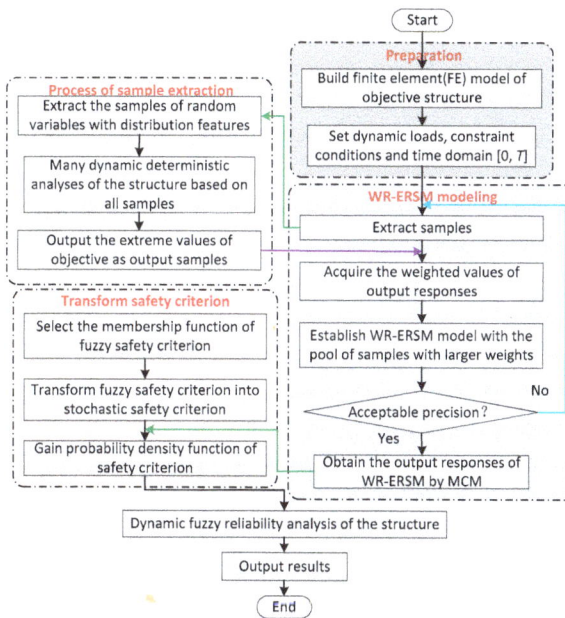

Figure 1. Flow chart of structural dynamic fuzzy reliability analysis with weighted regression extremum response surface method (WR-ERSM).

As revealed in Figure 1, the process of structural dynamic fuzzy reliability analysis comprises analytical preparation, WR-ERSM modeling, sample extraction, safety criterion transformation and reliability analysis. The analytical preparation is to structure finite model (FE) model and set all constraint conditions, workloads and time domain. The objective of the samples' extraction is to collect all input and output samples from dynamic deterministic analyses as one pool of samples for dynamic probabilistic analysis. Herein, the samples of random inputs are extracted by the full factorial design [29,30], and then the extrema of response processes are gained as new output responses based on dynamic deterministic analysis in the time domain of interest, and their weights are confirmed by a series of deterministic analyses with the acquired input samples and FE model. In the process of WR-ERSM modeling, the samples with larger weights, which include input samples and output samples, are chosen from the pool of samples as the fitting samples for the WR-ERSM modeling. When the fitting accuracy does not satisfy the requirements, the fitting samples are reselected to achieve the weighted values. Otherwise, the probability density functions (PDFs) of outputs are gained by MC method. The objective of safety criterion transformation is to transform the fuzzy safety criterion into a stochastic safety criterion based on the fuzzy entropy principle [31,32]. Lastly, structural dynamic fuzzy reliability analysis is performed to achieve the reliability degree considering the randomness of input variables and the fuzziness of safety criterion.

2.1. Weighted Regression Extremum Response Surface Method (WR-ERSM) Modeling

The ERSM was developed to evaluate structural dynamic reliability by considering the extremum values instead of all the output responses within the time domain of interest, and was proved to be efficient in terms of the efficiency improvement [18,33]. In other words, the random process of an output response in the time domain is transformed into a random variable as the ERSM is modeled. When $y(x)$ denotes the extremum of output response within the time domain $[0, T]$, corresponding to the input variables $x = [x_1, x_2, \ldots, x_k]^T$, where k is the number of inputs, the ERSM model can be expressed as:

$$y(x) = A + Bx + x^T C x \tag{1}$$

in which A, B and C indicate constant term, linear term and quadratic term. B and C are denoted as:

$$\begin{cases} B = [b_1, b_2, \cdots, b_k] \\ C = \begin{pmatrix} c_1 & & 0 \\ & \ddots & \\ 0 & & c_k \end{pmatrix} \end{cases} \tag{2}$$

here $i = 1, 2, \ldots, k$. Thus, the ERSM function can be rewritten as:

$$y(x) = a + \sum_{i=1}^{k} b_i x_i + \sum_{i=1}^{k} c_i x_i^2 \tag{3}$$

In this equation, the number of undetermined coefficients is $2k + 1$. To compute these coefficients, we extract a series of input samples by the full factorial design method in Equation (4).

$$\begin{aligned} E_1 &= (\mu_1, \mu_2, \cdots, \mu_k) \\ E_2 &= (\mu_1, \mu_2, \cdots, \mu_i \pm f\sigma_i, \cdots, \mu_k) \\ E_3 &= \left(\mu_1, \mu_2, \cdots, \mu_i \pm f\sigma_i, \cdots, \mu_j \pm f\sigma_j, \cdots, \mu_k\right) \\ &\vdots \\ E_s &= (\mu_1 \pm f\sigma_1, \mu_2 \pm f\sigma_2, \cdots, \mu_i \pm f\sigma_i, \cdots, \mu_k \pm f\sigma_k) \end{aligned} \tag{4}$$

where E_l ($l = 1, 2, \ldots, s$) is the l-th sampling category, namely experimental condition, which is the rule of generated sample set of random variable with respect to both the mean μ and standard deviation σ; the subscripts i, j indicate the i-th and j-th random variables; the subscript s expresses the number of sampling types; f denotes the empirical coefficient which is usually selected from 1 to 3.

Based on structural dynamic deterministic analysis and the least square method [1,19,34], the output responses are then acquired by Equation (5).

$$d = \left(v^T v\right)^{-1} v^T y$$

$$v = \begin{bmatrix} 1 & x_{11} & x_{12} & \cdots & x_{1k} & x_{11}^2 & x_{12}^2 & \cdots & x_{1k}^2 \\ 1 & x_{21} & x_{22} & \cdots & x_{2k} & x_{21}^2 & x_{22}^2 & \cdots & x_{2k}^2 \\ \vdots & \vdots & \vdots & \ddots & \vdots & \vdots & \vdots & \ddots & \vdots \\ 1 & x_{n1} & x_{n2} & \cdots & x_{nk} & x_{n1}^2 & x_{n2}^2 & \cdots & x_{nk}^2 \end{bmatrix} \tag{5}$$

$$d = [a, b_1, b_2, \cdots, b_k, c_1, c_2, \cdots, c_k]$$

The symbol v is the $n \times (k+1)$ matrix of input variables, in which n is the number of samples; d is the vector of undetermined coefficients in the ERSM model.

In respect of the ERSM, the computational burden is effectively reduced in structural dynamic reliability evaluation. However, the modeling precision is still unacceptable because of the limitations of quadratic polynomials in processing the high non-linearity problem and large-scale parameters. To resolve this issue, this study develops the WR-ERSM with respect to the ERSM and WR. We adopt the ERSM to compute the global extreme value rather than all the values for the dynamic output responses under different input parameters in the time domain $[0, T]$, and employ the WR to find the optimal parameters in the process of the ERSM modeling.

The comparison of the ERSM and the WR-ERSM are shown in Figure 2. The ERSM model (indicated by the red dotted curve) is established by all the samples based on the least square method. For the WR-ERSM modeling (denoted by the blue solid curve), we first apply the WR to select the efficient samples (annotated by the blue dots) with larger weights from the pool of n samples, to determine the undetermined coefficients and gain the WR-ERSM model. This method is termed the weighted least square method.

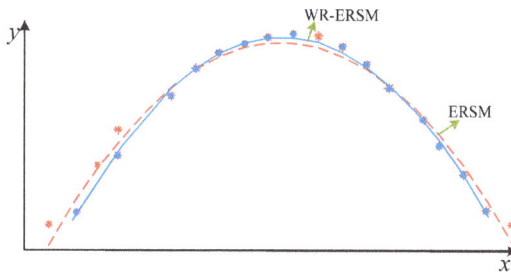

Figure 2. Basic thoughts of ERSM and WR-ERSM.

Regarding the ERSM model (Equation (1)), the WR-ERSM model $y_{WR}(x)$ is structured as:

$$y_{WR}(x) = A_{WR} + B_{WR}x + x^T C_{WR}x \tag{6}$$

where A_{WR}, B_{WR} and C_{WR} are the weighted constant, weighted linear vector and weighted quadratic matrix, respectively. B_{WR} and C_{WR} are denoted as

$$\begin{cases} B_{WR} = \left[b_{WR,1}, b_{WR,2}, \cdots, b_{WR,k}\right] \\ C_{WR} = \begin{bmatrix} c_{WR,1} & & & \\ & c_{WR,2} & & \\ & & \ddots & \\ & & & c_{WR,k} \end{bmatrix} \end{cases} \tag{7}$$

To intuitively express the WR-ERSM model, the Equation (6) is written as quadratic polynomial, i.e.,

$$y_{WR}(x) = a_{WR} + \sum_{i=1}^{k} b_{WR,i}x_i + \sum_{i=1}^{k} c_{WR,i}x_i^2 \tag{8}$$

in which a_{WR}, $b_{WR,i}$ and $c_{WR,i}$ are the weighted coefficient of A, B and C, respectively.

To determine these weighted coefficients in Equation (8), we first need to search the minimum of output responses $y_{obj}(x)$ with respect to n samples, the weighted values are then obtained by dividing the minimum value $y_{obj}(x)$ from all the output responses $y_{true}(x)$, which are ensured by dynamic deterministic analysis with the FE model. The m efficient samples with larger weights are selected from this pool of n samples, and the weighted matrix w is restructured. The related formulas are:

$$\begin{cases} y_{obj}(x) = \min\left|y_{true,j}(x)\right| \\ w_j = \dfrac{y_{obj}}{y_{true,j}(x)} \\ w = \begin{bmatrix} w_1 & & \\ & \ddots & \\ & & w_m \end{bmatrix} \end{cases} \tag{9}$$

here $j = 1, 2, \ldots, n$; $y_{true,j}(x)$ indicates the true value of the j-th output response; w_j is the weighted value of the j-th sample; m ($m \geq (2k + 1)$) is the number of efficient samples.

We confirm the undetermined coefficients of WR-ERSM model, i.e.,

$$d_{WR} = \left(v_{WR}^T w v_{WR}\right)^{-1} v_{WR}^T w y_{WR} \tag{10}$$

where d_{WR} denotes the vector of undetermined coefficients in the WR-ERSM model; v_{WR} is the matrix of efficient samples; y_{WR} is the output response corresponding to the efficient samples. d_{WR}, v_{WR} and y_{WR} are structured as:

$$\begin{cases} d_{WR} = \begin{bmatrix} a_{WR}, b_{WR,1}, b_{WR,2} \cdots, b_{WR,k}, c_{WR,1}, c_{WR,2} \cdots, c_{WR,k} \end{bmatrix} \\ v_{WR} = \begin{bmatrix} 1 & x_{WR,11} & x_{WR,12} & \cdots & x_{WR,1k} & x_{WR,11}^2 & x_{WR,12}^2 & \cdots & x_{WR,1k}^2 \\ 1 & x_{WR,21} & x_{WR,22} & \cdots & x_{WR,2k} & x_{WR,21}^2 & x_{WR,22}^2 & \cdots & x_{WR,2k}^2 \\ \vdots & \vdots & \vdots & \ddots & \vdots & \vdots & \vdots & \ddots & \vdots \\ 1 & x_{WR,m1} & x_{WR,m2} & \cdots & x_{WR,mk} & x_{WR,m1}^2 & x_{WR,m2}^2 & \cdots & x_{WR,mk}^2 \end{bmatrix} \\ y = [y(x_1), y(x_2), \cdots, y(x_n)]^T \end{cases} \tag{11}$$

Based on the above analysis, we can derive the WR-ERSM model.

2.2. Safety Criterion Transformation

In engineering practice, most factors have stochastic and fuzzy characteristics. For structural dynamic reliability analysis, various methods have been developed so far to process the effect of the random parameters. However, those methods are unable to resolve the influences of the randomness and fuzziness simultaneously. As typical fuzzy factors, the allowable values, e.g., deformation, stress, strain, and so forth, of the safety criterion generally depend on experimental statistics. Actually, these parameters always vary in small range in engineering. Hence, it is more reasonable to consider the randomness of inputs and the fuzziness of safety criterion in structural dynamic reliability analysis.

To address this issue, we transform the fuzzy safety criterion into a random safety criterion in the structural dynamic fuzzy reliability evaluation. This paper deals with the fuzzy safety criterion by the fuzzy entropy principle, which has been validated to be feasible [31,32,35,36]. For transforming the fuzzy safety criterion into random safety criterion, we first determine the membership function of safety criterion distribution feature, which is generally chosen as a triangular membership function in engineering practice [23,28,36]. Hence, we also select a triangular membership function to describe the information for the safety criterion as shown in Equation (12).

$$
u_{\tilde{y}}(x) = \begin{cases} \frac{x-a_1}{a-a_1}, a_1 < x \leq a \\ \frac{a_2-x}{a_2-a}, a < x \leq a_2 \\ 0, \text{otherwise} \end{cases}
\tag{12}
$$

We can compute the mean and standard deviation of stochastic safety criterion with a normal distribution based on the fuzzy entropy principle in Equation (13), and then acquire the PDF of safety criterion:

$$
\begin{aligned}
\mu_{eq} &= \left(\frac{1}{\int_{-\infty}^{+\infty} u_{\tilde{y}}(x)dx} \right) \int_{-\infty}^{+\infty} x u_{\tilde{y}}(x)dx \\
\sigma_{eq} &= \frac{1}{\sqrt{2\pi}} \exp(G_x - 0.5)
\end{aligned}
\tag{13}
$$

where μ_{eq} and σ_{eq} indicates the mean and standard deviation of equivalent random parameter; $u_{\tilde{y}}(x)$ is the membership function of fuzzy safety criterion; G_x is defined as:

$$
G_x = - \int_{-\infty}^{+\infty} u'_{\tilde{y}}(x) \ln u'_{\tilde{y}}(x)dx = - \int_{u_l}^{u_u} \frac{u_{\tilde{y}}(x)}{\int_{u_l}^{u_u} u_{\tilde{y}}(x)dx} \ln \frac{u_{\tilde{y}}(x)}{\int_{u_l}^{u_u} u_{\tilde{y}}(x)dx} dx
\tag{14}
$$

here u_u and u_l are the upper bound and lower bound of fuzzy variable interval, respectively.

Finally, the PDF of safety criterion is reshaped as:

$$
f_{eq}(z_{eq}) = \frac{1}{\sqrt{2\pi}\sigma_{eq}} \exp\left(-\frac{\left(z_{eq} - \mu_{eq}\right)^2}{2\sigma_{eq}^2} \right)
\tag{15}
$$

Through the above analysis, we can transform the fuzzy safety criterion into a stochastic safety criterion by using the fuzzy entropy principle.

2.3. Structural Dynamic Fuzzy Reliability Analysis

To accomplish structural dynamic reliability analysis, we need to build the PDF of output. In this case, we take structural stress as analytical object (output response). Based on the derived WR-ERSM model in Equation (8), we extract a large number of samples of the output based on MC method, and achieve the mean and standard deviation. The formula of PDF is then established, i.e.,

$$
f(z) = \frac{1}{\sqrt{2\pi}\sigma_z} \exp\left(-\frac{(z - \mu_z)^2}{2\sigma_z^2} \right)
\tag{16}
$$

where z expresses the structural stress; μ_z and σ_z are both mean value and standard deviation, respectively.

With the PDF of safety criterion, the fuzzy reliability index β and reliability degree Pr of the complex structure are:

$$
\begin{cases} \beta = \frac{\mu_z - \mu_{eq}}{\sqrt{\sigma_z^2 + \sigma_{eq}^2}} \\ Pr = \Phi(\beta) \end{cases}
\tag{17}
$$

where μ_z and σ_z present the mean value and standard deviation of output response; μ_{eq} and σ_{eq} indicate the mean value and standard deviation of the safety criterion.

3. Example Analysis

In this section, we regarded the dynamic fuzzy reliability analysis of an aeroengine turbine blisk as one case to verify the feasibility and effectiveness of the proposed WR-ERSM algorithm.

3.1. Deterministic Analysis for Turbine Blisk

Working in the extreme environment, turbine blisk endures high temperature, high pressure and high speed. To simulate the variation of turbine blisk stress under different operation status, the analytical range of start, idle, take off, climb and cruise is selected from the flight profile of the aeroengine in the time domain [0 s, 215 s] [37,38]. In this time domain [0 s, 215 s], 12 critical points of angular speed shown in Figure 3 are selected during the aeroengine operation of time domain. In this study, nickel-base alloy is selected as the material of the gas turbine blisk.

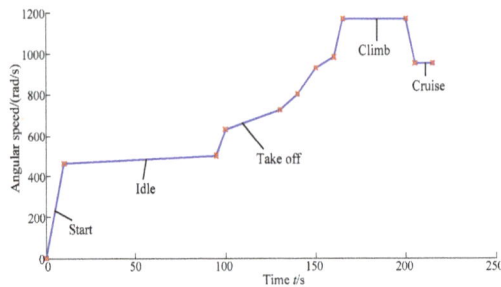

Figure 3. Change curve of angular speed in time domain [0 s, 215 s].

The turbine blisk is a typically cyclic symmetric structure comprising one disk and 40 blades, and is shown in Figure 4. To reduce the calculation burden, the 1/40 of the blisk model is regarded as the study object, besides the cooling holes on blisk are simplified. The FE models of the turbine blisk (29,332 elements and 47,933 nodes) and flow field (222,370 elements and 321,632 nodes) are shown in Figures 5 and 6, respectively.

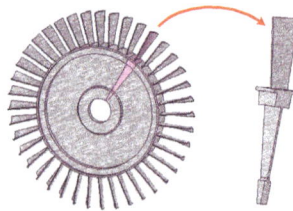

Figure 4. Geometric model of turbine blisk.

Figure 5. Finite element (FE) model of turbine blisk.

To simulate the variation of structural stress caused by fluid and structural loads within the time domain [0 s, 215 s], the dynamic deterministic analysis of the turbine blisk is fulfilled by both the close coupled analysis method and fluid–structure interaction [39–41]. The variation curve of turbine blisk stress in the time domain [0 s, 215 s] is displayed in Figure 7, in which σ is the stress of turbine blisk (similarly hereinafter).

Figure 6. FE model of flow field.

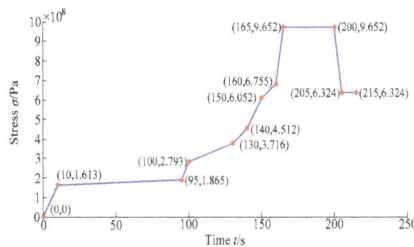

Figure 7. Stress variation curve of turbine blisk during [0 s, 215 s].

As revealed in Figure 7, the turbine blisk stress rises with the increasing angular speed. The maximum of stress emerges at $t = 165$ s in cruise phase. Thus, we regarded $t = 165$ s as the computational point of turbine blisk dynamic fuzzy reliability analysis. The nephograms of pressure and stress distributions on the fluid–structure coupling interface at $t = 165$ s are acquired in Figures 8 and 9, in which P is the pressure on the fluid–structure coupling interface. As revealed in Figure 9, the maximum stress is at the root of the turbine blade.

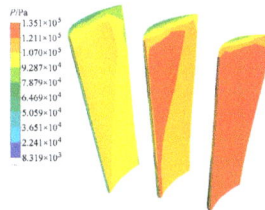

Figure 8. Pressure distribution on fluid–structure coupling interface.

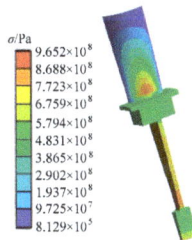

Figure 9. Stress distribution on turbine blisk.

3.2. The WR-ERSM Model of Turbine Blisk

To achieve the surrogate model of turbine blisk, the factors (parameters) impacting the analytical results are selected as the random input variables as listed Table 1, including inlet velocity, inlet pressure, material density and angular speed.

Table 1. The statistical characteristics of random input variables.

Parameters	Variable	Distribution	Mean, μ	St.Dev., δ
Inlet velocity (m·s^{-1})	v	Normal	168	5.04
Inlet pressure (Pa)	P	Normal	600,000	12,000
Material density (kg·m^{-3})	ρ	Normal	8210	246
Angular speed (rad·s^{-1})	w	Normal	1168	35

As revealed in Table 1, the mean μ of the random input variables is determined within the time domain by using the extremum selection method [42]. To establish the WR-ERSM model of turbine blisk, the samples are extracted from the random inputs and outputs at the selected calculation points by the full factorial design in Equation (4). The weighted values of output responses are calculated by Equation (5). The 40 samples are listed in Table 2.

Table 2. Weighted samples based on weighted regression analysis.

Parameters and Weighted Coefficient						Parameters and Weighted Coefficient					
v m·s^{-1}	P, ×10^5 Pa	ρ kg·m^{-3}	w rad·s^{-1}	σ × 10^8 Pa	W	v, m·s^{-1}	P, ×10^5 Pa	ρ, kg·m^{-3}	w, rad·s^{-1}	σ, ×10^8 Pa	W
168.00	6.00	8210	1168	9.687	0.9105	173.04	6.00	8210	1133	9.098	0.9694
162.96	6.00	8210	1168	9.693	0.9099	168.00	6.12	8210	1133	9.105	0.9687
168.00	5.88	8210	1168	9.686	0.9106	168.00	6.00	7964	1133	8.827	0.9992
168.00	6.00	7964	1168	9.392	0.9391	173.04	5.88	7964	1168	9.385	0.9398
168.00	6.00	8210	1133	9.105	0.9687	173.04	5.88	8210	1133	9.098	0.9694
173.04	6.00	8210	1168	9.686	0.9391	162.96	6.12	7964	1168	9.398	0.9385
168.00	6.12	8210	1168	9.687	0.9105	162.96	6.12	8210	1133	9.111	0.9681
168.00	6.12	8210	1203	10.29	0.8576	168.00	6.00	8456	1203	10.59	0.8575
162.96	5.88	8210	1168	9.687	0.9105	168.00	6.12	7964	1133	8.829	0.9989
162.96	6.00	7964	1168	9.391	0.9392	162.96	6.00	8456	1133	9.389	0.9394
162.96	6.00	8210	1133	9.105	0.9687	168.00	5.88	8456	1133	9.383	0.9400
168.00	5.88	7964	1168	9.391	0.9392	173.04	6.12	7964	1168	9.385	0.9398
168.00	5.88	8210	1133	9.105	0.9687	173.04	6.12	8210	1133	9.098	0.9694
168.00	6.00	7964	1133	8.827	0.9992	173.04	6.00	8456	1133	9.376	0.9407
173.04	6.12	8210	1168	9.687	0.9105	168.00	6.12	8456	1133	9.383	0.9400
162.96	6.12	8210	1168	9.693	0.9099	162.96	5.88	7964	1168	9.398	0.9385
168.00	6.00	8210	1203	10.28	0.8576	173.04	6.00	8210	1203	10.28	0.8576
173.04	5.88	8210	1168	9.681	0.9111	162.96	5.88	8210	1133	9.111	0.9681
173.04	6.00	7964	1168	9.385	0.9398	162.96	6.00	7964	1133	8.827	0.9992
168.00	6.12	7964	1168	9.391	0.9392	168.00	5.88	7964	1133	8.826	0.9993

Note: the symbols v, P, ρ and w are the inlet velocity, inlet pressure, material density and angular speed, respectively; σ presents the turbine blisk stress; W denotes the weighted value. Additionally, the underlined samples (20 samples) are used to establish the WR-ERSM model, and the underlined and bold samples (30 samples) are applied to derive the ERSM model.

Based on 20 groups of samples with larger weights underlined in Table 2, the coefficients of Equation (8) are acquired, and then the WR-ERSM model of turbine blisk is

$$y(x) = \begin{aligned} &-2.746 \times 10^9 + 2.258 \times 10^6 x_1 - 9.681 \times 10^2 x_2 + 1.063 \times 10^5 x_3 + 3.053 \times 10^6 x_4 - 7.036 \times 10^3 x_1^2 \\ &+ 8.078 \times 10^{-4} x_2^2 + 0.418 x_3^2 - 4.722 \times 10^2 x_4^2 \end{aligned} \quad (18)$$

Let the response $y(x)$ in Equation (18) obey a normal distribution, the dynamic fuzzy reliability analysis of turbine blisk is accomplished with the MC method. The simulation histories and stress histograms of turbine blisk are drawn in Figures 10 and 11, respectively.

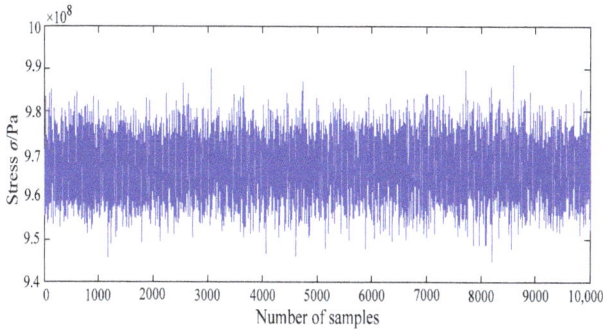

Figure 10. The simulation history of turbine blisk stress.

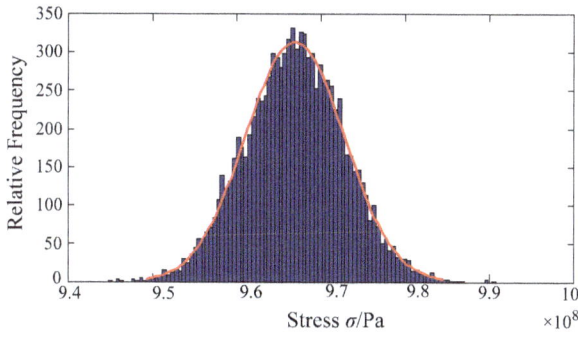

Figure 11. The histogram of turbine blisk stress.

As shown in Figures 11 and 12, the histogram of turbine blisk stress obeys a normal distribution with the mean value $\mu_z = 9.669 \times 10^8$ Pa and standard deviation $\sigma_z = 5.743 \times 10^6$ Pa. Moreover, in light of Equation (16), the built model in Equation (18) is rewritten as the PDF, i.e.,

$$f(z) = \frac{1}{\sqrt{2\pi} \times (5.743 \times 10^6)} \exp\left(-\frac{\left(z - 9.669 \times 10^8\right)^2}{2 \times (5.743 \times 10^6)^2}\right) \qquad (19)$$

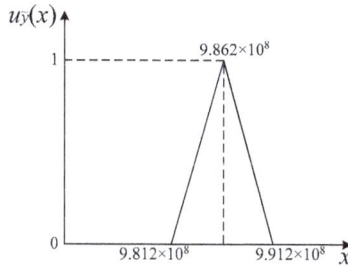

Figure 12. The triangular membership function of the turbine blisk.

3.3. Turbine Blisk Reliability Evaluation

The safety criterion is a typical fuzzy parameter because its allowable values are uncertain in practical engineering. To accomplish the dynamic fuzzy reliability analysis of the turbine blisk with the PDF, the fuzzy safety criterion needs to be transformed into a stochastic safety criterion by the fuzzy entropy principle. When the membership function of fuzzy safety criterion obeys a triangular distribution, the formula and distribution characteristics of the triangular membership function for the turbine blisk are obtained as shown in Equation (20) and Figure 12, respectively.

$$u_{\tilde{y}}(x) = \begin{cases} \frac{x - 9.812 \times 10^8}{5 \times 10^6}, & 9.812 \times 10^8 < x \le 9.862 \times 10^8 \\ \frac{9.912 \times 10^8 - x}{5 \times 10^6}, & 9.862 \times 10^8 < x \le 9.912 \times 10^8 \end{cases} \tag{20}$$

After confirming the membership function of fuzzy safety criterion, the triangular membership function is transferred into the PDF of safety criterion (namely a stochastic safety criterion) with a normal distribution by the fuzzy entropy theory. The mean value μ_{eq} and standard deviation σ_{eq} are consequently achieved. The PDF $f(z_{eq})$ can be expressed by:

$$f(z_{eq}) = \frac{1}{\sqrt{2\pi} \times (4.980 \times 10^6)} \exp\left(-\frac{(z_{eq} - 9.862 \times 10^8)^2}{2 \times (4.980 \times 10^6)^2}\right) \tag{21}$$

With respect to Equations (19) and (21), the probability density curves $f(z)$ and $f(z_{eq})$ of stress and the safety criterion of the turbine blisk are drawn in Figure 13, respectively.

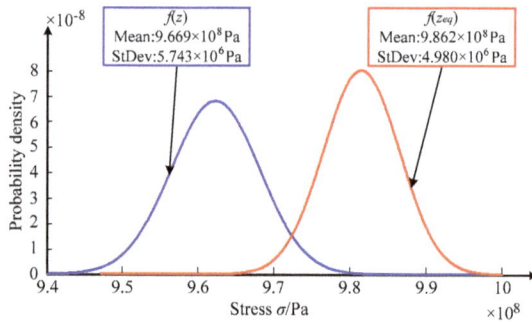

Figure 13. Distribution features of structural stress and safety criterion.

As demonstrated in Figure 13, the mean values (9.669×10^8 Pa and 9.862×10^8 Pa) and standard deviations (5.743×10^6 Pa and 4.980×10^6 Pa) of the two methods are acquired. In line with Equation (17), the structural reliability index and reliability degree are $\beta = 2.751$ and $Pr = 0.9970$, respectively.

4. WR-ERSM Verification Procedure

In this section, the proposed WR-ERSM is verified by the comparison with the MC method, ERSM based on least-square and equivalent stochastic transformation method (ESTM).

4.1. Model-Fitting Properties

By the 30 groups of underlined and bold experimental data in Table 2, the model of ERSM is established as:

$$\begin{aligned} y(x) = \ & -2.015 \times 10^8 + 3.734 \times 10^6 x_1 + 1.998 \times 10^3 x_2 + 9.989 \times 10^4 x_3 - 2.756 \times 10^6 x_4 - 1.137 \times 10^4 x_1^2 \\ & -1.669 \times 10^{-3} x_2^2 + 9.166 \times 10^{-1} x_3^2 + 1.907 \times 10^3 x_4^2 \end{aligned} \tag{22}$$

In this paper, we adopt the square-error r^2 and maximum absolute error r_{max} to test the fitting accuracy for the WR-ERSM and ERSM. The r^2 and r_{max} are illustrated as follows,

$$\begin{cases} r^2 = 1 - \dfrac{\sum\limits_{i=1}^{n}(y_i-\hat{y}_i)^2}{\sum\limits_{i=1}^{n}(y_i-\bar{y})^2} \\ r_{max} = \max\limits_{i=1}^{n}\left(\dfrac{|y_i-\hat{y}_i|}{S}\right) \end{cases} \tag{23}$$

in which n is the number of experimental data; y_i denotes the authentic output responses; $\hat{y}_i$ is the output responses gained by the mathematical models; $\bar{y}$ indicates the mean of the experimental data; S expresses the standard deviation of experimental data. If the square-error r^2 is close to 1 and the relative maximum absolute error r_{max} is close to 0, the fitting accuracy is high.

The remaining 10 groups of the experimental data in Table 2 are employed to test the fitting accuracy of the models (Equations (18) and (22)) with respect to r^2 and r_{max}. The results are listed in Table 3.

Table 3. The computational results of the ERSM model and WR-ERSM model.

Method	Fitting ERSM Model		Fitting Accuracy	
	Sample Number	Fitting Time, h	r^2	r_{max}
WR-ERSM	9	7.05	0.9984	0.0535
ERSM	29	22.39	0.9742	0.0834

As illustrated in Table 3, the proposed WR-ERSM only needs 9 samples for modeling, which is far less than 29 samples for the ERSM. Besides, the fitting time of WR-ERSM is 7.5 hours which is also far less than the 22.39 hours of the ERSM. As for the fitting accuracy, the square-error and maximum absolute error of the WR-ERSM and ERSM are 0.9984, 0.9742 and 0.0535, 0.0834, respectively. Because of the square error $0.9742 < 0.9984 \rightarrow 1$ and the relative maximum absolute error $0.0834 > 0.0535 \rightarrow 0$, the fitting accuracy of WR-ERSM is obviously higher than ERSM in modeling precision. Therefore, the WR-ERSM is superior to ERSM in fitting efficiency and accuracy. Because of the high computational accuracy of WR-ERSM, it is also demonstrated to be reasonable and efficient that WR is considered to select better samples to establish ERSM model.

4.2. Simulation Performances for Dynamic Fuzzy Reliability Analysis of Turbine Blisk

In this section, we compare the reliability degree assessments as using four methods, i.e., the MC method, ESTM, ERSM and WR-ERSM. The MC method with direction simulation is considered as the reference when the simulation precisions of other three methods are evaluated. The fuzzy reliability evaluation with the ESTM considers all calculations in the response process in the time domain [0 s, 215 s] without the simplification of the response process just like the ERSM. The ERSM is employed for the dynamic reliability analysis by simplifying the response process as a random parameter without the consideration of both the WR of the output responses in the sample selection for the modeling and the fuzziness of the safety criterion in the reliability analysis. When the WR-ERSM is applied to the dynamic fuzzy reliability estimation of turbine blisk, we completely regard the simplification of the response process, the WR of output responses and the fuzziness of safety criterion. All the calculations and simulation are completed based on the same input variables in Table 1 and computer environment. The computational results are shown in Table 4.

As revealed in Table 4, the WR-ERSM is closer to the MC method than both ESTM and ERSM for the reliability degree of turbine blisk. Besides, the proposed WR-ERSM has higher analytical accuracy than the ERSM and ESTM as the precision 0.9989 for the WR-ERSM is larger than the precision 0.9956 for the ERSM and the precision 0.9981. The result that the precision of the WR-ERSM is superior to the ERSM indicates that considering the fuzziness of the safety criterion besides the WR is efficient for the improvement of structural dynamic reliability analysis. The fact that the precision of the WR-ERSM is superior to the ESTM reveals that the ERSM is effective to deal with the transient in structural dynamic reliability analysis instead of the ESTM. In brief, the WR-ERSM, which considers both the WR of the parameters in the model and the fuzziness of the safety criterion in the analysis, is able to improve the computational accuracy of structural dynamic reliability analysis while keeping to a high degree of reliability. Besides, the structural dynamic fuzzy reliability analysis with the fuzziness of safety criterion is more reasonable and accurate than the structural dynamic reliability analysis without the fuzziness of the safety criterion.

Table 4. The simulation results of turbine blisk dynamic reliability analyses with four methods.

Methods	P_r	Errors	Precision, %
MC method	0.9981	-	-
ESTM	0.9962	0.0019	99.81
ERSM	0.9937	0.0044	99.56
WR-ERSM	0.9970	0.0011	99.89

Based on the above results, it is fully demonstrated that the developed WR-ERSM is able to improve computational efficiency and precision for structural dynamic reliability analysis while maintaining a high degree of reliability, by both introducing the WR method to find more effective samples for the ERSM modeling and considering the fuzziness of safety criterion in reliability analysis. The structural dynamic fuzzy reliability analysis with the WR-ERSM is effective and feasible for improving the designs of structures and mechanical system.

5. Conclusions

To improve the computational accuracy and efficiency of structural dynamic fuzzy reliability analysis, we present the weighted regression-based extremum response surface method (WR-ERSM) based on the extremum response surface method (ERSM) and weighted regression (WR), for structural dynamic fuzzy reliability analysis. Through this study, some conclusions are summarized as follows:

(1) The WR-ERSM is highly precise and efficient in structural dynamic reliability evaluation, since ERSM has the capacity of processing the transient problem;

(2) The WR approach can improve modeling accuracy so that the proposed WR-ERSM possesses high fitting efficiency and accuracy, due to the requirement of small samples;

(3) WR-ERSM possesses good simulation performance in structural dynamic fuzzy reliability evaluation, as the fuzzy safety criterion is considered to improve the precision;

(4) The change rule of turbine blisk structural stress from start to cruise for an aircraft is acquired with the maximum value of structural stress at $t = 165$ s and the reliability degree ($Pr = 0.997$) of the turbine blisk.

(5) The efforts of this study provide a promising method for the dynamic reliability analysis and evaluation of complex structures with respect to the working process.

Author Contributions: Conceptualization, C.-W.F. and Y.-W.F.; Methodology, C.L.; Software, C.L.; Validation, C.L., Y.-W.F. and C.-W.F.; Formal Analysis, C.-W.F.; Investigation, C.L.; Resources, Y.-W.F.; Data Curation, C.-W.F.; Writing-Original Draft Preparation, C.L.; Writing-Review & Editing, C.-W.F.; Visualization, C.L.; Supervision, C.-W.F.; Project Administration, Y.-W.F.; Funding Acquisition, Y.-W.F. and C.-W.F.

Funding: This research was funded by [National Natural Science Foundation of China] grant number [51875465 and 51605016], [Innovation Foundation for Doctor Dissertation of Northwestern Polytechnical University] grant number [CX201932], and [Research Start-up funding of Fudan University] grant number [FDU38341]. The APC was funded by [51875465].

Conflicts of Interest: The authors declare that there is no conflict of interests regarding the publication of this article.

References

1. Fei, C.W.; Choy, Y.S.; Hu, D.Y.; Bai, G.C.; Tang, W.Z. Dynamic probabilistic design approach of high-pressure turbine blade-tip radial running clearance. *Nonlinear Dyn.* **2016**, *86*, 205–223. [CrossRef]
2. Liu, Y.; Meng, L.L.; Liu, K.; Zhang, Y.M. Chatter reliability of milling system based on first-order second-moment method. *Int. J. Adv. Manuf. Technol.* **2016**, *87*, 801–809. [CrossRef]
3. Keshtegar, B. Conjugate finite-step length method for efficient and robust structural reliability analysis. *Struct. Eng. Mech.* **2018**, *65*, 415–422.
4. Zhang, J.F.; Du, X.P. A second-order reliability method with first-order efficiency. *J. Mech. Des.* **2010**, *132*, 101006. [CrossRef]
5. Huang, X.Z.; Li, Y.X.; Zhang, Y.M.; Zhang, X.F. A new direct second-order reliability analysis method. *Appl. Math. Model.* **2018**, *55*, 68–80. [CrossRef]
6. Hu, Z.L.; Du, X.P. Efficient reliability-based design with second order approximations. *Eng. Optim.* **2019**, *51*, 101–119. [CrossRef]
7. Nakamura, T.; Fujii, K. Probabilistic transient thermal analysis of an atmospheric reentry vehicle structure. *Aerosp. Sci. Technol.* **2006**, *10*, 346–354. [CrossRef]
8. Martinez-Velasco, J.A.; Guerra, G. Reliability analysis of distribution systems with photovoltaic generation using a power flow simulator and a parallel Monte Carlo approach. *Energies* **2016**, *9*, 537. [CrossRef]
9. Yang, X.J.; Yan, Y.L.; Xu, Z.Q.; Yan, X.Z. FEM simulation on the structural reliability of beam pumping unit based on the methods of Monte-Carlo. *Appl. Mech. Mater.* **2010**, *34–35*, 820–824. [CrossRef]
10. Yang, Y.H.; Liu, D.; He, Z.Y.; Luo, Z.J. Optimization of preform shapes by RSM and FEM to improve deformation homogeneity in aerospace forgings. *Chin. J. Aeronaut.* **2010**, *23*, 260–267.
11. Allaix, D.L.; Carbone, V.I. An improvement of the response surface method. *Struct. Saf.* **2011**, *32*, 165–172. [CrossRef]
12. Kaymaz, I. Application of kriging method to structural reliability problems. *Struct. Saf.* **2005**, *27*, 133–151. [CrossRef]
13. Zhao, Z.G.; Duan, X.J.; Wang, Z.M. A novel global method for reliability analysis with kriging. *Int. J. Uncertain. Quantif.* **2016**, *6*, 445–466. [CrossRef]
14. Kishore, R.A.; Mahajan, R.L.; Priya, S. Combinatory finite element and artificial neural network model for predicting performance of thermoelectric generator. *Energies* **2018**, *11*, 2216. [CrossRef]
15. Gondal, Z.U.R.; Lee, J. Reliability assessment using feed-forward neural network-based approximate meta-models. *Proc. Inst. Mech. Eng. Part O J. Risk Reliab.* **2012**, *226*, 448–454. [CrossRef]
16. Rocco, C.M.; Moreno, J.A. Fast Monte Carlo reliability evaluation using support vector machine. *Reliab. Eng. Syst. Saf.* **2002**, *76*, 237–243. [CrossRef]
17. Guo, Z.W.; Bai, G.C. Classification using least squares support vector machine for reliability analysis. *Appl. Math. Mech.* **2009**, *30*, 853–864. [CrossRef]
18. Zhang, C.Y.; Bai, G.C. Extremum response surface method of reliability analysis on two-link flexible rotor manipulator. *J. Cent. South Univ.* **2012**, *19*, 101–107. [CrossRef]
19. Fei, C.W.; Tang, W.Z.; Bai, G.C. Nonlinear dynamic probabilistic design of turbine disk-radial deformation using extremum response surface method-based support vector machine of regression. *Proc. Inst. Mech. Eng. Part G J. Aerosp. Eng.* **2015**, *229*, 290–300. [CrossRef]
20. Fei, C.W.; Tang, W.Z.; Bai, G.C. Dynamic probabilistic design for blade deformation with SVM-ERSM. *Aircr. Eng. Aerosp. Technol.* **2015**, *87*, 312–321. [CrossRef]
21. Broadie, M.; Du, Y. Risk estimation via weighted regression. In Proceedings of the 2011 Winter Simulation Conference, Phoenix, AZ, USA, 11–14 December 2011; pp. 3859–3870.
22. Kaymaz, I.; McMahon, C.A. A response surface method based on weighted regression for structural reliability analysis. *Probabilistic Eng. Mech.* **2005**, *20*, 11–17. [CrossRef]
23. Guo, S.X.; Lu, Z.Z. Procedure for computing the possibility and fuzzy probability of failure of structures. *Appl. Math. Mech.* **2003**, *24*, 338–343.
24. Hurtado, J.E.; Alvarez, D.A.; Ramirez, J. Fuzzy structural analysis based on fundamental reliability concepts. *Comput. Struct.* **2012**, *112*, 183–192. [CrossRef]
25. Zhu, S.P.; Liu, Q.; Zhou, J.; Yu, Z.Y. Fatigue reliability assessment of turbine discs under multi-source uncertainties. *Fatigue Fract. Eng. Mater. Struct.* **2018**, *41*, 1291–1305. [CrossRef]

26. Huang, H.Z. Structural reliability analysis using fuzzy sets theory. *Eksploat. I Niezawodn. Maint. Reliab.* **2012**, *4*, 284–294.

27. Wang, Z.L.; Li, Y.F.; Huang, H.Z.; Liu, Y. Reliability analysis of structure for fuzzy safety state. *Intell. Autom. Soft Comput.* **2012**, *18*, 215–224. [CrossRef]

28. Zhang, M.; Lu, S. A reliability model of blade to avoid resonance considering multiple fuzziness. *Proc. Inst. Mech. Eng. Part O J. Risk Reliab.* **2014**, *228*, 641–652. [CrossRef]

29. Zhu, S.P.; Liu, Q.; Peng, W.; Zhang, X.C. Computational-experimental approaches for fatigue reliability assessment of turbine bladed disks. *Int. J. Mech. Sci.* **2018**, *142–143*, 502–517. [CrossRef]

30. Lee, S.H.; Kwak, B.M. Response surface augmented moment method for efficient reliability analysis. *Struct. Saf.* **2006**, *28*, 261–272. [CrossRef]

31. Afsan, B.M.U.; Basu, C.K. Fuzzy topological entropy of fuzzy continuous functions on fuzzy topological spaces. *Appl. Math. Lett.* **2011**, *24*, 2030–2033. [CrossRef]

32. Markechova, D.; Riecan, B. Entropy of fuzzy partitions and entropy of fuzzy dynamical systems. *Entropy* **2016**, *18*, 19. [CrossRef]

33. Fei, C.W.; Tang, W.Z.; Bai, G.C. Novel method and model for dynamic reliability optimal design of turbine blade deformation. *Aerosp. Sci. Technol.* **2014**, *39*, 588–595. [CrossRef]

34. Zhang, C.Y.; Lu, C.; Fei, C.W.; Liu, L.J.; Choy, Y.Z.; Su, X.G. Multiobject reliability analysis of turbine blisk with multidiscipline under multiphysical field interaction. *Adv. Mater. Sci. Eng.* **2015**, *2015*, 649046. [CrossRef]

35. Li, L.Y.; Lu, Z.Z. Importance analysis for model with mixed uncertainties. *Fuzzy Sets Syst.* **2017**, *310*, 90–107. [CrossRef]

36. Sun, J.; Luo, Y. Reliability-optimization design based on fuzzy entropy for cylinder head bolts. *J. Appl. Sci.* **2013**, *13*, 5198–5203. [CrossRef]

37. Lattime, S.B.; Steinetz, B.M. Turbine engine clearance control systems: Current practices and future directions. *J. Propuls. Power* **2004**, *20*, 302–311. [CrossRef]

38. Fei, C.W.; Choy, Y.S.; Hu, D.Y.; Bai, G.C.; Tang, W.Z. Transient probabilistic analysis for turbine blade-tip radial clearance with multiple components and multi-physics fields based on DCERSM. *Aerosp. Sci. Technol.* **2016**, *50*, 62–70. [CrossRef]

39. Wang, H.B.; Wang, Z.G.; Sun, M.B.; Qin, N. Large eddy simulation of a hydrogen-fueled scramjet combustor with dual cavity. *Acta Astronaut.* **2015**, *108*, 119–128. [CrossRef]

40. Lu, C.; Feng, Y.W.; Fei, C.W.; Xue, X.F. Probabilistic analysis method of turbine blisk with multi-failure modes by two-way fluid-thermal-solid coupling. *Proc. Inst. Mech. Eng. Part C J. Mech. Eng. Sci.* **2018**, *232*, 2873–2886. [CrossRef]

41. Liu, J.; Nan, Z.; Yi, P. Validation and application of three-dimensional discontinuous deformation analysis with tetrahedron finite element meshed block. *Acta Mech. Sin.* **2012**, *28*, 1602–1616. [CrossRef]

42. Fei, C.W.; Bai, G.C. Extremum selection method of random variables for nonlinear dynamic reliability analysis of turbine blade deformation. *Propuls. Power Res.* **2012**, *1*, 58–63. [CrossRef]

energies

MDPI

Article

Initial Design Phase and Tender Designs of a Jacket Structure Converted into a Retrofitted Offshore Wind Turbine

Lorenzo Alessi [1], José A.F.O. Correia [2] and Nicholas Fantuzzi [1,*]

[1]   DICAM Department, University of Bologna, 40136 Bologna, Italy; alessilorenzo8@gmail.com
[2]   Faculty of Engineering (FEUP), University of Porto, PT-4200-465 Porto, Portugal; jacorreia@inegi.up.pt
*   Correspondence: nicholas.fantuzzi@unibo.it; Tel.: +39-051-209-3494

Received: 29 December 2018; Accepted: 14 February 2019; Published: 18 February 2019

Abstract: Jackets are the most common structures in the Adriatic Sea for extracting natural gas. These structural typologies are suitable for relative low water depths and flat sandy sea floors. Most of them have been built in the last 50 years. When the underground source finishes, these structures should be moved to another location or removed if they have reached their design life. Nevertheless, another solution might be considered: change the future working life of these platforms by involving renewable energy and transforming them into offshore wind towers. The present research proposal aims to investigate the possibility of converting actual structures for gas extraction into offshore platforms for wind turbine towers. This simplified analysis is useful for initial design phases and tender design, or generally when available information is limited. The model proposed is a new simplified tool used to study the structural analysis of the jacket structure, developed and summarized in 10 steps, firstly adopted to study the behavior of the oil and gas structure and then for the retrofitted wind tower configuration.

Keywords: offshore structures; oil and gas platforms; offshore wind turbines; retrofitting activities; renewable energy; dynamic analysis; wind and wave analysis; dynamic analysis of the structure; wave–structure interaction (WSI); probabilistic analyses of stochastic processes and frequency

---

1. Introduction

Policy support, technology advances, and a maturing supply chain are making offshore wind farms an increasingly viable option for renewable-based electricity generation, harnessing the more consistent and higher wind speeds available offshore instead the onshore solutions. Investment has picked up sharply in recent years, and with fewer restrictions on size and height with respect to their onshore counterparts, offshore wind turbines are becoming giants [1]. The growth of offshore wind creates potential synergies with the offshore hydrocarbons sector; integration could bring benefits in terms of reduced costs and improved environmental performance and utilization of infrastructure [2,3]. The possibility to electrify offshore oil and gas operations where there are wind farms nearby, or via floating turbines, reduces the need to run diesel or gas-fired generators on the platform and emissions of carbon dioxide (CO_2) and air pollutants [4]. The scope is to find new uses for existing offshore infrastructure once it reaches the end of its operational life or when the underground source finishes, in ways that might aid energy transitions [5,6]. The North Sea, a relatively mature oil and gas basin with a thriving renewable electricity industry, is already seeing some crossover between the sectors: some large oil and gas companies are major players in offshore wind; one former oil and gas company, Orsted in Denmark, has moved entirely to wind and other renewables [7,8]. A major problem in offshore wind energy design is the estimation of fatigue life and fatigue loads in an offshore environment [9,10]. The study of the best locations of offshore wind turbines is worth investigation as well [11]. However, many studies have been devoted to how the dynamic behaviors of wind turbines couple with jacket and pile-supported foundations [12–21].

This work aims to provide a tool for a general structural check analysis of a different retrofitting activities for the same offshore lattice structure. The research should allow the structural designer to evaluate the environmental loads, the dynamics of the structure, and the effects of changing the structural behavior of the platform over the different cases of study. All these analyses should be available in the early design phase when the data available about the site's environment and seabed conditions are limited. Furthermore, the tools shown should skip the use of the sophisticated and expensive software analysis that requires detailed data and a high level of understanding of the underlying physics. It follows that the calculation procedure will be faster than the long run-times of the software, which is based on time domain simulations of a high number of different load cases. MATLAB and Strand7 have been used as software programs in this research. In terms of accuracy, the most important requirement is that the approach should provide conservative results and at the same time avoid underestimation of loads and deformations. The issue of this project is to find a useful tool that is able to represent the behavior of a generic offshore structure subjected to environmental actions. The structure under investigation can be one of the oil and gas platforms sustained by jacket structures with a long pile foundation. The authors are aware of the complex problem that is presented here. Therefore, the present research is limited, but in agreement with the assumptions considered in this study. Further investigations can be performed in a future work such as tubular joint design [22] and fatigue assessment [23,24]. The influence of a different wind turbine such as vertical axis wind turbines [25] and wind turbines in other environments [26,27] is also worth investigating.

2. Theoretical Background

Offshore structures are complex systems subjected to severe dynamic loads due to wind, waves, currents, and mechanical loads. This makes the design of these structures a complex process that requires expertise from a wide range of fields. Foundations and substructures are specifically the key structural components that differentiate the design of wind towers from those of onshore ones. Due to the long service lifetime of offshore systems, the analysis of the performance of the substructure is a complex task that involves many steps including load analysis, dynamic analysis, evaluation of the fatigue life, as well as long-term deformations. This research presents a simplified methodology for the analysis of the ultimate and dynamics loads with a check on the most stressed pile of the different offshore retrofitted systems. The analysis was carried out such that the data required about the oil and gas platform, wind turbines, and the site are available in the early design phases. Thus, the methodology appears to obtain a conservative estimation. An apparent limitation in the research is the modeling of first the oil and gas platform and second of the offshore wind turbine called 6DOF with all the retrofitting configurations, which was realized in general form, and it found adaptation in different specific applications. The model proposed considers a discrete system of masses and stiffness, which is just a first rough representation of the real offshore lattice structure with pile foundation. Another inherent limitation in the methodology, which is a natural consequence of the approach taken, is the load estimation based on linear analysis and where they are applied. In the model proposed, all the environmental actions are considered at each node of the cantilever beam, while in the real world, all the steel elements are subjected to the wave and wind force. Another issue of the loads is that in this research is that they have not been calculated using partial safety factors regarding the load combination. The present analysis is performed only over one direction with the assumption that the wind and the wave act in the same direction at the same time. Fatigue state analysis is out of the scope of the present research; nevertheless, this is central to the structural analysis, the earthquake analysis, and the checking of the state of the structure like welding, bolts, and grouting, which for sake of simplicity's, have been bypassed. The simplified analysis proposed in this research can be used in the fatigue design and evaluation including the effects of welding, bolts, and the stiffness of tubular connections [28,29]. The next stage of the current research study will address fatigue analysis using current (global rules) and advanced approaches (local rules) for the critical structural detail considering the wave and wind loads as indicated in the DNVGLstandards.

The equation of motion associated with a generic structure is a second order linear differential equation with constant coefficients, and it changes form in relation to the structural properties that have to be modeled. The general set of equations representing structural motion can be defined by the following matrix form:

$$M\ddot{\xi} + C\dot{\xi} + K\xi = p \tag{1}$$

that represents a finite set of N ordinary, linear differential equations in N independent coordinates that contain the dominant features of the structural motion. These coordinates are the elements of the column vector:

$$\xi = [\xi_1, \xi_2, \dots, \xi_N]^T \tag{2}$$

The loading vector $p = p(t)$ is comprised of N loads p_i, $i = 1, 2, \dots, N$ where the load $p_i = p_i(t)$ is located at the respective node i. The vector representation is:

$$p = [p_1, p_2, \dots, p_n, \dots, p_N]^T \tag{3}$$

In this analysis, the mass matrix M is assumed to be a diagonal matrix with elements $m_i > 0$, $i = 1, 2, \dots, N$, in which the element's subscript is its associated node point. It has been said that each generalized coordinate represents either the displacement or the rotation of a portion of the structure's mass at a specific node point. The mass lumping method is probably the most popular method of discretizing the supporting framework and the rigid body portions of an offshore structure. In this research, the mass lumping method will be applied where all the offshore mass structure is modeled as lumped at the top of a cantilever beam. The stiffness matrix K for an N degree of freedom structure is a symmetric matrix of $N \times N$ elements. The stiffness is defined as the force applied to the structure in order to produce a unitary displacement. The constant k_{ij} in other terms is that force that is required at node i to counteract a unit elastic displacement $\xi_{sj} = 1$ imposed at node j, under the condition that all displacements $\xi_{si} = 0$ for $i \neq j$. If a displacement condition is applied sequentially to each node, then the net force at each node j can be obtained by superposition. For a structure with N degrees of freedom, the damping matrix C is defined as a symmetric array of $N \times N$ constants c_{ij}. Damping is an influence within or upon an oscillatory system that has the effect of reducing, restricting, or preventing its oscillations. In physical systems, damping is the capacity of the system to dissipate the energy stored in the oscillation within itself without damage. In this analysis, the damping force q_{Di} for the structural mode coordinate ξ is assumed to be a linear combination of the generalized coordinate velocities $\dot{\xi}_i$, $i = 1, 2, \dots, N$:

$$q_D = C\dot{\xi} \tag{4}$$

The damping matrix can be cast in several different specialized forms, each of which has the advantage of easily utilizing available experimental data to determine the elements c_{ij}. One such form is Rayleigh damping in which C is proportional to the system's mass and also the system's stiffness:

$$C - a_1 M + a_2 K \tag{5}$$

in which a_1 and a_2 are Rayleigh constants, and they are fixed for a given dynamic system. Rayleigh constants are determined using a standardized procedure [30]. The latter equation is pre- and post-multiplied by the matrix of the shape functions as $X^T C X = a_1 X^T K X + a_2 X^T M X$ with the orthogonal properties $X^T K X = diag(\omega_n)$ and $X^T M X = I$. Introducing the nth modal damping factor ζ_n, the last result becomes $X^T C X = diag(a_1 \omega_n^2 + a_2) = diag(2\zeta_n \omega_n)$. After equating the nth diagonal terms above, it follows that $\zeta_n = (\omega_n/2)a_1 + (0.5\omega_n)a_2$. This last result shows that, for an N degree of freedom system for which the N frequencies $\omega_1, \omega_2, \dots, \omega_N$ are known, the two Rayleigh constants a_1 and a_2 are uniquely determined if any two of ζ_n are specified. For instance, if ζ_k and ζ_m are specified, then the equation yields the following two simultaneous equations from which a_1 and a_2 can be calculated $\zeta_k = (\omega_k/2)a_1 + (0.5\omega_k)a_2$ and $\zeta_m = (\omega_m/2)a_1 + (0.5\omega_m)a_2$. Then, the remaining $N - 2$ values of ζ_n for all $n \neq m$ can be determined.

3. Reference Model

Let us suppose that the oil and gas platform studied in this detection is part of a wide context of a gas field, in a water depth of 87 m. The platform sub-structure is a steel jacket, installed by lifting, supporting a three generic level deck for the drilling and extraction activities. The jacket configuration is represented by a four-legged tubular structure, and the jacket base dimensions are approximately 32 m by 20 m (excluding the sleeves). The jacket dimensions at the top interface are 8 m × 8 m. Jacket planes are located at the following elevations from top the bottom with respect to the still water level: +6 m, −8.5 m, −24.5 m, −43.5 m, −64.5 m. The leg diameter varies from 1200 mm by 40 mm at the top to 1600 mm by 50 mm at the bottom. For the bracing elements, the diameter varies from 600 mm by 10 mm–800 mm by 15 mm. Figure 1 is a possible representation of the jacket substructure, the subject of this analysis, realized using Strand7 and MATLAB software.

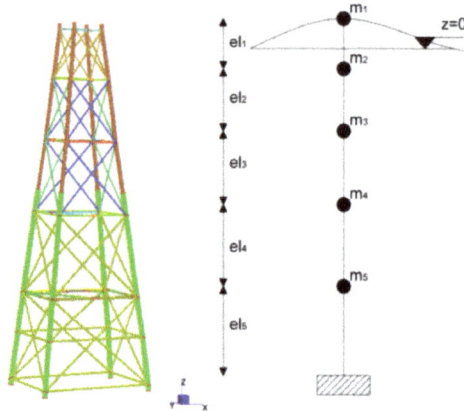

Figure 1. Jacket modeled by Strand7 and MATLAB.

All analysis steps necessary for the simplified design of an offshore structure have been done introducing a 10-step modeling approach:

1. Structural and environmental parameters
2. Undamped motion of the structure
3. Damped frequencies' definition
4. Response to environmental loads
5. Maximum load pile applied
6. Time domain solution
7. Transfer function definition
8. Spectral density function of the generic load
9. Response spectrum
10. 3σ approach

In order to test if the activity model conforms to the real response of the structure, two methods of analysis will be tested for the same structure. One has been developed using MATLAB software and the other one using the Strand7 software. If the results are coherent, the modeling could be said to be efficient and suitable for the representation of the behavior of a generic jacket steel structure.

Platform properties and environmental parameters are described below. The steel jacket is modeled with five lumped masses placed at different heights, one at each horizontal plane and clamped at the bottom. The generic lumped mass is the sum of all the tubular masses surrounding the center of the horizontal plane, in particular half above and half below the generic horizontal plane. A real complex steel jacket structure can be studied using an FE model in Strand7. All the masses have been calculated considering the specific weight of steel 7850 kg/m³ multiplied by the generic cross-section and multiplied by the length of the tubular members. The diagonal mass matrix M of the 5DOF model expressed in kg developed in MATLAB:

$$M = \begin{bmatrix} m_1 & 0 & 0 & 0 & 0 \\ 0 & m_2 & 0 & 0 & 0 \\ 0 & 0 & m_3 & 0 & 0 \\ 0 & 0 & 0 & m_4 & 0 \\ 0 & 0 & 0 & 0 & m_5 \end{bmatrix} = \mathrm{diag}[581{,}889, 122{,}432, 147{,}074, 242{,}301, 206{,}739] \tag{6}$$

The stiffness matrix K has been determined along the direction x. Once having applied the definition of stiffness to all the system, it is possible to define (using MATLAB) the matrix K in N/m as:

$$K = \begin{bmatrix} k_{11} & k_{21} & k_{31} & k_{41} & k_{51} \\ k_{12} & k_{22} & k_{32} & k_{42} & k_{52} \\ k_{13} & k_{23} & k_{33} & k_{43} & k_{53} \\ k_{14} & k_{24} & k_{34} & k_{44} & k_{54} \\ k_{15} & k_{25} & k_{35} & k_{45} & k_{55} \end{bmatrix} = \begin{bmatrix} 0.6568 & -0.6568 & 0 & 0 & 0 \\ -0.6568 & 1.1290 & -0.4722 & 0 & 0 \\ 0 & -0.4722 & 0.9560 & -0.4838 & 0 \\ 0 & 0 & -0.4838 & 1.0523 & -0.5685 \\ 0 & 0 & 0 & -0.5685 & 1.0390 \end{bmatrix} \cdot 10^9 \tag{7}$$

The sea state analysis has been simplified, and sample values have been chosen through Pierson–Moskowitz spectra. Such a distribution is an empirical equation for deep water waves that defines how the energy is distributed with respect to the wave frequency. It is based on the assumption of the fully-developed sea in which the wind blows constantly for a long time over an unlimited fetch area. Another wave spectrum (termed JONSWAP) might be taken into account. It is typical of the North Sea characterized by a limited fetch area. The aim of this project is not to reproduce a specific study of a particular case, but to find a generic tool to reproduce an easier and faster way expressed in terms of analysis cost of the dynamic behavior of a generic lattice structure. The quality of this model is also promoted by the flexibility, the adaptability of the frame, which represents a valid approach to consider different cases. The reader is free to reproduce the analysis using the JONSWAP spectra.

These data characterize the definition of the wave loads defined by Morison's equation against the structure. The following sea state conditions have been chosen as wave height $H = 11.6$ m, the wave period $T = 12$ s, the wavelength $\lambda = 1.56T^2 = 226$ m, the wave frequency $\omega = 2\pi/T = 0.5236$ rad/s, and the salt water density $\rho = 1031$ kg/m³.

The free undamped motion of the structure is investigated solving the eigenvalue problem:

$$\det(K - \omega^2 M) = 0 \tag{8}$$

Thus, the five undamped natural structural frequencies for free vibration in rad/s of the 5DOF model in rad/s are:

$$\omega_1 = 11.4498, \quad \omega_2 = 35.8886, \quad \omega_3 = 68.9503, \quad \omega_4 = 87.9880, \quad \omega_5 = 110.9217 \tag{9}$$

Modal shape matrix X is also identified for each natural structural frequency ω_n. Each modal vector is identified by $\hat{\xi}_n = [\hat{\xi}_{1n}, \hat{\xi}_{2n}, \dots, \hat{\xi}_{Nn}]^T = [1, \hat{\xi}_{2n}, \dots, \hat{\xi}_{Nn}]^T$. Those vectors are normalized with respect to the mass matrix to form the new modal vectors $x_n = \hat{\xi}_n / e_n$, where e_n is a set of positive and real constants computed from $\hat{\xi}_n^T M \hat{\xi}_n = e_n^2$. It follows that the modal shape matrix X of the 5DOF model is defined as the assembly of the normalized modal vectors x_n:

$$X = [x_1, x_2, \ldots, x_5] = \begin{bmatrix} 1.1 & 0.6 & 0.3 & -0.2 & 0.2 \\ 1.0 & -0.1 & -1.0 & 1.1 & -2.2 \\ 0.8 & -0.9 & -1.7 & 0.7 & 1.5 \\ 0.5 & -1.4 & 0.1 & -1.3 & -0.4 \\ 0.3 & -1.0 & 1.4 & 1.3 & 0.2 \end{bmatrix} \cdot 10^{-3} \tag{10}$$

Associated with five undamped natural structural frequencies for free vibration ω_n with $n = 1, 2, \ldots, 5$ are the five corresponding vibration modes of the 5DOF model.

A benchmark test with Strand7 software is carried out for verification purposes using natural frequency analysis. Forty vibration modes have been considered, because the FE model is made of several 1D beam elements. The five vibration modes given by the MATLAB code are compared to the ones obtained via FE in Figure 2. The undamped natural frequencies in rad/s are: $\omega_1 = 14.33$, $\omega_2 = 35.00$, $\omega_3 = 57.93$, $\omega_4 = 76.97$, and $\omega_5 = 104.62$.

It can be noted that the mode shapes obtained by MATLAB are close to the ones obtained by FE simulation, as well as the corresponding natural undamped frequencies.

Subsequently, damped natural frequencies should be computed because they better represent the structural behavior of real structures. Rayleigh damping is considered for this purpose. Damped frequencies are characterized by the two Rayleigh constants a_1 and a_2, which are uniquely determined if any two of the modal damping factors ζ_n are specified. Since the first few modes will dominate the motion, it is reasonable to choose $\zeta_1 = \zeta_2 = 0.05$.

Thus, the five damped natural structural frequencies for free vibration in rad/s of the 5DOF model are:

$$\omega_{d1} = 11.4355, \quad \omega_{d2} = 35.8437, \quad \omega_{d3} = 68.7227, \quad \omega_{d4} = 87.5516, \quad \omega_{d5} = 110.0883 \tag{11}$$

The fourth step, of the present procedure, regards the response of the structure to the harmonic wave. Airy theory is considered, and it is assumed that the motion of the structure is much smaller than the motion of the wave; thus, Morison's equation can be applied. The inertia counterpart of the flow is represented through the inertia coefficient $C_M = 2$, while the drag counterpart by the drag coefficient $C_D = 0.8$. The structure has been modeled with four vertical legs $N_l = 4$ plus two horizontal cross braces $N_c = 2$, which are normal to the flow. The calculation of the wave forces follows the guidelines reported by classical books of offshore structural modeling [30]. Wave loads for the present problem in kN are:

$$p_1 = 580.53, \quad p_2 = 1132, \quad p_3 = 1125.3 \quad p_4 = 1092, \quad p_5 = 1111.1 \tag{12}$$

The fifth step is related to the pile stress analysis at the foundation level. Global bending moment at the foundation level is considered due to all the forces of the model. The final stress on the pile is determined by the bending moment and the weight of the structure (Figure 3).

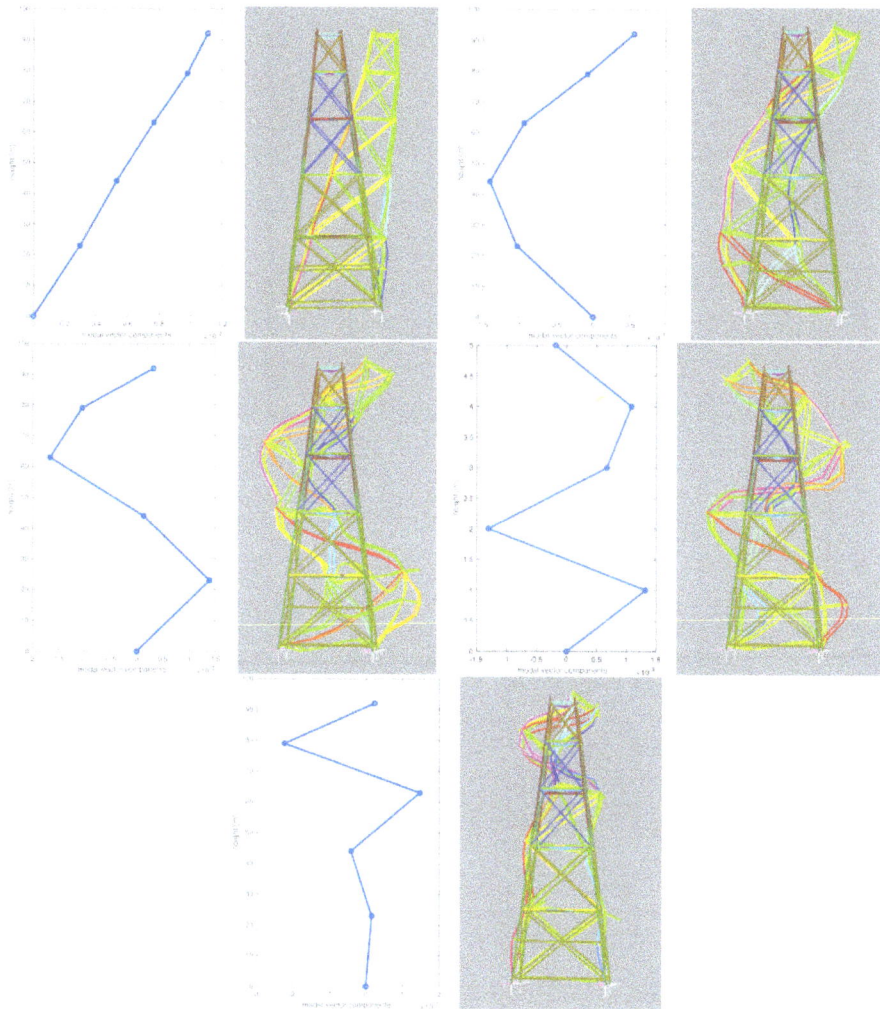

Figure 2. Mode shape comparison between 5DOF and the FE model.

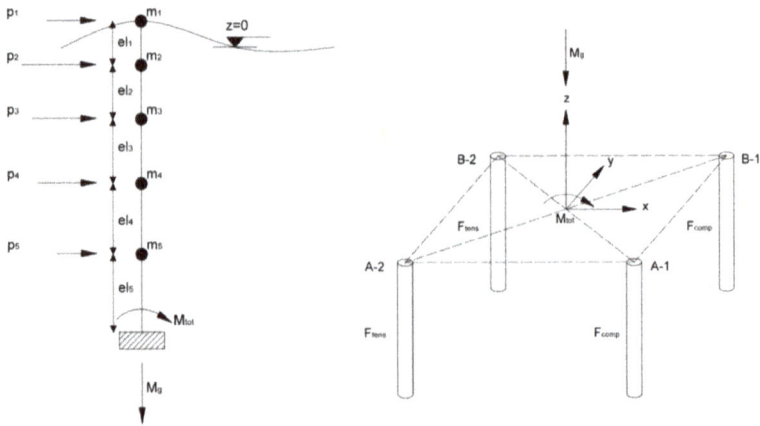

Figure 3. Pile stress analysis of the 5DOF model.

Moreover, a wind force F_{wind}= 410 kN is considered acting on the superstructure z_{wind} = +7.5 m. Therefore, the force applied on the pile B-1 is $F_{B1} = -8316.28$ kN. As the numerical benchmark, the same analysis is performed in the FE model. The maximum compression value of the stress in piles given by Strand7 is in the pile B1 and is equal to $F_{B1} = -9922$ kN, while the MATLAB calculation has given $F_{B1} = -8316$ kN. There is a high correlation between the two values; indeed, they differ by only of 16%. It follows that the 5DOF model proposed in this research is working properly.

The sixth step is based on the time domain solution. These five structural steady state displacements over a time of 30 s, or in other words, the response of the structure subjected to five cycles of wave loading, are depicted in Figure 4.

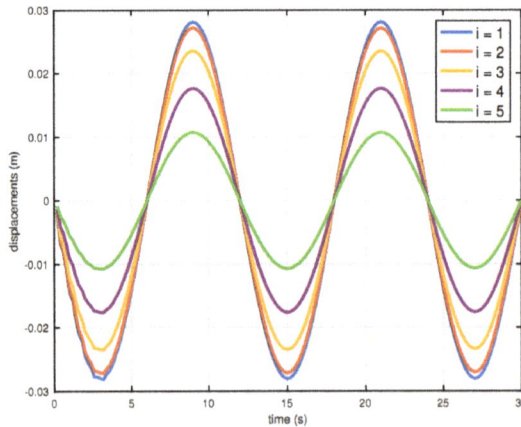

Figure 4. Time domain plot of the 5DOF model.

The absolute value of the maximum peaks in m are: $\xi_1 = 0.02813$, $\xi_2 = 0.027214$, $\xi_3 = 0.02356$, $\xi_4 = 0.017664$, $\xi_5 = 0.01073$. These values are then compared with the results obtained with the 3σ approach that defines structural safety according to the stochastic approach. The simulation is verified via FE analysis, leading to $\xi_1 = 0.0384$, $\xi_2 = 0.0348$, $\xi_3 = 0.0278$, $\xi_4 = 0.0187$, $\xi_5 = 0.0101$.

The maximum displacements obtained with MATLAB and FE are of the same magnitude. The ratios between the displacement obtained with MATLAB and Strand7 are listed in Table 1.

Table 1. Comparison between MATLAB and Straus7 displacements.

	MATLAB	FE	Ratio
ζ_1	0.0281	0.0384	0.73
ζ_2	0.0272	0.0348	0.78
ζ_3	0.0236	0.0278	0.85
ζ_4	0.0177	0.0187	0.95
ζ_5	0.0107	0.0101	1.06

Small deviations can be observed. However, this is due to the simplifications introduced; nevertheless, as a preliminary design phase, the results can be considered worthy for proceeding in the present study. It follows that the model fully satisfies the requirements of this research.

The seventh step defines the wave transfer functions for the model. The transfer function $G(\omega)$ is a function that relates the wave height H of the incident wave to the load imparted to the structural component. It is generally defined in harmonic form: $G(\omega) = G_0 e^{j\omega t}$ with $j = \sqrt{-1}$, and G_0 is a complex number independent of time. The transfer function depends on the flow regimes and the structural component. Transfer functions are computed for all components and assembled for the structure. The transfer function can be applied only to linear analysis (as in the present case), and the effect of multiple wave excitation can be considered (which is not possible in the nonlinear regime). The calculation of the wave transfer function follows the indication provided in [30], and the plot of the modulus square of the five transfer functions $|x_k^T G(p,\omega)|^2 = |G(\bar{q}_k,\omega)|^2$ is given in Figure 5.

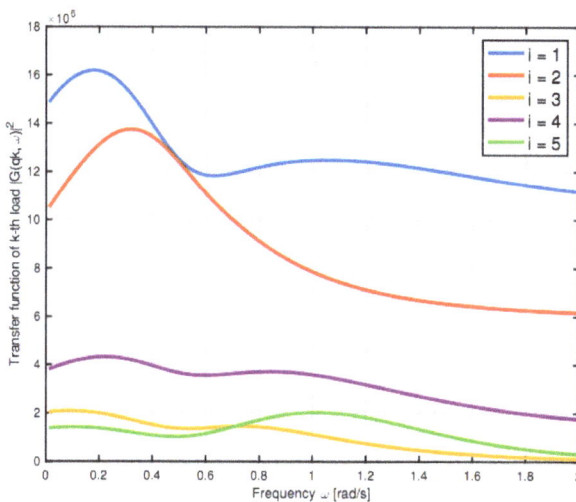

Figure 5. Wave transfer functions for the 5DOF model.

Spectral Density Function of the Generalized Force Component in Modal Coordinates

The eighth step is based on the definition of the spectral density function of the generalized force component in modal coordinates. It is necessary to determine first the Pierson–Moskowitz wave spectrum [30]:

$$S_\eta(\omega) = 0.0081 \frac{g^2}{\omega^5} e^{-B/\omega^4}$$ (13)

with $B = 0.74(g/V)^4$ where v is the wind speed in m/s at a height of 19.5 m above the still water level and ω is expressed in rad/s. Multiplying the transfer function of the load in modal coordinates by the Pierson–Moskowitz spectrum, it achieved the spectral density function of the generalized force component in modal coordinates (Figure 6).

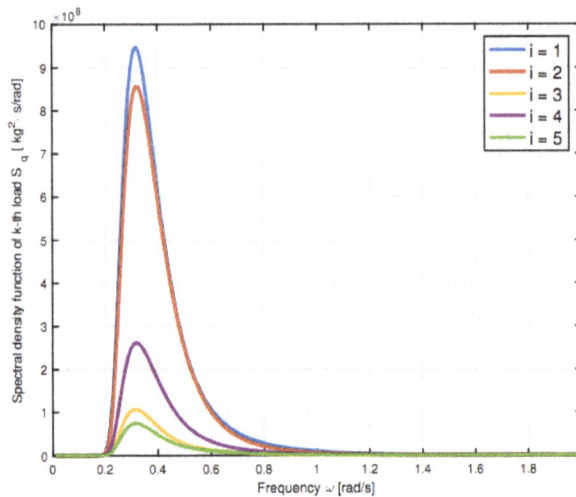

Figure 6. Spectral density functions of the generalized force component in modal coordinates for the 5DOF model.

The ninth step is focused on the spectral density in terms of modal coordinates, and it is given by $S(y_k, \omega) = |H_k(\omega)|^2 S(\bar{q}_k, \omega)$ with $k = 1, 2, \ldots, 5$. $S(\bar{q}_k, \omega)$ is the spectral density function of the generalized force component in modal coordinates, and $|H_k(\omega)| = [(\omega_k^2 - \omega^2)^2 + (2\zeta_K \omega_k \omega)^2]^{-1/2}$ is the modulus of the harmonic response function for the kth mode. Once the spectral density in terms of modal coordinates $S(y_k, \omega)$ has been determined, is possible to define the spectral density in terms of physical coordinates $S(\xi_k, \omega) = \sum_{n=1}^{N} x_{kn}^2 S(y_k, \omega)$ applying the inverse of the modal analysis, as shown in [30] (Figure 7).

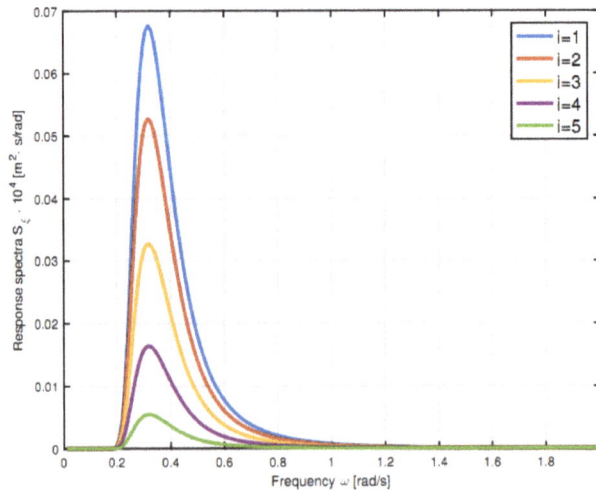

Figure 7. Response spectra for the displacement in physical coordinates of the fixed-leg platform.

The last step includes the stochastic approach design using the 3σ approach. By computing the area covered by the response displacement spectra, the variance of such displacement can be carried out $\sigma^2(\xi_k)$. The extreme limits of ξ_k are $\pm 3\sigma(\xi_k)$ with $k = 1, 2, \ldots, 5$. If the static stresses and deflection of the members are within these extreme limit values, then the structure is assumed safe. Assuming a Gaussian process, there is only a 0.026% chance that each response exceeds the $\pm 3\sigma(\xi_k)$ limits.

The extreme displacement limit values in m are:

$$3\sigma(\xi_1) = 0.3577, \quad 3\sigma(\xi_1) = 0.3158, \quad 3\sigma(\xi_1) = 0.2491, \quad 3\sigma(\xi_1) = 0.1763, \quad 3\sigma(\xi_1) = 0.1014 \quad (14)$$

After comparing these results with the maximum peak displacements (Figure 4), the reader will easily state that the structure can be said to be safe.

4. Jacket Supporting the Wind Turbine

The aim of the present paper is to find a useful tool able to model the behavior accurately of a new offshore structure that supports the wind turbine subjected to environmental actions. The structure under investigation presents the same jacket structure with a long pile foundation placed in the same position with the same characteristics. All analysis steps necessary for the simplified design for the new offshore structure have been introduced in the previous section. The introduced 10-step modeling approach will be considered also in the following.

The wind tower is made of steel with a length of 80 m. The nacelle mass is 240 tons. The rotor mass is 110 tons. The turbine diameter is 126 m (Figure 8). The wind tower has a truncated cone shape; at the bottom, the diameter is 4 m and the thickness is 0.18 m; at the top, the diameter is 3.5 m and the thickness is 0.10 m. At half tower height, the diameter is 3.75 m and the thickness is 0.14 m. The hub is at z = +90 m. The wind tower produces a rated power of 5 MW. It is attached to the substructure through a concrete (density 1600 kg/m^3) transition piece with 9.6 × 9.6 × 4 m^3 of volume. The whole structure will be treated as a discrete system with 5 + 1 lumped masses.

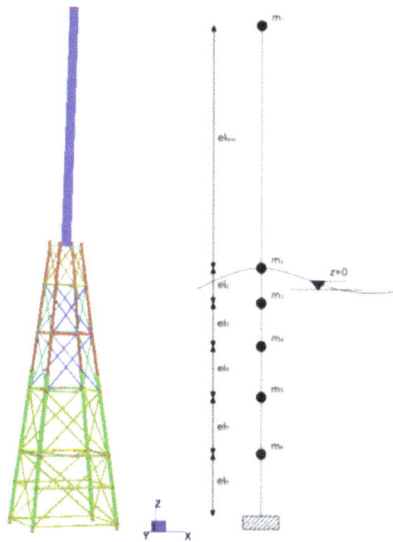

Figure 8. Wind tower model.

The lumped masses of the jacket are calculated using the same procedure illustrated above. It follows that the mass matrix M of the 6DOF model in kg is:

$$M = \begin{bmatrix} 937{,}000 & 0 & 0 & 0 & 0 & 0 \\ 0 & 1{,}587{,}300 & 0 & 0 & 0 & 0 \\ 0 & 0 & 122{,}400 & 0 & 0 & 0 \\ 0 & 0 & 0 & 147{,}100 & 0 & 0 \\ 0 & 0 & 0 & 0 & 242{,}300 & 0 \\ 0 & 0 & 0 & 0 & 0 & 206{,}700 \end{bmatrix} \tag{15}$$

The stiffness matrix K has been determined along the x direction using the proper definition. Thus, matrix K in N/m is:

$$K = \begin{bmatrix} 0.121 & -0.121 & 0 & 0 & 0 & 0 \\ -0.121 & 0.6689 & -0.6568 & 0 & 0 & 0 \\ 0 & -0.6568 & 1.1290 & -0.4722 & 0 & 0 \\ 0 & 0 & -0.4722 & 1.0345 & -0.5622 & 0 \\ 0 & 0 & 0 & -0.5622 & 1.1307 & -0.5685 \\ 0 & 0 & 0 & 0 & -0.5685 & 1.2073 \end{bmatrix} \cdot 10^9 \tag{16}$$

The same sea state of the previous section has been considered.

The free undamped motions of the present structure can be determined solving the corresponding eigenvalue problem. The six undamped natural frequencies computed using MATLAB and the FE model are listed in Table 2 in rad/s.

Table 2. Undamped natural frequencies computed using MATLAB and the FE model.

	MATLAB	FE	Ratio
ω_1	3.38	3.39	≈ 1
ω_2	8.53	8.54	≈ 1
ω_3	36.02	36.00	≈ 1
ω_4	71.90	71.88	≈ 1
ω_5	90.62	90.60	≈ 1
ω_6	111.50	111.53	≈ 1

$\omega_1 = 3.3879$, $\omega_2 = 8.5393$, $\omega_3 = 36.0211$, $\omega_4 = 71.8985$, $\omega_5 = 90.6182$, $\omega_6 = 111.5014$.

The modal shape matrix X of the 6DOF model is defined as the assembly of the normalized modal vectors x_n as:

$$X = [x_1, x_2, \ldots, x_6] = \begin{bmatrix} 1.0 & -0.2 & -0.0 & 0.0 & -0.0 & -0.0 \\ 0.1 & 0.7 & 0.2 & -0.1 & 0.1 & 0.1 \\ 0.1 & 0.6 & -0.5 & 1.3 & -1.3 & -2.1 \\ 0.1 & 0.5 & -1.2 & 1.5 & -0.4 & 1.6 \\ 0.0 & 0.3 & -1.5 & -0.3 & 1.2 & -6.0 \\ 0.0 & 0.1 & -0.9 & -1.4 & -1.4 & 0.2 \end{bmatrix} \cdot 10^{-3} \tag{17}$$

and its vibrational modes of shape are compared to the ones obtained by FE analysis in Figure 9. As follows, all the frequencies computed via FE analysis in rad/s are: $\omega_1 = 2.74$, $\omega_2 = 7.16$, $\omega_3 = 26.13$, $\omega_4 = 50$, $\omega_5 = 63.81$, and $\omega_6 = 80.39$. The modes of shape obtained by MATLAB calculation and the ones obtained by Strand7 are quite similar, as well as the corresponding natural undamped frequencies. The correlation between the two approaches shows the quality of the model.

As done for the previous case, damped frequencies of the 6DOF model in rad/s are computed as: $\omega_{d1} = 3.3836$, $\omega_{d2} = 8.5286$, $\omega_{d3} = 35.2135$, $\omega_{d4} = 67.8688$, $\omega_{d5} = 82.9486$, $\omega_{d6} = 97.4530$.

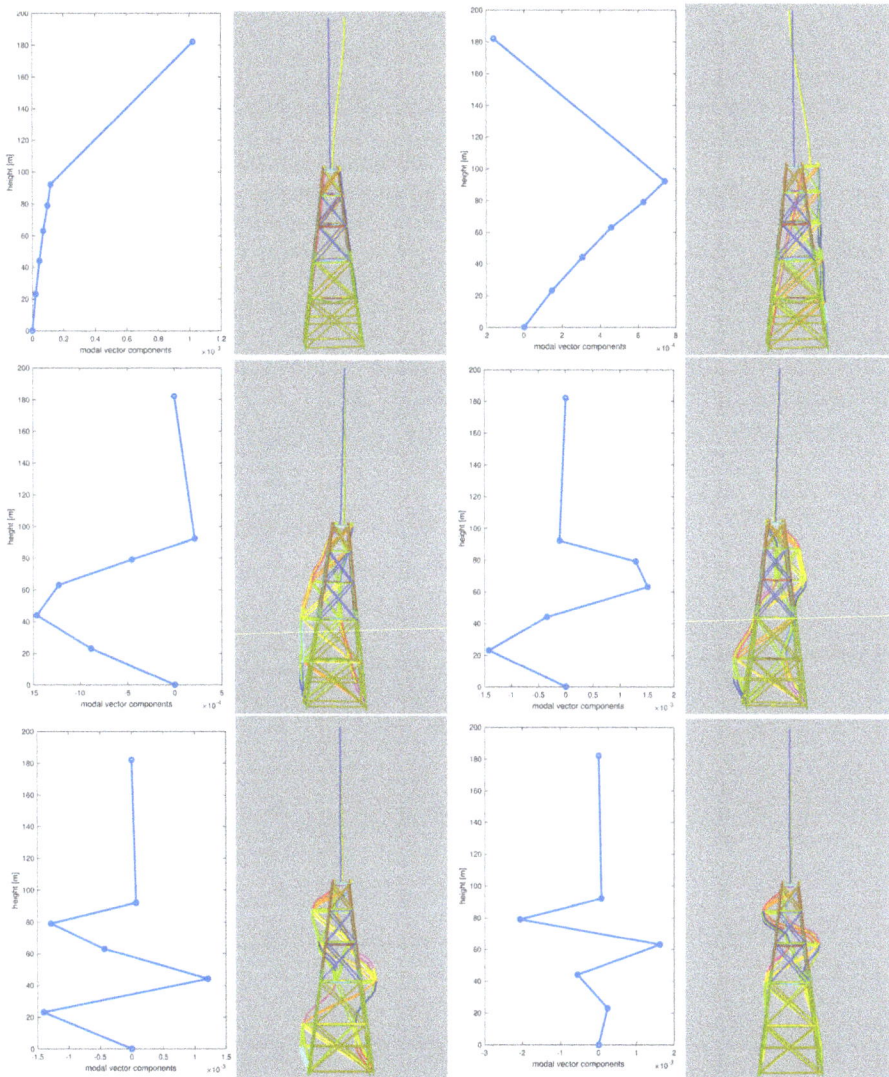

Figure 9. Vibrational mode comparison between the 6DOF and FE models.

The harmonic wave of the previous case is considered below. Regarding the wind load, the correspondent force is calculated at z = +90 m. The wind force is given by two contributions: one from the wind acting against the swept rotor area and the other from the wind acting against the tower. The force considered at the top node of the model $p_1 = 3055$ kN will take into account the wind on the swept rotor area and the wind on the half tower. The remaining half part of the wind on the tower will be add to the first wave load. The wind force is given by two contributions: one from the wind acting against the swept rotor area and the other from the wind acting against the tower. The force considered at the top node of the model p_1 will take into account the wind on the swept rotor area and the wind on the half tower. The remaining half part of the wind on the tower will be added to the first wave load. The counterpart wind on the tower is calculated by integrating the wind velocity over the tower height, and it is equal to $f_w = \int u(z)dz = 191.58$ N with $u(z) = 1.5(z/3)^{1/7}$

and with $0 \leq z \leq 90$ m. The counterpart wind on the rotor is $f_{rot} = 0.5\rho u_{rot} A = 3054.9$ kN, where the wind velocity at the rotor hub is $u_{rot} = 20$ m/s. It follows that the wind load applied at the top of the wind turbine is $p_1 = f_w/2 + f_{rot} = 3055$ kN.

The pile foundation check is carried out as previously described with reference to Figure 10. For this new configuration, the maximum compression value in the piles given in B-1 is equal to $FB1 = -21{,}131$ kN. The same pile in the 5DOF configuration withstood a compression equal to $FB1 = -8316$ kN. In the new configuration, the pile B-1 carries more than two and a half-times the stress carried in the previous one. Supposing that the compression threshold pile limit is $Fmax = -18{,}000$ kN, it is quite clear that the new load produced by the 6DOF configuration is no longer bearable for the pile B-1.

The generic long steel pile foundation in this structure presents the following characteristics: diameter of 2.13 m, wall thickness of 50 mm, pile length of 96 m, of which 75 m are embedded. The corresponding axial pile capacity is $Q_r = 28{,}800$ kN. In order to take into account the differences between the structure with the cantilever model, a safety factor equal to $\gamma_r = 1.60$ has been chosen. It follows that the maximum bearing load for the pile is $F_{max} = Q_r/\gamma_r = 18{,}000$ kN.

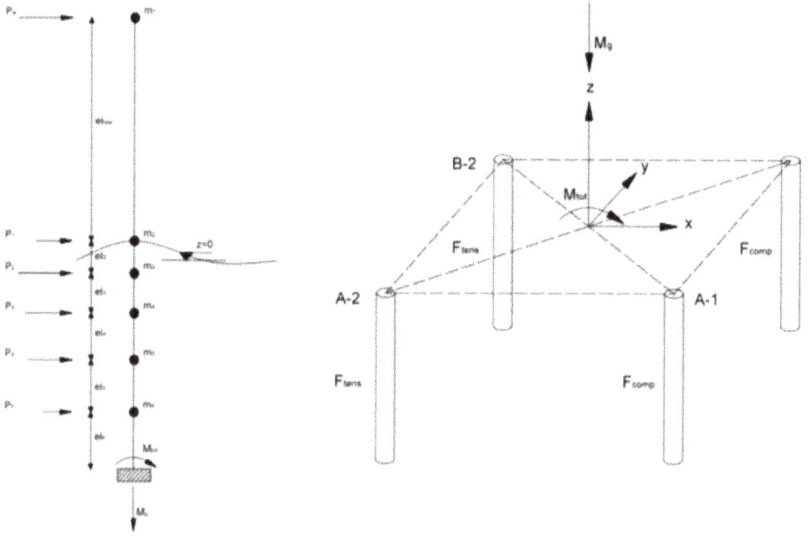

Figure 10. Pile stress analysis of the 6DOF model.

Performing the static analysis with Strand7 of the 6DOF model, the maximum displacement for each node is: $\xi_1 = 1.3893$, $\xi_2 = 0.1012$, $\xi_3 = 0.712$, $\xi_4 = 0.434$, $\xi_5 = 0.0244$, $\xi_6 = 0.0103$.

These values will be compared with the $3\sigma(\xi_k)$ approach below.

Although the jacket substructure has the same configuration as the 5DOF model, the wave transfer functions are not identical, but similar to the previous case. This is due to the mode shape matrix X of the 6DOF model. It follows that the equations are the same as the 5DOF configuration, but the functions of the five wave transfer functions are slightly different.

In order to define the spectral density function of the generalized force component, it is requested to define first the wave and wind spectra. For the wind analysis, the Kaimal spectrum is used. The parameters chosen for the Kaimal spectrum are:

$$S_u(\omega) = \sigma_u^2 \frac{6.868 \frac{L_u}{U_{mean}}}{(1 + 10.32 \frac{\omega L_u}{U_{mean}})^{5/3}} \tag{18}$$

where the mean wind velocity $U_{mean} = 6$ m/s, the variance of the wind speed $\sigma_U^2 = \lambda_w^2 \cdot (\Gamma(1 + \frac{2}{k_w}) - (\Gamma(1 + \frac{1}{k_w}))^2 = 12.81$ (m/s)^2 with $\lambda_w = 6.7$ and $k_w = 1.72$ Weibull parameters of probability density of gamma function Γ, roughness parameter $z_0 = 0.001$, the integral length scale $L_u = 300(\frac{el_1}{300})^{0.46+0.074log(z_0)} = 321$ m, and the air density $\rho_{air} = 1.225$ kg/m^3 (Figure 11).

Wave modeling is carried out using the Pierson–Moskowitz spectrum. Multiplying the wind transfer function of the load $H_{uf} = C_T A \rho_{air} U_{mean}$ in the modal coordinates with the thrust coefficient $C_T = 0.8$ by the Kaimal spectrum S_u for the wind and the wave transfer functions $|x_k^T G(p, w)|^2$ by the Pierson–Moskowitz S_η, the spectral density function of the wind force in modal coordinates $S_f = H_{uf}^2 \cdot S_u$ and the spectral density of the generalized wave force in modal coordinates S_{qi}, respectively, are achieved (Figure 12).

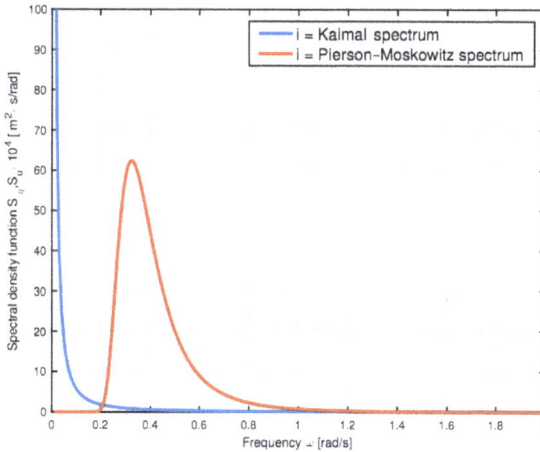

Figure 11. Wave and wind spectral density function.

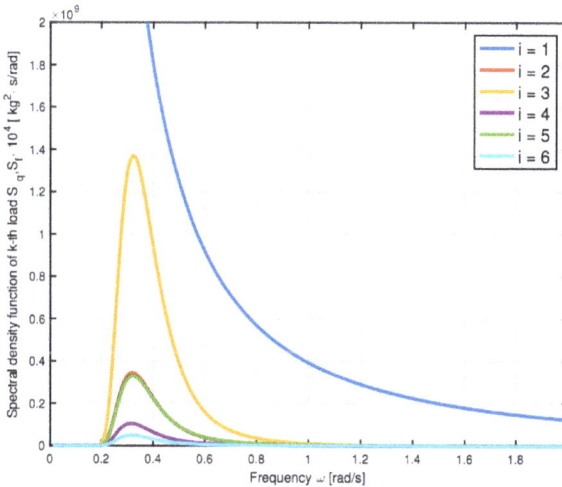

Figure 12. Spectral density of the generic environmental load in modal coordinates of the 6DOF model.

The spectral density of the wind load in modal coordinates $i = 1$ has values much higher than the wave loads; in particular, it tends to infinity when ω tends to zero. For the sake of simplicity, the spectral density function of the wind force in modal coordinates will be limited. The wind transfer function of the load H_{uf} in modal coordinates, which converts the wind speed spectrum to the wind load spectrum, is based on the assumption that $u(t)$ has the same value over the area, which in general is not true.

Once the spectral density function of the wind load and generalized wave component are defined in modal coordinates S_f and S_{qi}, the wind transfer function from the spectral density of wind force into spectral density in modal coordinates is carried out as:

$$|H_{wind}(\omega)|^2 = \frac{1}{m_i \cdot 4\pi\omega_i^4} \frac{1}{\{1 - (\frac{\omega}{\omega_i})^2\}^2 + 4\zeta_i^2(\frac{\omega}{\omega_i})^2} \tag{19}$$

and the wave transfer functions from spectral density of the generalized wave force component into spectral density in modal coordinates $|H_k(\omega)|^2$. The spectral density function in terms of modal coordinates $S(y_k, \omega)$ is given by multiplying the spectral density function of the load in modal coordinates with the correspondent transfer function in modal coordinates. Let J_1 be the joint acceptance of the mode, which accounts for the correlation of wind loading along the rotor, considering the mode shape.

$$J_1 = \frac{1}{X^4}|H_{wind}(\omega)|^2 \tag{20}$$

Once the spectral density in terms of modal coordinates $S(y, \omega)$ has been determined, it is possible to define the spectral density in terms of physical coordinates $S(\xi_k, \omega)$ (Figure 13).

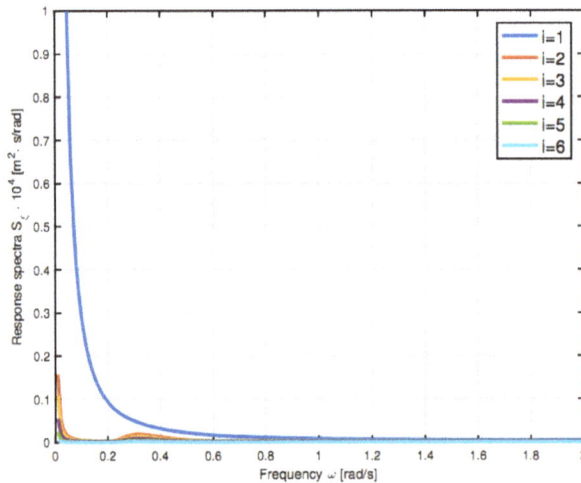

Figure 13. Response spectra for the displacement in physical coordinates of the offshore wind turbine plot.

The last 10-step methodology phase defines the variance of the structural displacements in physical coordinates $\sigma^2(\xi_k)$. The extreme limits of ξ_k are $\pm 3\sigma(\xi_k)$ with $k = 1, 2, \ldots, 6$. If the static stresses and deflection of the members are within these extreme limit values, then the structure is assumed safe. Considering a Gaussian process, there is only a 0.026% chance that each response exceeds the $\pm 3\sigma(\xi_k)$ limits. The extreme displacement values of the 6DOF configuration in m are: $3\sigma(\xi_1) = 1.3822$, $3\sigma(\xi_2) = 0.3471$, $3\sigma(\xi_3) = 0.2917$, $3\sigma(\xi_4) = 0.2154$, $3\sigma(\xi_5) = 0.1505$, $3\sigma(\xi_6) = 0.0748$. After comparing these results with the ones given above, it can be stated that the structure cannot be safe for the first node at the top of the wind tower, in particular $\xi_1 = 1.40$ m, whereas the limit is $3\sigma(\xi_1) = 1.38$ m. The 6DOF model built from the adaptation and extension of the 5DOF model has to be modified in order to find the safety condition. In order to do that, hereafter, different retrofitting examples will be presented.

5. Improved Retrofitting Models

Retrofitting refers to the addition of new technology or features to older systems, extending the service life. In this case, different retrofitting typologies are illustrated. The retrofitting models proposed in this section are analyzed following the same methodology previously shown. The approach is suitable for tender design and early design steps of jacket structures. The aim is to identify a useful tool that is able to represent the behavior of the new offshore structure subjected to environmental actions. The system under investigation presents the same jacket and a big wind tower installed on top of it. The whole system will be analyzed with MATLAB calculation and supported by Strand7 software. The aim is to find the right intervention that will ensure the modified offshore structure without causing stability trouble for the system. This research proposes three main retrofitting activities that can be adopted for the offshore wind turbine. They are: crown pile, mooring lines, and 2-MW wind turbine.

5.1. Crown Pile Configuration

The crown pile solution represents the same configuration of the 6DOF model except for the piles at the base of the jacket. In addition to the four corner piles, here there are an extra four piles placed at the midspan of each side at $z = -d$ connected each to the other piles with a rectangular ring (Figure 14). Regarding Figure 14 and all the following configurations, $z = -d$ represents the sea bottom level when the vertical axis z is zero at the Still Water Line (SWL) pointing upward. The reference system shown in Figure 14 of the FE model considers the reference system at the sea bottom, so $z = 0$. However, all the present MATLAB models consider the z axis pointing upward with the sea bottom at $z = -d$.

Figure 14. Crown pile model detail.

The difference from the previous model is in the definition of the mass matrix M, where the mass $m_6 = 19{,}202$ kg. The six undamped natural structural frequencies for free vibration in rad/s are: $\omega_1 = 3.3879$, $\omega_2 = 8.5391$, $\omega_3 = 35.7613$, $\omega_4 = 70.5186$, $\omega_5 = 89.1430$, $\omega_6 = 111.4539$. For the nth frequency ω_n exists a modal vector ξ^n, which can be computed solving the linear eigenvalue problem (see [30] p. 226).

The crown pile mode shapes are represented in Figure 15.

The damped frequencies of the 6DOF model in rad/s are: $\omega_{d1} = 3.3837$, $\omega_{d2} = 8.5284$, $\omega_{d3} = 34.9825$, $\omega_{d4} = 66.6939$, $\omega_{d5} = 81.8239$, $\omega_{d6} = 97.4230$.

Wave and wind loads do not change with respect to the previous 6DOF case.

The pile foundation check leads to a maximum compression in the pile B-1 equal to $F_{B1} = -17{,}178$ kN lower than the pile compression limit $F_{max} = -18{,}000$ kN. The reader can easily state that this first configuration helps the wind turbine in terms of stability. The maximum node displacements of the crown pile model in m are: $\xi_1 = 1.3893$, $\xi_2 = 0.1011$, $\xi_3 = 0.711$, $\xi_4 = 0.433$, $\xi_5 = 0.0243$, $\xi_6 = 0.0101$. These values will be compared with the results obtained with the $3\sigma(\xi_k)$ approach to ensure structural safety.

The five wave transfer function are rather similar to the ones obtained for the 6DOF.

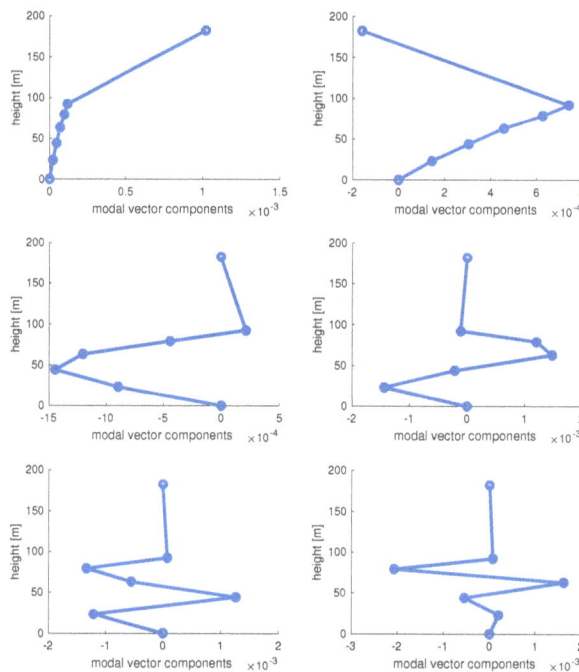

Figure 15. Crown pile vibration modes.

Multiplying the wind transfer function of the load H_{uf} in modal coordinates by the Kaimal spectrum S_u for the wind and the wave transfer functions $|x_k^T G(p,\omega)|^2$ by the Pierson–Moskowitz S_η, the spectral density function of the wind force in modal coordinates S_f and the spectral density of the generalized wave force in modal coordinates S_{qi}, respectively, are achieved. Such spectral densities are depicted in Figure 16.

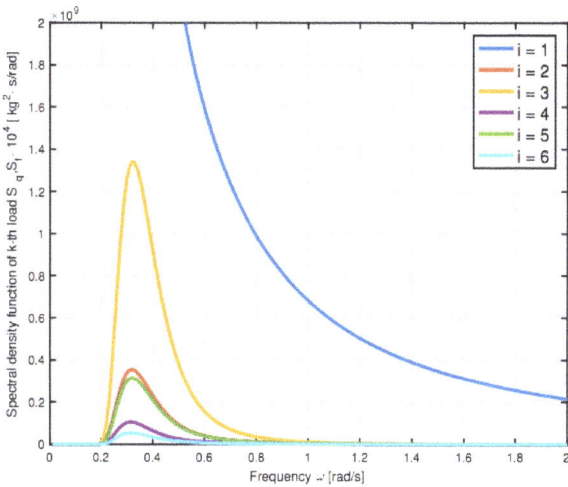

Figure 16. Crown pile spectral density of the generic environmental load in modal coordinates.

The response spectra (Figure 17) for the displacements in physical coordinates of the offshore wind turbine crown pile configuration have been determined.

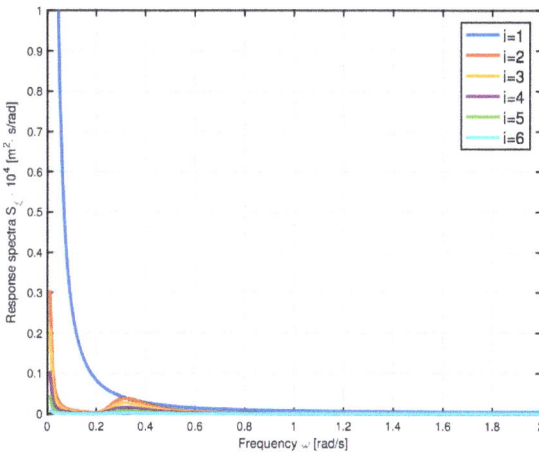

Figure 17. Response spectra for the displacements in physical coordinates of the offshore wind turbine crown pile configuration.

The area beneath these spectral density functions represents the variance of the generic displacement in physical coordinates $\sigma^2(\xi_k)$. The extreme limits of ξ_k are $\pm 3\sigma(\xi_k)$ with $k = 1, 2, \ldots, 6$. If the static stresses and deflection of the members are within these extreme limit values, then the structure is assumed safe. Considering a Gaussian process, there is only a 0.026% chance that each response exceeds the $\pm 3\sigma(\xi_k)$ limits.

The extreme displacement values of the crown pile configuration in m are: $3\sigma(\xi_1) = 1.3821$, $3\sigma(\xi_2) = 0.3470$, $3\sigma(\xi_3) = 0.2916$, $3\sigma(\xi_4) = 0.2152$, $3\sigma(\xi_5) = 0.1503$, $3\sigma(\xi_6) = 0.0751$.

Comparing these results with the ones given by Figure 15, the reader will easily state that the structure cannot be safe for the first node at the top of the wind tower, in particular $\xi_1 = 1.3893$ m and

$3\sigma(\xi_1) = 1.3821$. The crown pile configuration has obtained higher values of safety expressed by the $3\sigma(\xi)$ approach results, but this is not enough yet for the global stability.

5.2. Mooring Line Configuration

The mooring line configuration presents the same configuration of the 6DOF model, which in addition has a mooring line system that links the midspan second horizontal plane point of the jacket to the seabed, limiting the movement (Figure 18). The mooring line has been modeled as a steel truss element subjected to tension action given by the environmental loads. It has a diameter cross-section of 0.1 m and a length of 98 m, and it creates an angle of 60° with the seabed. For this proposal, only the x direction will be considered, but for real cases, it could be extended and applied for each direction.

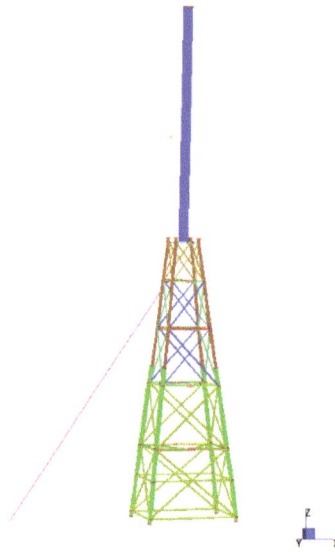

Figure 18. Mooring line configuration model detail.

The mooring line masses are calculated following the same approach shown previously. Here, the mooring pile configuration mass matrix M in kg:

$$M = \begin{bmatrix} 0.9370 & 0 & 0 & 0 & 0 & 0 \\ 0 & 1.5873 & 0 & 0 & 0 & 0 \\ 0 & 0 & 0.1224 & 0 & 0 & 0 \\ 0 & 0 & 0 & 0.1471 & 0 & 0 \\ 0 & 0 & 0 & 0 & 0.2423 & 0 \\ 0 & 0 & 0 & 0 & 0 & 0.2067 \end{bmatrix} \cdot 10^6 \quad (21)$$

The stiffness matrix K has been determined over the x direction. Once having applied the definition of stiffness to all the system, it is possible to define the matrix K in N/m as:

$$K = \begin{bmatrix} 0.121 & -0.121 & 0 & 0 & 0 & 0 \\ -0.121 & 0.6689 & -0.6568 & 0 & 0 & 0 \\ 0 & -0.6568 & 1.1342 & -0.4722 & 0 & 0 \\ 0 & 0 & -0.4722 & 0.9560 & -0.4838 & 0 \\ 0 & 0 & 0 & -0.4838 & 1.0523 & -0.5685 \\ 0 & 0 & 0 & 0 & -0.5685 & 1.0390 \end{bmatrix} \cdot 10^9 \qquad (22)$$

The sea state is unchanged with respect to the previous cases.

The six undamped natural structural frequencies for free vibration in rad/s are: $\omega_1 = 3.3737$, $\omega_2 = 8.2907$, $\omega_3 = 33.7583$, $\omega_4 = 67.6558$, $\omega_5 = 87.5002$, $\omega_6 = 111.0619$. Natural frequencies of the present case are depicted in Figure 19.

The response of the mooring line configuration to the harmonic waves has been considered using the same hypothesis for the 6DOF model. The wave loads and the wind load present the same values for the 6DOF system.

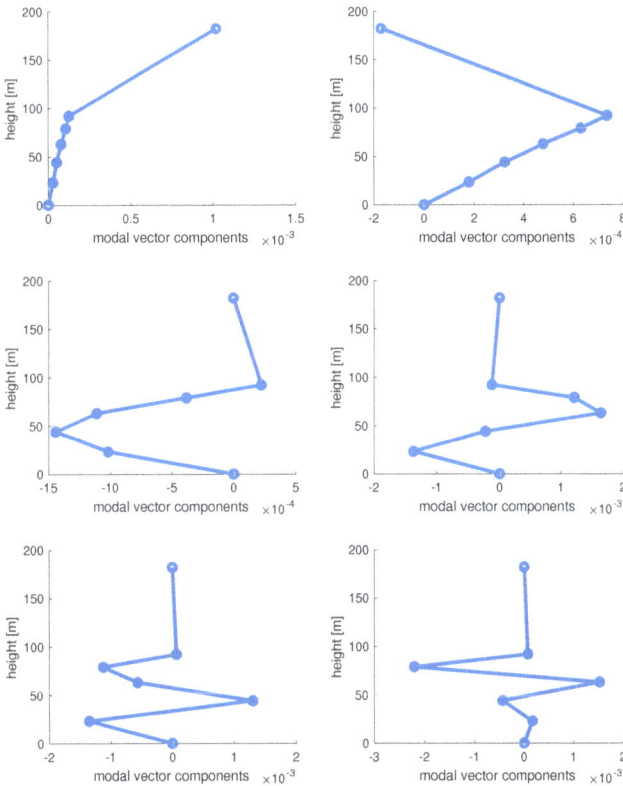

Figure 19. Mooring line vibrational modes.

A foundation check is performed for both the piles and the mooring line anchoring. The tension in the mooring line from Strand7 calculation is $F_{ml} = 427.5$ kN. The calculation of the stress applied on the pile B-1 considers all the environmental loads p_i with the horizontal and vertical component of F_{ml} that are applied to the structure. The compression in the pile B-1 is equal to $F_{B1} = -20{,}959$ kN, which is still too high for the compression pile limit capacity.

The damped frequencies of the 6DOF model in rad/s are: $\omega_{d1} = 3.3695$, $\omega_{d2} = 8.2803$, $\omega_{d3} = 33.0309$, $\omega_{d4} = 64.0793$, $\omega_{d5} = 80.2370$, $\omega_{d6} = 95.8081$.

The maximum nodal displacements of the mooring model in m are: $\xi_1 = 1.3854$, $\xi_2 = 0.0983$, $\xi_3 = 0.0685$, $\xi_4 = 0.0416$, $\xi_5 = 0.0235$, $\xi_6 = 0.0099$. These values will be then compared with the results obtained with the $3\sigma(\xi_k)$ for checking structural safety.

The five wave transfer functions are similar to the ones obtained for the previous case.

Multiplying the wind transfer function of the load H_{uf} in modal coordinates by the Kaimal spectrum S_u for the wind and the wave transfer functions $|x_k^T G(p,\omega)|^2$ by the Pierson–Moskowitz S_η, the spectral density function of the wind force in modal coordinates S_f and the spectral density of the generalized wave force in modal coordinates S_{qi}, respectively, are achieved. These spectra are illustrated by Figure 20.

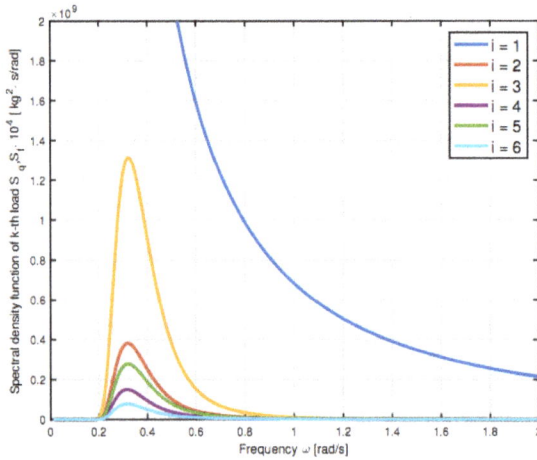

Figure 20. Mooring line spectral density of the generic environmental load in modal coordinates.

The response spectra (Figure 21) for the displacements in physical coordinates of the offshore wind turbine crown pile configuration have been determined.

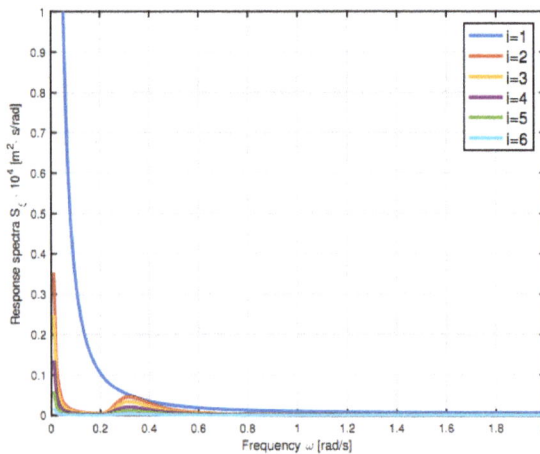

Figure 21. Response spectra for the displacements in physical coordinates of the offshore wind turbine mooring line configuration.

The area beneath these spectral density functions represents the variance of the generic displacement in physical coordinates $\sigma^2(\xi_k)$. The extreme limits of ξ_k are $\pm 3\sigma(\xi_k)$ with $k = 1, 2, \ldots, 6$. If the static stresses and deflection of the members are within these extreme limit values, then the structure is assumed safe. Considering a Gaussian process, there is only a 0.026% chance that each response exceeds the $\pm 3\sigma(\xi_k)$ limits. The extreme displacement values of the crown pile configuration in m are: $3\sigma(\xi_1) = 1.3965$, $3\sigma(\xi_2) = 0.3768$, $3\sigma(\xi_3) = 0.3204$, $3\sigma(\xi_4) = 0.2459$, $3\sigma(\xi_5) = 0.1724$, $3\sigma(\xi_6) = 0.0995$.

Comparing these results with the ones previously given, the reader will easily state that the structure cannot be safe for the first node at the top of the wind tower, in particular $\xi_1 = 1.3854$ m and $3\sigma(\xi_1) = 1.3965$. The mooring configuration has obtained higher values of safety expressed by the $3\sigma(\xi)$ approach results, but it is not enough yet for the global stability.

5.3. The 2-MW Wind Turbine Configuration

The modeling of the 2-MW configuration is characterized by the same jacket sub-structure, but with a smaller wind tower. Tower length is 60 m. The nacelle mass is 57 tons. The rotor mass is 23 tons. The diameter is 66 m. The wind tower has a truncated cone shape. At the bottom, the diameter is 4 m and the thickness is 0.18 m. At the top, the diameter is 3.5 m and the thickness is 0.10 m. At half the tower height, the diameter is 3.75 m and the thickness is 0.14 m. The hub is at z = +70 m. The wind tower produces a rated power of 2 MW. As for the 6DOF model, it is attached to the sub-structure through the same concrete transition piece. The sketch of the 2-MW offshore wind turbine configuration is given in Figure 22.

Figure 22. The 2-MW model detail.

All the masses have been calculated considering the specific weight of the steel 7850 kg/m^3 multiplied by the generic cross-section and multiplied by the length of the tubular member. It follows that the mass matrix M of the 6DOF model in kg is:

$$M = \begin{bmatrix} 0.5203 & 0 & 0 & 0 & 0 & 0 \\ 0 & 1.4834 & 0 & 0 & 0 & 0 \\ 0 & 0 & 0.1224 & 0 & 0 & 0 \\ 0 & 0 & 0 & 0.1471 & 0 & 0 \\ 0 & 0 & 0 & 0 & 0.2423 & 0 \\ 0 & 0 & 0 & 0 & 0 & 0.2067 \end{bmatrix} \cdot 10^6 \qquad (23)$$

The stiffness matrix K has been determined over the x direction. Once having applied the definition of stiffness to all the system, it is possible to define the matrix K in N/m as:

$$K = \begin{bmatrix} 0.121 & -0.121 & 0 & 0 & 0 & 0 \\ -0.121 & 0.6689 & -0.6568 & 0 & 0 & 0 \\ 0 & -0.6568 & 1.1290 & -0.4722 & 0 & 0 \\ 0 & 0 & -0.4722 & 1.0345 & -0.5622 & 0 \\ 0 & 0 & 0 & -0.5622 & 1.1307 & -0.5685 \\ 0 & 0 & 0 & 0 & -0.5685 & 1.2073 \end{bmatrix} \cdot 10^9 \qquad (24)$$

The sea state remains unchanged with respect to the other configurations.

The six undamped natural structural frequencies for free vibration in rad/s are: $\omega_1 = 4.5039$, $\omega_2 = 8.8834$, $\omega_3 = 36.1127$, $\omega_4 = 71.9487$, $\omega_5 = 90.6417$, $\omega_6 = 111.5328$. They are also graphically represented in Figure 23.

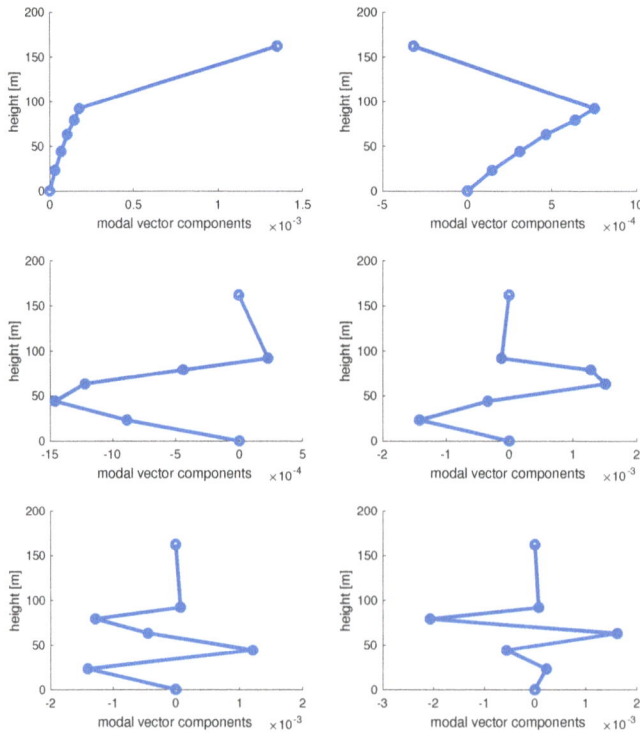

Figure 23. The 2-MW configuration vibrational modes.

The damped frequencies of the 6DOF model in rad/s are: $\omega_{d1} = 4.4983$, $\omega_{d2} = 8.8723$, $\omega_{d3} = 35.5149$, $\omega_{d4} = 68.8136$, $\omega_{d5} = 84.6690$, $\omega_{d6} = 100.6118$.

The response of the harmonic wave has been considered the same as the 6DOF model. The wind load due to the different wind towers will be lower in particular if it is applied at z = +70 m and is equal to $P_w = 838$ kN.

The compression in the pile B1 is equal to $F_{B1} = -13,288$ kN, lower than the compression limit value of the pile.

The maximum node displacements of the mooring model in m are: $\xi_1 = 0.2989$, $\xi_2 = 0.0504$, $\xi_3 = 0.0413$, $\xi_4 = 0.0302$, $\xi_5 = 0.0.0194$, $\xi_6 = 0.0099$. These values will be then compared with the results obtained with the $3\sigma(\xi_k)$ approach that will define the structural safety.

The five wave transfer functions are similar to the ones obtained for the previous cases.

Multiplying the wind transfer function of the load H_{uf} in modal coordinates by the Kaimal spectrum S_u for the wind and the wave transfer functions $|x_k^T G(p,\omega)|^2$ by the Pierson–Moskowitz S_η, the spectral density function of the wind force in modal coordinates S_f and the spectral density of the generalized wave force in modal coordinates S_{qi}, respectively, are achieved (Figure 24).

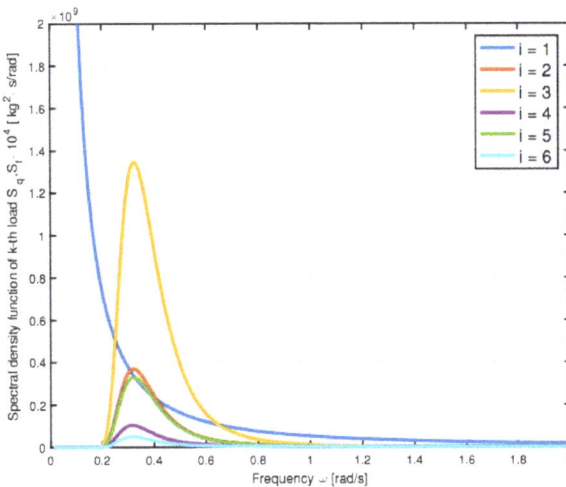

Figure 24. The 2-MW configuration spectral density of the generic environmental load in modal coordinates.

For the spectral density of the wind load in modal coordinates, the same consideration is made for the 6DOF case.

As done for the previous model, the response spectra for the displacements in physical coordinates of the offshore wind turbine crown pile configuration have been determined and are shown in Figure 25.

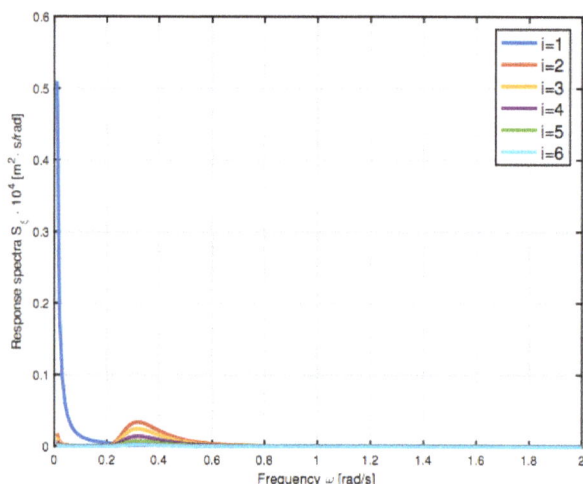

Figure 25. Response spectra for the displacements in physical coordinates of the 2-MW offshore wind turbine configuration.

The area beneath these spectral density functions represents the variance of the generic displacement in physical coordinates $\sigma^2(\xi_k)$. The extreme limits of ξ_k are $\pm 3\sigma(\xi_k)$ with $k = 1, 2, \ldots, 6$. If the static stresses and deflection of the members are within these extreme limit values, then the structure is assumed safe. Considering a Gaussian process, there is only a 0.026% chance that each response exceeds the $\pm 3\sigma(\xi_k)$ limits.

The extreme displacement values of the 2-MW wind turbine configuration in m are: $3\sigma(\xi_1) = 0.4208$, $3\sigma(\xi_2) = 0.1847$, $3\sigma(\xi_3) = 0.1572$, $3\sigma(\xi_4) = 0.1234$, $3\sigma(\xi_5) = 0.0950$, $3\sigma(\xi_6) = 0.0503$.

Comparing these results with the ones given above, the reader will easily state that the structure cannot be safe for the first node at the top of the wind tower, in particular $\xi_1 = 0.2989$m and $3\sigma(\xi_1) = 0.4208$. The 2-MW configuration has obtained lower values of safety expressed by the $3\sigma(\xi)$ approach results and together with the pile check analysis completes the global stability of the 2-MW retrofitted configuration.

6. Conclusions

Due to the long service lifetime of offshore systems, the analysis of the performance of the substructure is a complex task that involves many steps including load analysis, dynamic analysis, evaluation of the fatigue life, as well as long-term deformations.

This work presents a simplified methodology for the analysis of the ultimate and dynamic loads with a check on the most stressed pile of the different offshore retrofitted systems. The analysis was carried out such that the data required about the oil and gas platform, wind turbines, and the site are available in the early design phases. Thus, the methodology appears to obtain a conservative estimation. It is obvious that the present study is not free from limitations, but it can be improved in further investigations.

The first limitation is due to the simplified modeling made of discrete masses, one for each jacket level. Nevertheless, high correlations have been shown with the example proposed by the references.

The present work proposes five retrofitted solutions: four related to the substructure operations and one regarding the wind turbine. They are respectively the crown pile configuration, long pile, mooring line, stirrups, and the 2-MW wind turbine configuration. The safety concept of the retrofitted models is expressed in terms of pile check analysis and the 3σ approach.

Nevertheless, the final values obtained by the different configuration typologies are slightly higher or lower than the threshold limit, and the check is slightly fulfilled or not; here, the goal is just to show a simple method to investigate an offshore structure. The present research can be considered as the starting point for a reliability analysis regarding decommissioned jacket platforms for wind energy generations in an offshore environment.

Author Contributions: Data curation, L.A.; Formal analysis, L.A.; Investigation, L.A.; Methodology, L.A. and N.F.; Project administration, N.F.; Resources, J.A.F.O.C. and N.F.; Software, N.F.; Supervision, N.F.; Validation, L.A. and J.A.F.O.C.; Visualization, N.F.; Writing—original draft, L.A.; Writing—review & editing, J.A.F.O.C. and N.F.

Funding: This research received no external funding

Acknowledgments: The authors acknowledge "Fondazione Flaminia" (Ravenna, Italy) for supporting the present research. The research topic is one of the subjects of the Center for Off-shore and Marine Systems Engineering (COMSE), Department of Civil Chemical Environmental and Materials Engineering (DICAM), University of Bologna. This work was also supported by: UID/ECI/04708/2019—CONSTRUCT—Instituto de I&D em Estruturas e Construções funded by national funds through the FCT/MCTES (PIDDAC).

Conflicts of Interest: The authors declare no conflict of interest.

References

1. Muskulus, M.; Schafhirt, S. Design Optimization of Wind Turbine Support Structures—A Review. *J. Ocean Wind Energy* **2014**, *1*, 12–22.
2. International Energy Agency. *Offshore Energy Outlook*; International Energy Agency: Paris, France, 2018.
3. Global Wind Energy Council (GWEC). *Global Wind Report 2015*; Technical Report; Global Wind Energy Council: Brussels, Belgium, 2016.
4. Huang, W.; Liu, J.; Zhao, Z. The state of the art of study on offshore wind turbine structures and its development. *Ocean Eng.* **2009**, *2*. Available online: http://en.cnki.com.cn/Article_en/CJFDTOTAL-HYGC200902021.htm (accessed on 10 February 2018).
5. Ho, A.; Mbistrova, A. *The European Offshore Wind Industry: Key Trends and Statistics 2016*; Technical Report; WindEurope: Brussels, Belgium, 2017.
6. ORE Catapult. *Cost Reduction Monitoring Framework 2016: Summary Report to the Offshore Wind Programme Board*; Technical Report; Offshore Renewable Energy Catapult: Glasgow, UK, 2017.
7. Mone, C.; Stehly, T.; Maples, B.; Settle, E. *2014 Cost of Wind Energy Review*; National Renewable Energy Laboratory: Oak Ridge, TN, USA, 2015.
8. Musial, W.D.; Sheppard, R.E.; Dolan, D.; Naughton, B. *Development of Offshore Wind Recommended Practice for U.S. Waters*; Offshore Technology Conference Houston, Texas, TX, USA, 6–9 May 2013.
9. Veldkamp, D. A probabilistic evaluation of wind turbine fatigue design rules. *Wind Energy* **2008**, *11*, 655–672. [CrossRef]
10. DNV GL. *DNVGL-ST-0437: Loads and Site Conditions for Wind Turbines*. Available online: https://rules.dnvgl.com/docs/pdf/DNVGL/ST/2016-11/DNVGL-ST-0437.pdf (accessed on 10 February 2018).
11. Negro, V.; Lopez-Gutierrez, J.-S.; Esteban, M.D.; Matutano, C. Uncertainties in the design of support structures and foundations for offshore wind turbines. *Renew. Energy* **2014**, *63*, 125–132. [CrossRef]
12. Seidel, M. Jacket substructures for the REpower 5M wind turbine. In Proceedings of the Conference Proceedings European Offshore Wind, Berlin, Germeny, 4–6 December 2007. Available online: http://www.marc-seidel.de/Papers/Seidel_EOW_2007.pdf (accessed on 10 February 2018).
13. Lozano-Minguez, E.; Kolios, A.J.; Brennan, F.P. Multi-criteria assessment of offshore wind turbine support structures. *Renew. Energy* **2011**, *36*, 2831–2837. [CrossRef]
14. Popko, W.; Vorpahl, F.; Zuga, A.; Kohlmeier, M.; Jonkman, J.; Robertson, A.; Larsen, T.J.; Yde, A.; Sætertrø, K.; Okstad, K.M.; et al. Offshore Code Comparison Collaboration Continuation (OC4), Phase I—Results of Coupled Simulations of an Offshore Wind Turbine with Jacket Support Structure. In Proceedings of the Twenty-second International Offshore and Polar Engineering Conference, Rhodes, Greece, 17–22 June 2012.
15. Saha, N.;Gao, Z.; Moan, T.; Naess, A. Short-term extreme response analysis of a jacket supporting an offshore wind turbine. *Wind Energy* **2014**, *17*, 87–104. [CrossRef]
16. Luo, T.; Tian, D.; Wang, R.; Liao, C. Stochastic Dynamic Response Analysis of a 10 MW Tension Leg Platform Floating Horizontal Axis Wind Turbine. *Energies* **2018**, *11*, 3341. [CrossRef]

17. Sivalingam, K.; Martin, S.; Wala, A.A.S. Numerical Validation of Floating Offshore Wind Turbine Scaled Rotors for Surge Motion. *Energies* **2018**, *11*, 2578. [CrossRef]
18. Ku, C.-Y.; Chien, L.-K. Modeling of Load Bearing Characteristics of Jacket Foundation Piles for Offshore Wind Turbines in Taiwan. *Energies* **2016**, *9*, 625. [CrossRef]
19. Chen, I-W., Wong, B.-L.; Lin, Y.-H.; Chau, S.-W.; Huang, H.-H. Design and Analysis of Jacket Substructures for Offshore Wind Turbines. *Energies* **2016**, *9*, 264. [CrossRef]
20. Chew, K.H.; Ng, E.Y.K.; Tai, K. Offshore Wind Turbine Jacket Substructure: A Comparison Study Between Four-Legged and Three-Legged Designs. *J. Ocean Wind Energy* **2014**, *1*, 74–81.
21. Chew, K.H.; Ng, E.Y.K.; Tai, K.; Muskulus, M.; Zwick, D. Structural Optimization and Parametric Study of Offshore Wind Turbine Jacket Substructure. In Proceedings of the Twenty-Third International Offshore and Polar Engineering Conference, Anchorage, Alaska, 30 June–5 July 2013.
22. Dubois, J.;Muskulus, M.; Schaumann, P. Advanced Representation of Tubular Joints in Jacket Models for Offshore Wind Turbine Simulation. *Energy Procedia* **2013**, *35*, 234–243. [CrossRef]
23. Kang, J.; Liu, H.; Fu, D. Fatigue Life and Strength Analysis of a Main Shaft-to-Hub Bolted Connection in a Wind Turbine. *Energies* **2019**, *12*, 7. [CrossRef]
24. Liu, F.; Li, H.; Li, W.; Wang, B. Experimental study of improved modal strain energy method for damage localisation in jacket-type offshore wind turbines. *Renew. Energy* **2014**, *72*, 174–181. [CrossRef]
25. Abkar, M. Theoretical Modeling of Vertical-Axis Wind Turbine Wakes. *Energies* **2019**, *12*, 10. [CrossRef]
26. Palmieri, M.; Bozzella, S.; Cascella, G.L.; Bronzini, M.; Torresi, M.; Cupertino, F. Wind Micro-Turbine Networks for Urban Areas: Optimal Design and Power Scalability of Permanent Magnet Generators. *Energies* **2018**, *11*, 2759. [CrossRef]
27. Wicaksono, Y.A.; Prija Tjahjana, D.D.D.; Hadi, S. Influence of omni-directional guide vane on the performance of cross-flow rotor for urban wind energy. *AIP Conf. Proc.* **2018**, *1931*, 030040. [CrossRef]
28. Farhan, M.; Mohammadi, M.R.S.; Correia, J.A.; Rebelo, C. Transition piece design for an onshore hybrid wind turbine with multiaxial fatigue life estimation. *Wind Eng.* **2018**, *42*, 286–303. [CrossRef]
29. Mourao, A.; Correia, J.A.F.O.; Castro, J.M.; Correia, M.; Leziuk, G.; Fantuzzi, N.; De jesus, A.M.P.; Calcada, R.A.B. Fatigue Damage Analysis of Offshore Structures using Hot-Spot stress and Notch Strain Approaches. *Mater. Res. Forum* **2019**, in press.
30. Wilson, J.F. *Dynamics of Offshore Structures*; Wiley: Hoboken, NJ, USA, 2002.

energies

MDPI

Article

Dynamic Study of a Rooftop Vertical Axis Wind Turbine Tower Based on an Automated Vibration Data Processing Algorithm

Ying Wang [1], Wensheng Lu [1], Kaoshan Dai [2,1,*], Miaomiao Yuan [1] and Shen-En Chen [3]

[1] State Key Laboratory of Disaster Reduction in Civil Engineering, Tongji University, Shanghai 200092, China; tj_wangying@foxmail.com (Y.W.); wally@tongji.edu.cn (W.L.); 209jgsys@tongji.edu.cn (M.Y.)
[2] Department of Civil Engineering, Sichuan University, Chengdu 610065, China
[3] Department of Civil & Environmental Engineering, University of North Carolina, Charlotte, NC 28223, USA; schen12@uncc.edu
* Correspondence: kdai@scu.edu.cn; Tel.: +86-28-8599-6688

Received: 3 July 2018; Accepted: 6 August 2018; Published: 13 November 2018

Abstract: When constructed on tall building rooftops, the vertical axis wind turbine (VAWT) has the potential of power generation in highly urbanized areas. In this paper, the ambient dynamic responses of a rooftop VAWT were investigated. The dynamic analysis was based on ambient measurements of the structural vibration of the VAWT (including the supporting structure), which resides on the top of a 24-story building. To help process the ambient vibration data, an automated algorithm based on stochastic subspace identification (SSI) with a fast clustering procedure was developed. The algorithm was applied to the vibration data for mode identification, and the results indicate interesting modal responses that may be affected by the building vibration, which have significant implications for the condition monitoring strategy for the VAWT. The environmental effects on the ambient vibration data were also investigated. It was found that the blade rotation speed contributes the most to the vibration responses.

Keywords: vertical axis wind turbine; structural health monitoring; operational modal analysis; stochastic subspace identification; vibration test

1. Introduction

In commercial wind-power generation, most attention is placed on the deployment of horizontal axis wind turbines (HAWTs) as they are thought to be more efficient than the vertical axis wind turbines (VAWTs) [1]. However, recent studies on VAWTs found that they can be driven by gentler winds and have no need of yaw. These advantages make VAWTs ideal for urban power generation [1,2].

VAWTs can be generally divided into two types: drag-type and lift-type. Lift-type VAWTs have a bigger tip-speed ratio; thus, they can convert the highest amount of energy. Lift-type VAWTs, such as the Darrieus rotor with egg-beater blades (Φ-type) and the Darrieus rotor with straight blades (H-type), are the most widely used VAWTs [3,4]. Furthermore, their designs were improved with computer fluid dynamics (CFD) and finite element analysis (FEA), making them more efficient in energy conversion [5–7]. However, studies on VAWTs are limited when compared to those on horizontal wind turbines. Although several laboratory vibration tests were performed to study the dynamic behavior of VAWTs [8–11], only a limited number of field vibration measurements were conducted. Some studies showed that the vibration of the VAWT support structure may be influenced by the wind turbine's operation [9]. Although there are some research reports about the development of rooftop VAWTs which can make better use of high-altitude winds [12,13], very few instances of on-site monitoring are published.

The implementation of structural health monitoring (SHM) systems in a structure can help ensure optimal performances of the structure. This is achieved via real-time vibration measurements and a better understanding of the dynamic loads [14–16]. The SHM of wind turbine towers was proposed for system monitoring and was integrated into the Supervisory Control and Data Acquisition (SCADA) systems [17–20]. The SHM of HAWT towers concerns problems including resonance responses caused by operation frequency or blade passing frequency [21–23], foundation scouring [24], blade flange conditions [25], and structural performances under extreme events [26,27].

Understanding the dynamics of VAWTs can be helpful for improving the design of VAWTs and the implementation of SHM in the VAWT systems. In this paper, along with the installation of an H-type VAWT on the roof of a 24-story dormitory building, a structural health monitoring system was implemented, and the dynamic responses of the VAWT were studied and reported. The objective of this study was to understand how the VAWT behaves dynamically under normal operation conditions. Ambient vibration measurements were obtained for both the building and the VAWT, including the support tower. The data were analyzed to find the interactions between the building and the VAWT structure, as well as the environmental effects on the vibration responses. To facilitate modal parameter identification, an automated algorithm based on stochastic subspace identification (SSI) and a fast clustering algorithm was developed. The results are presented herein along with discussions on the interactions between the building and the tower and their significance for condition monitoring.

The paper is organized as follows: in Section 2, the VAWT is firstly introduced and the vibration measurement campaign is described. Data analyses and a discussion of the results are provided in Sections 3 and 4, respectively. The findings are summarized in Section 5.

2. The VAWT Wind Turbine and Vibration Measurements

An H-type VAWT was installed on the roof of a 24-story dormitory building belonging to the Tongji University, Shanghai, China (referred to as the #6 building at the Zhangwu campus).

Vibration measurements of the #6 building were conducted prior to the VAWT installation. Several electromagnetic accelerometers were placed on the building's rooftop, and the lateral movements of the building along both the X-direction (longer axis) and the Y-direction (shorter axis) were measured. These electromagnetic accelerometers, with a moving coil as the sensing unit, have good performances in the low-frequency range. Details of the electromagnetic accelerometers can be found in Table 1. Figure 1 shows the #6 building, the relative location of the VAWT, and the deployment of the vibration sensors. Accelerations of the roof of the #6 building were measured with a sampling frequency of 128 Hz, and the duration of the building acceleration data was around 9 min.

Table 1. Details of accelerometers used for building and tower testing.

Item	Piezoelectric Sensors (Tower)	Electromagnetic Sensors (Building)
Sensitivity	9.83 mV/(ms^2)	329 mV/(ms^2)
Resolution	0.001 ms^2	3 × 10^{-6} ms^2
Frequency range	1.0 to 5000 Hz (±5%)	0.25 to 100 Hz (+1 dB to −3 dB)
	0.5 to 7000 Hz (±10%)	
Acceleration range	500 ms^2	200 ms^2
Weight	4.4 g	800 g

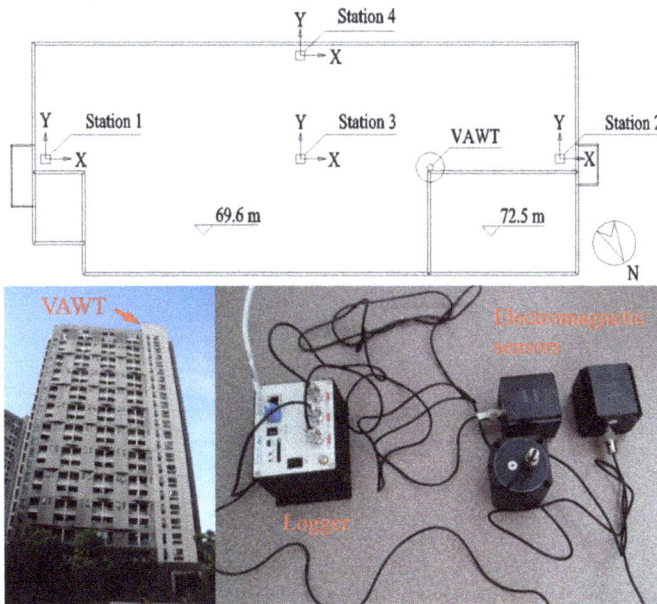

Figure 1. Sensor deployment on the #6 building's roof.

Both the vibration-time histories and the power spectrum density (PSD) of the roof were obtained along the X- and Y-directions. The building vibrations obtained are shown in Figure 2. Based on the PSD diagram, the first few modes of natural frequencies of the #6 building are shown to be less than 12 Hz (Table 2). The peaks of higher vibration modes are not as obvious as the lower modes; thus, high-frequency modes are more difficult to determine (Figure 2b,d). The frequencies shown in Table 2 were the measured natural frequencies of the #6 building. The roof vibrations, shown in Figure 3, include (1) translational motion along the X-direction, (2) translational motion along the Y-direction, and (3) torsional motion. Since the long axis of the building is along the X-direction, the torsional motion (corresponding to 1.3 Hz) can be observed for the Y-direction measurements of Station 1 and Station 2, indicating possibly more torsional vibration for a substructure (e.g., a VAWT) if it is installed on the edge of a building than for one installed at the center of the building.

Table 2. Natural frequencies of the #6 building.

Direction	Frequency (Hz)			
	1st	2nd	3rd	4th
X	0.94	3.60	6.90	10.60
Y	0.71	1.30	4.10	9.60

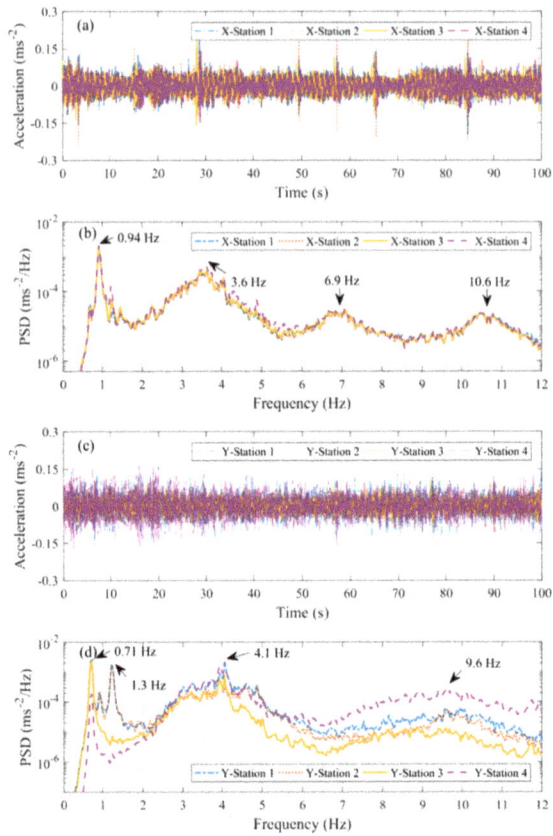

Figure 2. Ambient responses of the #6 building's roof: (**a**) time histories along the X-direction; (**b**) power spectrum density (PSD) along the X-direction; (**c**) time histories along the Y-direction; (**d**) PSD along the Y-direction.

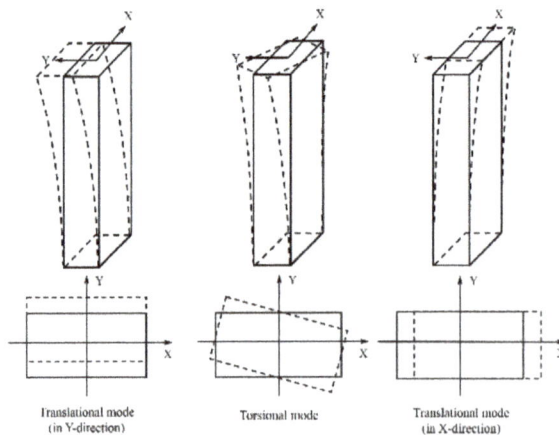

Figure 3. Building mode shape sketch.

132

The VAWT consists of a power generation component that includes five turbine blades, a deflector, a generator, and a hollow shaft. Figure 4 shows the wind turbine, which is attached via metal-strip hoops to the exterior of an extension structure for the elevator shaft on the rooftop. The design details of the VAWT are listed in Table 3. The metal-strip hoops were designed to laterally constrain the wind turbine tower and were embedded into the wall of the #6 building with expansion bolts. To reduce vibration of the tower, rubber gaskets were placed between the hoops and the tower. There were three hoops attached to the structure with the last hoop very close to the base block. Details of the hoops can be seen in Figure 4. The heavy concrete base block, laid on the roof, was used to weigh down the tower. Bolts were embedded into the concrete base block (foundation) and were connected to the bottom flange of the tower. The bolt size and anchor depth are shown in Table 4.

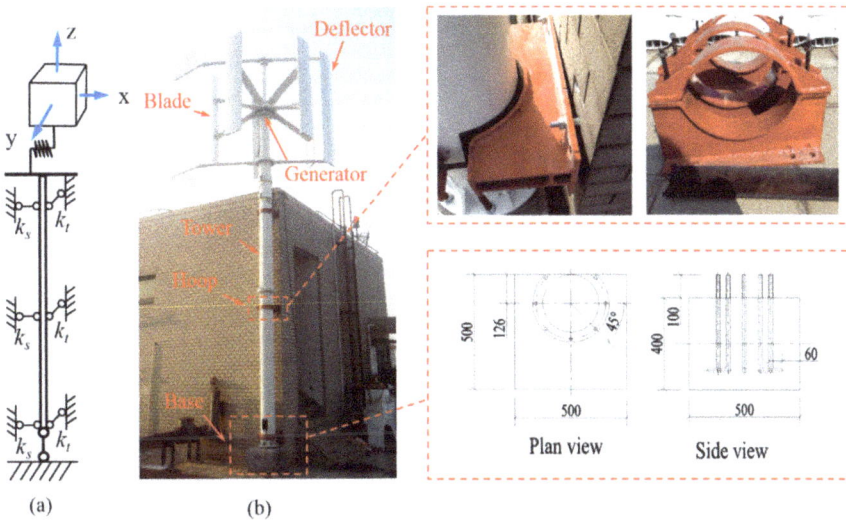

Figure 4. The H-type vertical axis wind turbine (VAWT): (**a**) modeling of the vibration system; (**b**) the VAWT and the details of its supports (including schematics; unit: mm).

Table 3. Details of the wind turbine.

Parameter	Value
Rated power	2 kW
Rated spin speed	250 rpm
Rated wind speed	25 m/s
Maximum design wind speed	32 m/s
Blade height	2 m
Blade span	2.2 m
Weight (generator + blades)	240 kg

Table 4. Details of the supports of the wind turbine.

Hoop	
Thickness	6 mm
Material	Q235 steel
Bolts	M10
Expansion bolts	M12
Anchorage depth	≥75 mm
Rubber gasket thickness	5 mm
Tower	
Thickness	4 mm
Material	Q235 steel
External diameter	250 mm
Base	
Bolts	M18

The theoretical frequencies of the VAWT structure were also calculated before the installation to avoid the resonance issue, considering building vibrations. The vibration responses of interest for the wind turbine are the lateral movements of the power generation component as described in the simplified model shown in Figure 4a [28]. Other than the rigid foundation at the base, the bending behaviors of the tower are critically limited by the stiffness of the three hoop supports on the tower shaft. As shown in Figure 4, the hoops consist of half-rings connected to the hoop base using the bolt connectors and rubber gaskets. Hence, the constraining actions of the hoops can be described as shear in the perpendicular direction to the bolts (X-direction) and tension in the bolt connection direction (Y-direction). Also shown in Figure 4 is the simplified dynamic model of the monitoring system, which includes the wind generation component (rigid mass) and the connection to the top hoop support (rod). Different stiffness models can be constructed for the hoop support. The tensile stiffness of a single hoop, k_t, can be described as

$$k_t = \frac{1}{\frac{1}{2k_{M10}} + \frac{1}{4k_{M12}} + \frac{1}{2k_R}}, \tag{1}$$

where k_{M10}, k_{M12}, and k_R are the tensile stiffness of M10 bolts and M12 bolts, and the compression stiffness of the rubber gaskets, respectively. The shear stiffness of a single hoop k_s can be described as

$$k_s = \frac{4k_{M12,s}k_R}{4k_{M12,s} + k_R}, \tag{2}$$

where $k_{M12,s}$ is the shear stiffness of the M12 bolts. In the following discussion, the vibration responses of the wind turbine are separated into the shear (X) direction and the tension (Y) direction. Using this very simplistic dynamic model, the fundamental frequencies of the tower were derived as 7.2 Hz in the shear direction (X-direction of the tower shaft and X-direction of the building) and 6.0 Hz in the tension direction (Y-direction of the tower shaft and Y-direction of the building).

A monitoring system was designed for the data collection, storage, and processing. Parameters associated with environment, power generation, and structural performances were collected continuously. Measurements were stored in a computer installed on the roof. The system was also designed to allow remote monitoring of data. Figure 5 shows the monitoring setup of the VAWT tower. In this case, a third-generation (3G) network was used for the wireless data transmission. This paper focuses on vibration data collected from this condition monitoring system.

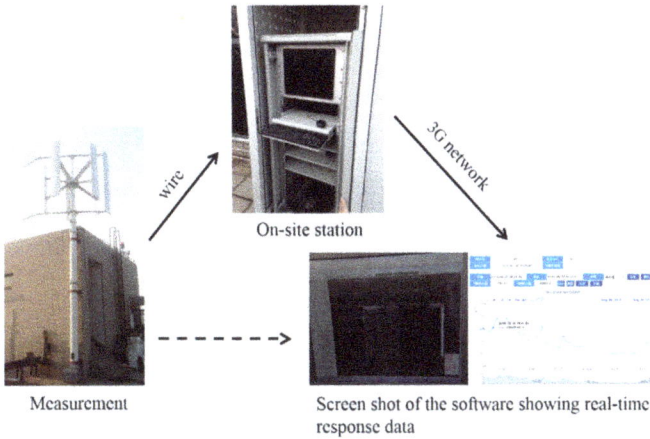

Figure 5. Monitoring setup of the VAWT tower.

To measure the vibration modes of the wind turbine, along with the installation of the VAWT, three pairs of piezoelectric accelerometers were mounted along the tower at three sections with different heights; each pair had two sensors, deployed in two directions perpendicular to each other at each section: (a) the X-direction which runs perpendicular to the hoop bolts (shear direction), and (b) the Y-direction which parallels the tensioning of the hoop bolts. Figure 6 shows the sensor locations, which were designated as cross Sections 1–3. Sensors in cross Section 1 (the generator) were installed inside the tower shaft; thus, the set-up was not influenced by the rotating blades. The other section sensors were installed on the outside surface of the tower. Different from electromagnetic sensors, a piezoelectric sensor adopts a piezoelectric crystal as the sensing unit, and it can be made into a smaller size compared to that of the electromagnetic ones. Details of the piezoelectric accelerometers deployed for tower testing can be found in Table 1.

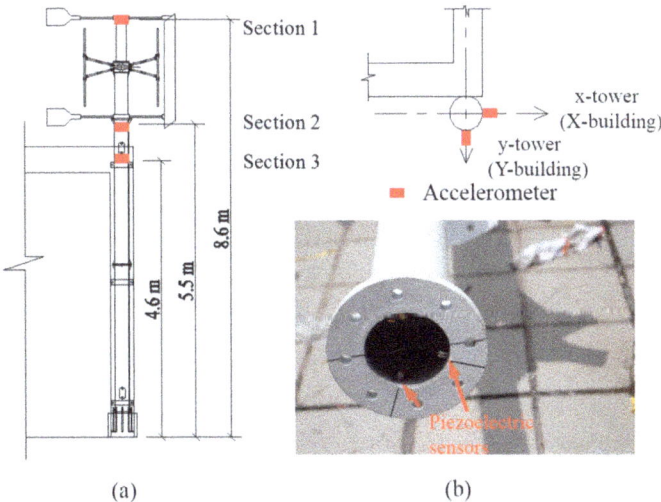

Figure 6. Deployment of sensors: (**a**) schematics of the VAWT wind turbine; (**b**) the sensor attachments.

To monitor the environmental conditions, a tachometer, a thermometer, and a voltmeter were also installed. The tachometer was used to measure the rotation speed of the wind turbine. The thermometer kept track of the daily environmental temperature changes near the structure. The power generation of the wind turbine in Watts could be translated from the measurements obtained by the voltmeter. The sampling frequencies of all the measured data were set at 1024 Hz. Typical acceleration time histories for a time window of 10 min are shown in Figure 7. Accelerations labeled x1, x2, and x3, and y1, y2, and y3 correspond to cross Sections 1–3 along the X-direction and the Y-direction of the tower, respectively. Typical time histories in a day (24 h) of rotation speed, temperature, and voltage (power generation) recorded on 28 August 2016 are presented in Figure 8.

A cup anemometer with a wind-direction indicator was installed near the VAWT, which measured the speed and direction of the incoming wind. The anemometer was the F220S model manufactured by the Nanhua Electronics Company, Shanghai, China. Measurements of wind speed and wind direction were recorded about every 10 s. The definition of wind direction was as follows: 0° corresponded to the north direction, and 90° corresponded to the east direction. Since the wind field in urban areas can be quite complex, both the wind speed and direction measurements fluctuated constantly. Only the moving averages with a step length of 10 min were used in current study. Figure 9 shows a typical recorded wind scenario throughout a day (28 August 2016) and its moving average. It can be seen from Figures 8 and 9 that both the maximum wind speed and the maximum temperature occurred at noon, and the temperature and the wind speed were strongly correlated. The rotation speed, which should be linear to the quadratic of the wind speed, only had a medium correlation (the correlation coefficient was 0.581) with the raw wind speed, and was strongly correlated (the correlation coefficient was 0.888) with the average wind speed. This happened since the rotation of the VAWT calls for a continuous wind force rather than a gust. The correlation coefficients between the rotation speed, wind speed, and temperature calculated with both raw wind-speed data and averaged wind-speed data are shown in Table 5. The relationship between the power and the wind speed is also provided in Figure 10.

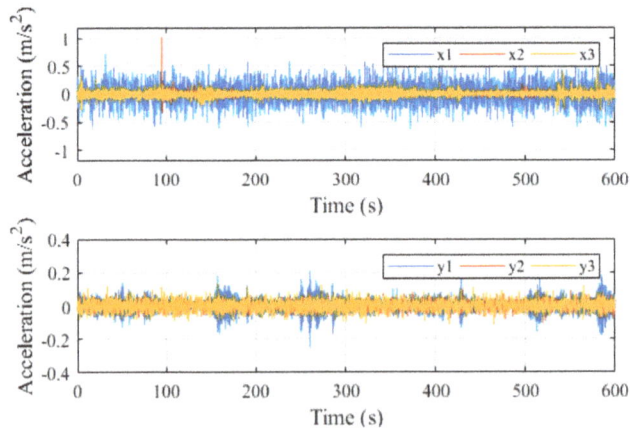

Figure 7. VAWT accelerations at 12 a.m. on 28 August 2016.

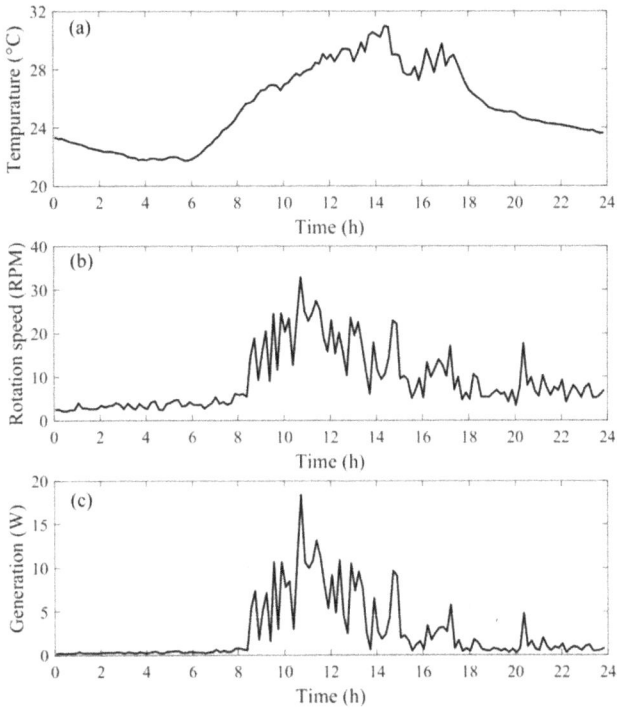

Figure 8. Typical monitoring data, recorded on on 28 August 2016: (**a**) wind turbine rotational speed; (**b**) temperature; (**c**) power generation distribution.

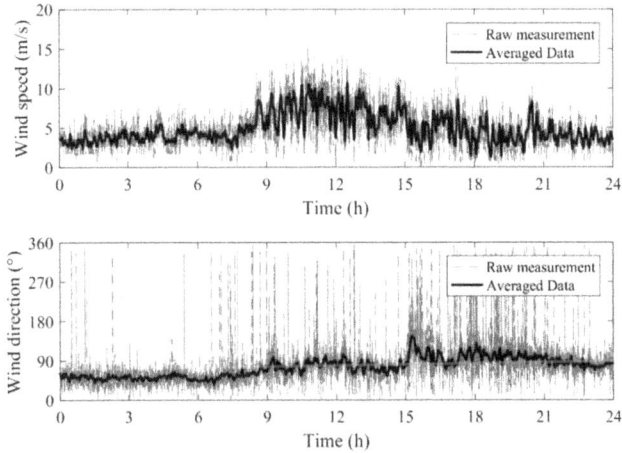

Figure 9. Wind speed and wind direction on 28 August 2016.

Table 5. Correlation coefficients between rotation speed, wind speed, and temperature.

Correlation Coefficients	Rotation Speed$^{1/2}$	Wind Speed (Raw Data)	Temperature
Rotation speed	1	0.581	0.757
Wind speed	0.581	1	0.516
Temperature	0.757	0.516	1
Correlation Coefficients	**Rotation Speed$^{1/2}$**	**Wind Speed (Moving Average)**	**Temperature**
Rotation speed	1	0.888	0.757
Wind speed	0.888	1	0.677
Temperature	0.757	0.677	1

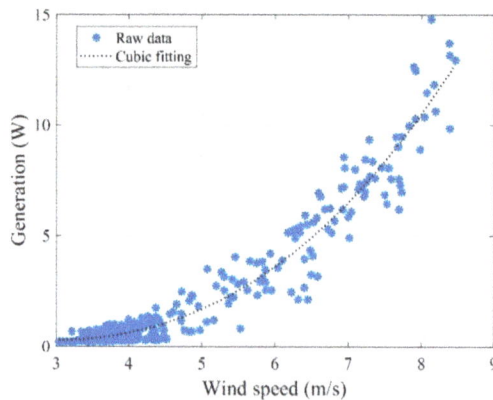

Figure 10. Power curve of the rooftop VAWT on 28 August 2016.

3. Data Analysis

3.1. Frequency Domain Analyses

The frequency domain analyses were conducted using the power spectrum density (PSD). The PSD of the ambient response of a structure can be written as

$$\mathbf{P}_{XX}(f) \propto [\mathbf{T}(f)]\mathbf{I}[\mathbf{T}(f)]^{\mathrm{T}}, \tag{3}$$

where $\mathbf{P}_{XX}(f)$ is the PSD of the response of a structure, $\mathbf{T}(f)$ is the transfer function, which contains the information of a structure, and $\mathbf{I}$ is a unit matrix since the ambient excitation can be regarded as white noise. $\mathbf{P}_{XX}(f)$ represents the system characteristics of a structure, and the period gram method was used to calculate the PSDs. The period gram method uses windows to divide data into sections, and averages the PSDs of each section; hence, it can help accentuate the modal amplitudes and suppress the noise base in the frequency domain [29,30].

3.2. SSI Method

The classical SSI method is based on the state-space model of a system. A discrete-time state-space model of the system under ambient excitation without feedback signals can be described as

$$\begin{cases} \mathbf{x}_{k+1} = \mathbf{A}\mathbf{x}_k + \mathbf{w}_k \\ \mathbf{y}_k = \mathbf{C}\mathbf{x}_k + \mathbf{v}_k \end{cases}, \tag{4}$$

where the subscript k is the time, $\mathbf{x}_k$ is the state vector, $\mathbf{y}_k$ is a measurement vector, $\mathbf{w}_k$ is an excitation vector that is considered as white noise, $\mathbf{v}_k$ is a measurement noise vector, $\mathbf{A}$ is a state transition matrix, and $\mathbf{C}$ is an output location matrix.

The key to obtaining the modal parameters of the system is to solve for the state transition matrix $\mathbf{A}$, and to calculate its eigenvalues and eigenvectors. The procedure of the classical SSI method to obtain the modal parameters of a system is explained below [31,32].

Suppose the number of measurement is m. A Hankel matrix with $2i$ row blocks and j columns can be constructed as

$$
\mathbf{H}_{0|2i-1} =
\begin{bmatrix}
\mathbf{y}_0 & \mathbf{y}_1 & \mathbf{y}_2 & \cdots & \mathbf{y}_{j-1} \\
\mathbf{y}_1 & \mathbf{y}_2 & \mathbf{y}_3 & \cdots & \mathbf{y}_j \\
\cdots & \cdots & \cdots & \cdots & \cdots \\
\mathbf{y}_{i-1} & \mathbf{y}_i & \mathbf{y}_{i+1} & \cdots & \mathbf{y}_{i+j-2} \\
\mathbf{y}_i & \mathbf{y}_{i+1} & \mathbf{y}_{i+2} & \cdots & \mathbf{y}_{i+j-1} \\
\mathbf{y}_{i+1} & \mathbf{y}_{i+2} & \mathbf{y}_{i+3} & \cdots & \mathbf{y}_{i+j} \\
\cdots & \cdots & \cdots & \cdots & \cdots \\
\mathbf{y}_{2i-1} & \mathbf{y}_{2i} & \mathbf{y}_{2i-1} & \cdots & \mathbf{y}_{2i+2j-2}
\end{bmatrix}_{2mi \times j}
=
\begin{bmatrix}
\mathbf{H}_{0|i-1} \\
\mathbf{H}_{i|2i-1}
\end{bmatrix}
=
\begin{bmatrix}
\mathbf{H}_p \\
\mathbf{H}_f
\end{bmatrix},
\tag{5}
$$

where $\mathbf{H}_{0|i-1}$ and $\mathbf{H}_{i|2i-1}$ are matrices representing the first mi rows and the second mi rows of $\mathbf{H}_{0|2i-1}$, respectively; and $\mathbf{H}_p$ and $\mathbf{H}_f$ are equal to $\mathbf{H}_{0|i-1}$ and $\mathbf{H}_{i|2i-1}$, respectively, with subscripts p and f standing for past and future times.

With the orthogonal projection of $\mathbf{H}_f$ to $\mathbf{H}_p$, the projection matrix $\mathbf{O}_i$ is obtained as

$$
\mathbf{O}_i = \mathbf{H}_f / \mathbf{H}_p = \mathbf{H}_f \mathbf{H}_p^{\mathrm{T}} \left(\mathbf{H}_p \mathbf{H}_p^{\mathrm{T}} \right)^+ \mathbf{H}_p,
\tag{6}
$$

where $(*)^+$ denotes the pseudo-inverse of a matrix.

The projection matrix $\mathbf{O}_i$ can be decomposed further using the singular-value decomposition, which gives

$$
\mathbf{O}_i = \mathbf{U}_i \mathbf{\Sigma}_i \mathbf{V}_i^{\mathrm{T}} =
\begin{pmatrix} \mathbf{U}_{i1} & \mathbf{U}_{i2} \end{pmatrix}
\begin{pmatrix} \mathbf{\Sigma}_{i1} & \\ & \mathbf{\Sigma}_{i2} = 0 \end{pmatrix}
\begin{pmatrix} \mathbf{V}_{i1}^{\mathrm{T}} \\ \mathbf{V}_{i2}^{\mathrm{T}} \end{pmatrix}
= \mathbf{U}_{i1} \mathbf{\Sigma}_{i1} \mathbf{V}_{i1}^{\mathrm{T}},
\tag{7}
$$

where $\mathbf{\Sigma}_i$ is a diagonal matrix of dimension $mi \times j$; $\mathbf{U}_i$ and $\mathbf{V}_i$ are orthogonal matrices of dimensions $mi \times mi$ and $mi \times j$, respectively; $\mathbf{\Sigma}_{i1}$ and $\mathbf{\Sigma}_{i2}$ are submatrices of $\mathbf{\Sigma}_i$ that have dimensions $2n \times 2n$ and $(mi - 2n) \times (j - 2n)$, respectively, in which n is the degree of freedom of the system; $\mathbf{U}_{i1}$ and $\mathbf{U}_{i2}$ are submatrices of $\mathbf{U}_i$; and $\mathbf{V}_{i1}$ and $\mathbf{V}_{i2}$ are the submatrices of $\mathbf{V}_i$. Finally, the observability matrix $\mathbf{\Gamma}_i$ and the Kalman filter state sequence $\hat{\mathbf{X}}_i$ can be obtained as

$$
\mathbf{\Gamma}_i = \mathbf{U}_{i1} \mathbf{\Sigma}_{i1}^{1/2};
\tag{8}
$$

$$
\hat{\mathbf{X}}_i = \mathbf{\Sigma}_{i1}^{1/2} \mathbf{V}_{i1}^{\mathrm{T}}.
\tag{9}
$$

The state transition matrix $\mathbf{A}$ can then be obtained as

$$
\mathbf{A} = \mathbf{\Gamma}_1^+ \mathbf{\Gamma}_2,
\tag{10}
$$

where $\mathbf{\Gamma}_1$ and $\mathbf{\Gamma}_2$ are formed as the first $m \times (i-1)$ and last $m \times (i-1)$ rows of $\mathbf{\Gamma}_i$, respectively. Eigenvalues and eigenvectors of $\mathbf{A}$ can be obtained using eigenvalue decomposition:

$$
\mathbf{A} = \mathbf{\Psi} \mathbf{\Lambda} \mathbf{\Psi}^{-1},
\tag{11}
$$

where $\Lambda = diag(\lambda_1 \lambda_2 \dots \lambda_r \dots \lambda_{2n})$, in which λ_r is the r-th eigenvalue of $\mathbf{A}$ and $\Psi = \begin{pmatrix} \psi_1 & \psi_2 & \dots & \psi_r & \dots & \psi_{2n} \end{pmatrix}$. The r-th natural frequency of the system in Hz can be calculated as

$$f_r = \frac{abs(\ln(\lambda_r) \times f_s)}{2\pi}, \tag{12}$$

where f_s is the sampling frequency in Hz, and the r-th damping-ratio of the system can be calculated as

$$\xi_r = \frac{-real(\ln(\lambda_r) \times f_s)}{abs(\ln(\lambda_r) \times f_s)}. \tag{13}$$

The associated mode shape can be calculated as

$$\varphi_r = \mathbf{C}\psi_r, \tag{14}$$

where the measurement matrix $\mathbf{C}$ is formed as the first m row of Γ_i.

In this study, the stability diagrams were used to assist the identification of the "stable" modes. In a stability diagram, there are user-defined stability criteria for natural frequencies and for damping ratios. When an identified modal parameter corresponding to the calculation order of $2n$ is different from that corresponding to the order $2n - 2$, which is lower than the threshold of a criterion for the identified parameter, then the criterion is considered to meet at the order $2n$. When the two criteria (pertaining to natural frequency and damping ratios) are simultaneously met at the order $2n$, the identified modal parameters are considered "stable", and a marker is drawn at the point in the stability diagram, whose horizontal and vertical positions correspond to $2n$ and the identified natural frequency, respectively. In this work, the criteria for frequencies and damping ratios were set as 1% and 5%, respectively, and was set as 50. Figure 11 shows a typical stability diagram for the current study using the Y-direction results. Figure 11a shows the calculation order as a function of frequency, which shows three vibration modes. Figure 11b shows the damping ratio as a function of the vibration frequencies indicating less obvious vibration modes. This observation has important implications for the automation of the mode identification process, which is discussed below.

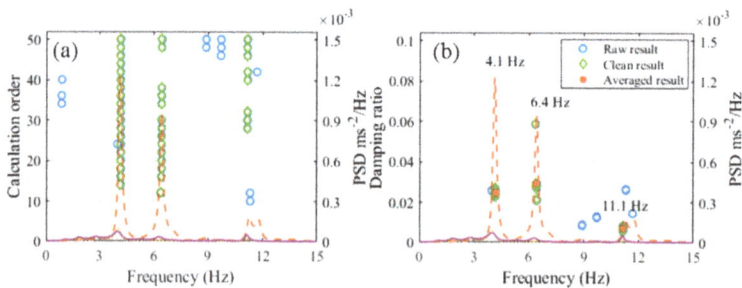

Figure 11. Stabilization diagram using the stochastic subspace identification (SSI) method and identified vibration parameters: (**a**) calculation order as a function of frequency; (**b**) damping ratio as a function of frequency (raw results—prior to pole-picking algorithm; clean results—after application of pole-picking algorithm).

3.3. Automatic Pole-Picking Procedure

Since long term monitoring produces a significant amount of data, to facilitate data analysis, the vibration parameters need to be identified using automated procedures. Automated modal analysis procedures based on the stochastic subspace identification (SSI) method were used [33–35]. Methods based on clustering adopt a stability diagram; therefore, they can be used concurrently with commercial modal analysis software [33,35]. In this study, a new cluster method was adopted for the automated process. The cluster method does not require iteration and costs less calculation time.

With the SSI method, poles corresponding to different calculation orders can be obtained. Each pole contains the information on frequency, damping ratio, and mode shape. The similarity between different orders of calculated mode shapes is evaluated by a parameter called the modal assurance criterion (MAC). In this work, the number of poles clustered within a certain frequency and MAC range was called a local density ρ, and a value was assigned to the cluster. The pole-picking procedure uses the local density and its distance to the other poles to define the cluster. It can be presented as follows [36]:

(1) Calculate frequency distances d_{ij} and MAC_{ij} between each pole set, i and j:

$$d_{ij} = \left| \frac{f_i - f_j}{f_j} \right|;$$ (15)

$$MAC_{ij} = \frac{\left(\boldsymbol{\varphi}_i^T \boldsymbol{\varphi}_j \right)^2}{\left(\boldsymbol{\varphi}_i^T \boldsymbol{\varphi}_i \right)\left(\boldsymbol{\varphi}_j^T \boldsymbol{\varphi}_j \right)}$$ (16)

(2) If d_{ij} and MAC_{ij} satisfy the following criteria, then add one to the local density value ρ_i of pole i:

- $d_{ij} \times 100\% < 1\%$;
- $(1 - MAC_{ij}) \times 100\% < 1\%$.

(3) Pick the poles with relatively high local densities.

In this work, only poles with a local density ρ_i higher than five were used. Averages of clean poles corresponding to different modes were then identified as final results.

Figure 12 summarizes the pole-picking algorithm for mode identification. The values defined in the two selection processes should be determined based on a proper understanding of the behaviors of the particular structure and the serviceability (limits on vibration amplitudes) requirements. The selection of the criteria range will not critically impact the number of "stable" modes that can be identified.

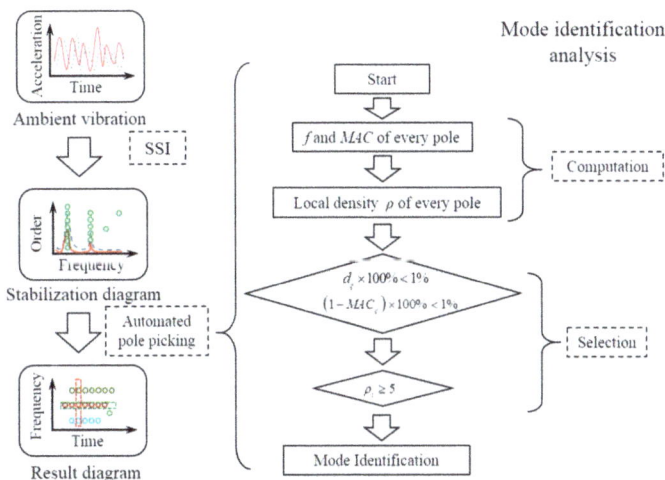

Figure 12. Flow chart of the proposed pole-picking approach.

4. Results and Discussion

4.1. Vibration Composition Analysis

The VAWT vibration is fundamentally caused by wind force, but the vibration amplitudes did not show a strong correlation with wind speed, since the measured excitations and responses were not of the same phase. To determine how the wind turbine vibration signature changed, the PSD diagrams of the turbine vibration accelerations along both the X- and Y-directions of cross Section 1 were presented along with the turbine rotation speed (transferred into blade passing frequencies to indicate whether there were excitations caused by blade rotation) in contour plots, and are shown in Figure 13. Measurements on 28 August 2016 were used for the analysis. It was first noted that there were little changes in the frequency values associated with the rotation speeds; thus, it was concluded that the rotation of blades did not cause significant disturbances on the global structural mode identification, which is unlike the HAWTs [16,18]. However, it was also noted that the PSD amplitude of this VAWT vibration increased with an increase in rotation speed.

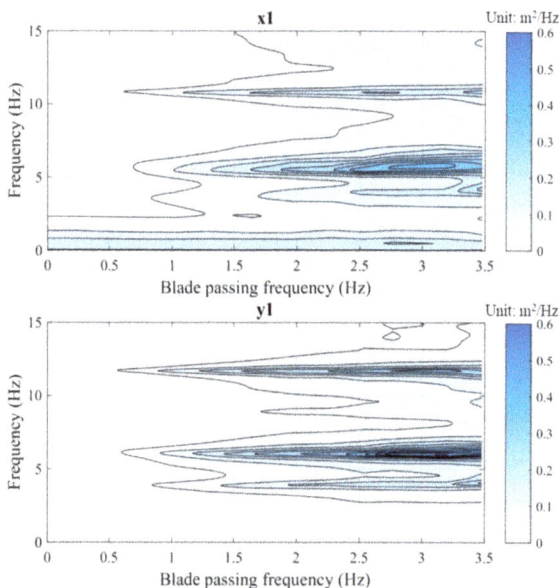

Figure 13. Vibration amplitude versus turbine blade passing frequency.

PSDs from the acceleration time histories were obtained from the #6 building roof (Station 2, which was close to the VAWT) and the wind turbine tower. Figure 14 shows the vibration PSDs of the building and the tower superimposed. Measurements of cross Section 1 along both the X-direction and the Y-direction were taken as references since their vibration amplitudes were the biggest and led to higher signal-to-noise ratios. Because the sensors deployed on the tower were less sensitive toward low-frequency responses, the relationships of the dynamic responses between the building and the tower under 2 Hz was ignored. Three peaks at 3.6, 5.6, and 11.7 Hz can be observed in Figure 14 along the X-direction. For the Y-direction, peaks at 3.8, 6.0, and 11.7 Hz can be seen in Figure 14.

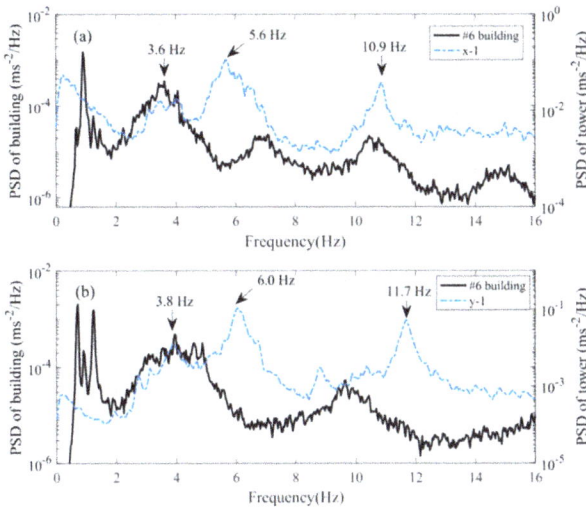

Figure 14. Comparison of acceleration PSDs measured from the roof of the #6 building and the VAWT:
(a) X-direction; (b) Y-direction.

A closer observation of Figure 14 shows that some modes are closely associated with the building
vibration frequencies; hence, they can be identified as building-associated modes. Specifically, the 3.6
and 3.8 Hz modes of the tower along both the X- and Y-directions have corresponding peaks that
match the vibration mode of the #6 building (see also Table 2). On the other hand, the peaks at 5.8 Hz
and 6.0 Hz along the X- and Y-directions, respectively, are the fundamental frequencies of the wind
turbine tower, and they are unique as the corresponding PSD values of the #6 building do not exist.
This indicates that the building contributed to the vibration of the tower.

Table 6 summarizes the vibration frequencies identified along both the X- and Y-directions. Due to
the rotation of the deflector and the H-shape blades, frequencies of some modes can fluctuate within a
narrow range.

Table 6. Vibration frequencies of the tower.

Direction	Frequency (Hz)			
	1st	2nd	3rd	4th
x	3.6	5.8	10.7	
y	3.8	6.0	9.7	11.7

According to Figures 13 and 14, 5.6 Hz and 6.0 Hz were the domain frequencies of the VAWT
vibration along the X-direction and Y-direction, respectively. The predominant building-associated
VAWT vibration modes were 3.6 Hz along the X-direction and 3.8 Hz along the Y-direction since peaks
with the same frequencies can be found in the PSD curves of the building vibrations. The third mode
frequency of the building along the X-direction was 6.9 Hz, and it influenced the VAWT vibration,
since the contour plot was sparser above 5.6 Hz than below 5.6 Hz, as shown in in Figure 13 (x1 plot).

Although the influence cannot be seen clearly in Figure 14, a 9.6-Hz mode, which is the fourth
peak along the Y-direction building PSD curve, shows an obvious contour in Figure 14 (y1 plot).
Furthermore, a 10.9 Hz mode along the X-direction and 11.7 Hz along the Y-direction are actual VAWT
vibration modes. This 10.9-Hz (along the X-direction) VAWT vibration is very close to the fourth
building vibration mode along the X-direction with a frequency of 10.6 Hz.

4.2. Mode Identification

Data from 26 August to 4 September 2016 were used to generate the scatter plots for mode identification with the automated algorithm introduced in Section 3, which are shown in Figure 15. Because most of the modes are shown as scattered data points, a line fit was used to help determine the specific frequencies (colored scatters) of interest. It can be found that, generally, the concerned frequencies decreased as the temperature or blade passing speed increased. Usually, the high temperature and severe vibration of the structure may reduce the rigidness of the system due to steel material characteristics and the boundary feature of the VAWT. Wind direction has little influence on the vibration frequencies as the VAWT is insensitive to the incoming flow. The influence of the blade passing speed on the vibration frequency is not as strong as that of the wind speed. The reason can be that the change of rotation speed lags behind the airflow velocity change.

The identified mode shapes along both the X-direction (hoop bolt shear) and the Y-direction (hoop bolt tension) of the tower are shown in Figure 16. Superimposed in the amplitude plots (Figure 16) are the idealized mode shapes. According to Reference [8], the mode shapes of the Φ-type VAWT shaft are combinations of a simply supported beam and a cantilever. However, this Φ-type VAWT was attached with cables, which is different from the H-type VAWT we studied. A laboratory experimental study was conducted with an H-type VAWT in Reference [9]. However, it does provide mode shapes of the VAWT. A stereo vision technique was used to measure a Φ-type VAWT vibration in Reference [11], and the mode shapes of the Φ-type VAWT shaft were also cantilevers. However, the mode shapes of the H-type VAWT shaft in our study are not completely cantilever shapes since the shaft was connected with two bearings, which cannot be simplified as totally rigid boundary conditions.

It is important to point out that the mode shape corresponding to 9.7 Hz along the Y-direction was different from the others, with the vibration amplitude of the topmost section (cross Section 1) smaller than that of the bottom section (cross Section 3). This mode is a building-associated mode, but the vibration of the shaft is at an opposite phase with the building; thus, the vibration on the top is counteracted. Also shown in Figure 16 are the views from the top of the vibration modes.

The differentiation between building-associated and non-building-associated modes implies that, for future condition monitoring, these two types of modes can be separately studied to identify the influence of building vibration to the tower vibration and to study the tower-alone vibration. The building-associated modes may indicate the PSD or amplitude effects on the VAWT vibration, and the non-building-associated modes would indicate the connections and overall integrity of the VAWT.

Finally, the fundamental frequency along the Y-direction (tension direction of the hoop bolts) was consistent with the calculated values, which was 6.0 Hz. However, the frequency corresponding to the X-direction (shear direction of the hoop bolts) was lower than the calculated value, which was 7.2 Hz. This may be because of an overestimation of the connection stiffness in the theoretical model.

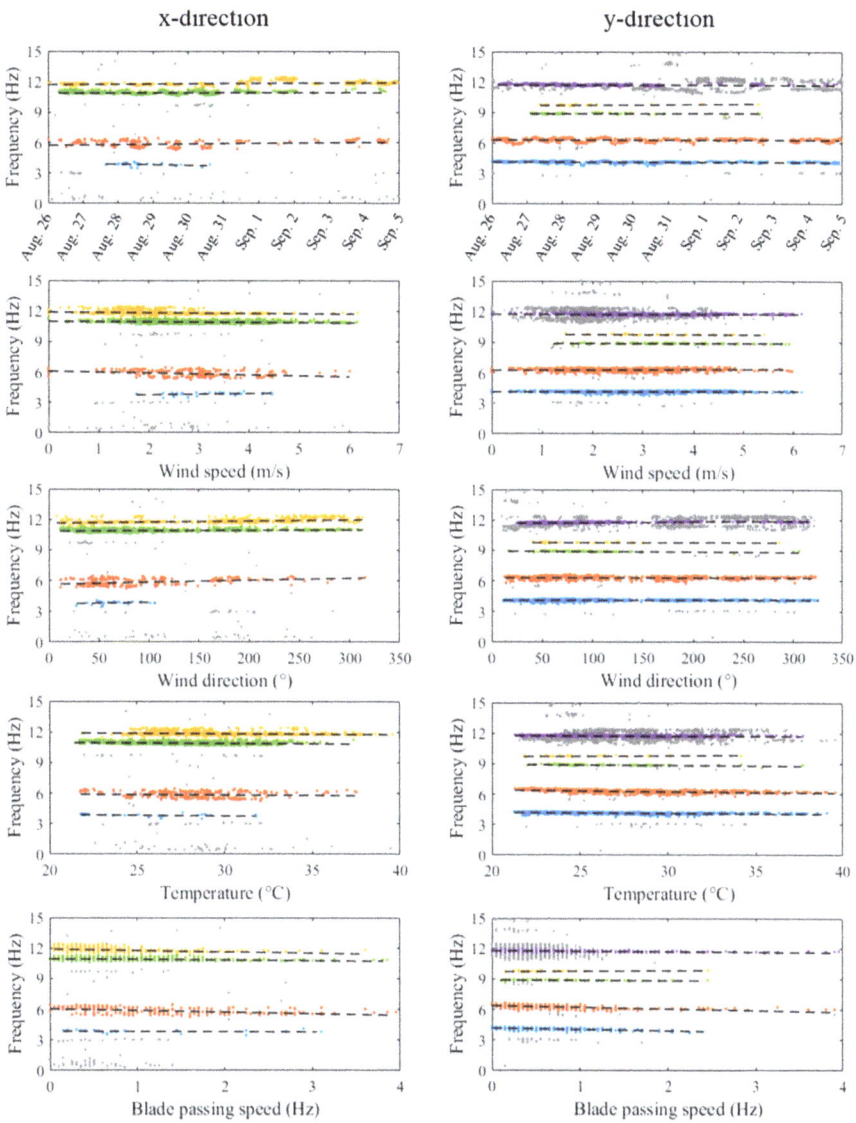

Figure 15. Ambient parameter effects on tower vibrations (26 August to 4 September).

(a) x-direction

(b) y-direction

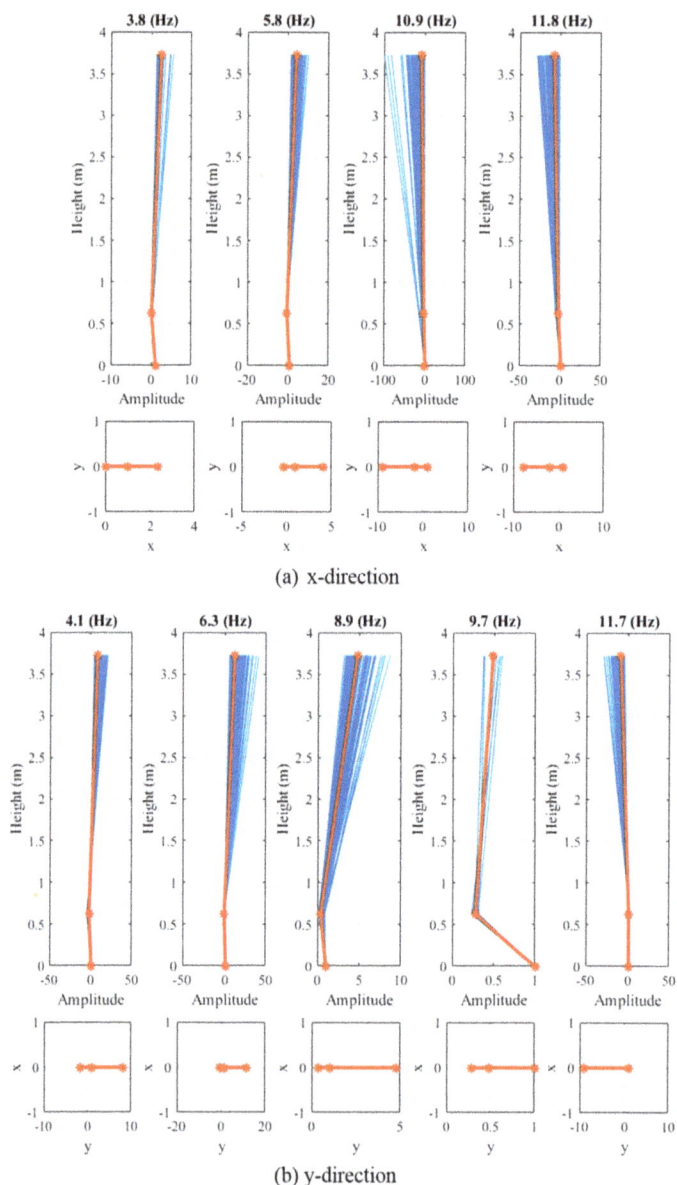

Figure 16. Identified mode shapes and the associated schematics of the four dominant modes along (a) the X-direction, and (b) the Y-direction (26 August to 4 September).

5. Conclusions

This paper presents a field vibration study of an H-type VAWT on a rooftop. To understand the vibration performances of the VAWT installed on the rooftop of the #6 building at Tongji University, a health monitoring system was implemented, and the dynamic behaviors of the VAWT were studied using vibration measurements under ambient conditions. To process the vibration data, an automated algorithm based on stochastic subspace identification (SSI) and a fast clustering approach was developed and presented to show how modes could be determined. The modified method helps accommodate the frequency and mode shape requirements for mode identification, and the results successfully identified several modes of the VAWT tower vibration along the tension direction and shear direction of the hoop connection.

The results showed that some of the vibration responses of the VAWT tower (e.g., 3.6 Hz along the X-direction and 3.8 Hz along the Y-direction) may be affected by the vibrations of the #6 building. Hence, the modal behaviors can be differentiated into building-associated and non-building-associated modes. Among the vibration frequencies identified from both the building and tower measurements, the first mode vibration of the building (0.94 Hz along the X-direction, 0.71 Hz along the Y-direction, and 1.3 Hz in torsion) has little influence on tower vibration. The second bending modes of the building along both directions (3.6 Hz along the X-direction and 4.1 along the Y-direction), however, can affect tower vibration, since the frequencies of the second mode of the building are relatively close to those fundamental frequencies of the tower. This observation is important for the structural health monitoring and life-cycle condition maintenance strategy of the VAWT. The understanding of the effect of building-associated and non-building-associated modes on the VAWT tower responses may also help improve the design of the VAWT tower.

The environmental effects on the ambient vibration data were also investigated, and it was found that the blade rotation speed had a significant effect on the VAWT vibration PSD amplitudes, as shown in Figure 13, indicating that higher blade rotation speeds can result in higher vibration amplitudes of the wind turbine. On the other hand, temperature and wind direction, in general, had little effect on the PSD amplitudes.

Author Contributions: Y.W. analyzed the monitoring data and prepared the manuscript. W.L. is the principal investigator of this project. K.D. is the leading author of this manuscript and prepared the original draft together with the first author. M.Y. helped collect data and S.-E.C. helped with paper writing, reviewing & editing.

Acknowledgments: This research was funded by the International Collaboration Program of Science and Technology Commission of Ministry of Science and Technology, China (Grant No. 2016YFE0105600), the International Collaboration Program of Science and Technology Commission of Shanghai Municipality and Sichuan Province (Grant Nos. 16510711300 and 18GJHZ0111), the National Natural Science Foundation of China (Grant No. U1710111), the Fundamental Research Funds for Central Universities of China, and the China Scholarship Council.

Conflicts of Interest: The authors declare no conflicts of interest.

Abbreviations

HAWT	horizontal axis wind turbine;
VAWT	vertical axis wind turbine;
SHM	structural health monitoring;
SSI	stochastic subspace identification;
PSD	power spectrum density;
MAC	modal assurance criterion.

Symbols

k_t	tensile stiffness of a single hoop;
k_s	shear stiffness of a single hoop;
k_{M10}, k_{M12}	tensile stiffness of M10 bolt and M12 bolt, respectively;
k_R	compression stiffness of the rubber gasket;
$k_{M12,s}$	shear stiffness of the M12 bolt;
$\mathbf{P_{XX}}(f)$	PSD of the response of a structure;
$\mathbf{T}(f)$	transfer function;
$\mathbf{I}$	unit matrix;
x_k	state vector;
y_k	measurement vector;
w_k	excitation vector;

v_k	measurement noise vector;
A	state transition matrix;
C	output location matrix;
k_t	number of measurement;
H	Hankel matrix;
O	projection matrix;
Γ	observability matrix;
X̂	Kalman filter state sequence;
λ	eigenvalue of **A**;
Ψ	eigenvector of **A**;
f_s	sampling frequency;
$2n$	maximum calculation order;
f	natural frequency;
ξ	damping ratio;
φ	mode shape;
ρ	local density of a result point;
d	frequency distance between two result points;
MAC	mode shape similarity (modal assurance criterion) between two result points.

References

1. Riegler, H. HAWT versus VAWT: Small VAWTs find a clear niche. *Refocus* **2003**, *4*, 44–46.
2. Han, D.; Heo, Y.G.; Choi, N.J.; Nam, S.H.; Choi, K.H.; Kim, K.C. Design, Fabrication, and Performance Test of a 100-W Helical-Blade Vertical-Axis Wind Turbine at Low Tip-Speed Ratio. *Energies* **2018**, *11*, 1517. [CrossRef]
3. Marie, D.G.J. Turbine Having its Rotating Shaft Transverse to the Flow of the Current. U.S. Patent No. 1,835,018, 8 December 1931.
4. Bhutta, M.M.A.; Hayat, N.; Farooq, A.U.; Ali, Z.; Jamil, S.R.; Hussain, Z. Vertical axis wind turbine—A review of various configurations and design techniques. *Renew. Sustain. Energy Rev.* **2012**, *16*, 1926–1939. [CrossRef]
5. Jin, X.; Zhao, G.; Gao, K.J.; Ju, W. Darrieus vertical axis wind turbine: Basic research methods. *Renew. Sustain. Energy Rev.* **2015**, *42*, 212–225. [CrossRef]
6. Yang, Y.; Guo, Z.; Zhang, Y.; Jinyama, H.; Li, Q. Numerical Investigation of the Tip Vortex of a Straight-Bladed Vertical Axis Wind Turbine with Double-Blades. *Energies* **2017**, *10*, 1721. [CrossRef]
7. Mabrouk, I.B.; Hami, A.E.; Walha, L.; Zghal, B.; Haddar, M. Dynamic vibrations in wind energy systems: Application to vertical axis wind turbine. *Mech. Syst. Signal Process.* **2017**, *85*, 396–414. [CrossRef]
8. James, G.H.I.; Carne, T.G.; Lauffer, J.P. *The Natural Excitation Technique (NExT) for Modal Parameter Extraction from Operating Wind Turbines*; Nasa STI/Recon Technical Report N; 1993; Volume 93, pp. 260–277. Available online: https://prod.sandia.gov/techlib-noauth/access-control.cgi/1992/921666.pdf (accessed on 7 June 2018).
9. Mclaren, K.; Tullis, S.; Ziada, S. Measurement of high solidity vertical axis wind turbine aerodynamic loads under high vibration response conditions. *J. Fluids Struct.* **2012**, *32*, 12–26. [CrossRef]
10. Malge, A.; Pawar, P. Wind tunnel and numerical performance analysis of multi-storey vertical axis wind turbines. *J. Renew. Sustain. Energy* **2015**, *7*, 053121. [CrossRef]
11. Najafi, N.; Paulsen, U.S. Operational modal analysis on a VAWT in a large wind tunnel using stereo vision technique. *Energy* **2017**, *125*, 405–416. [CrossRef]
12. Rankine, R.K.; Chick, J.P.; Harrison, G.P. Energy and carbon audit of a rooftop wind turbine. *Proc. Inst. Mech. Eng. A J. Power Energy* **2006**, *220*, 643–654. [CrossRef]
13. Balduzzi, F.; Bianchini, A.; Carnevale, E.A.; Ferrari, L.; Magnani, S. Feasibility analysis of a Darrieus vertical-axis wind turbine installation in the rooftop of a building. *Appl. Energy* **2012**, *97*, 921–929. [CrossRef]
14. Brownjohn, J.M. Structural health monitoring of civil infrastructure. *Philos. Trans. R. Soc. Lond. A Math. Phys. Eng. Sci.* **2007**, *365*, 589–622. [CrossRef] [PubMed]

15. Ou, J.; Li, H. Structural health monitoring in mainland china: Review and future trends. *Struct. Health Monit.* **2010**, *9*, 219–231.
16. Lan, C.; Li, H.; Ou, J. Traffic load modelling based on structural health monitoring data. *Struct. Infrastruct. Eng.* **2011**, *7*, 379–386. [CrossRef]
17. Rohrmann, R.G.; Thöns, S.; Rücker, W. Integrated monitoring of offshore wind turbines—Requirements, concepts and experiences. *Struct. Infrastruct. Eng.* **2010**, *6*, 575–591. [CrossRef]
18. Smarsly, K.; Hartmann, D.; Law, K.H. A computational framework for life-cycle management of wind turbines incorporating structural health monitoring. *Struct. Health Monit.* **2013**, *12*, 359–376. [CrossRef]
19. Shirzadeh, R.; Devriendt, C.; Bidakhvidi, M.A.; Guillaume, P. Experimental and computational damping estimation of an offshore wind turbine on a monopile foundation. *J. Wind Eng. Ind. Aerodyn.* **2013**, *120*, 96–106. [CrossRef]
20. Devriendt, C.; Weijtjens, W.; El-Kafafy, M.; De Sitter, G. Monitoring resonant frequencies and damping values of an offshore wind turbine in parked conditions. *IET Renew. Power Gener.* **2014**, *8*, 433–441. [CrossRef]
21. Hu, W.H.; Thöns, S.; Rohrmann, R.G.; Said, S.; Rücker, W. Vibration-based structural health monitoring of a wind turbine system. Part I: Resonance phenomenon. *Eng. Struct.* **2015**, *89*, 260–272. [CrossRef]
22. Hu, W.H.; Thöns, S.; Rohrmann, R.G.; Said, S.; Rücker, W. Vibration-based structural health monitoring of a wind turbine system Part II: Environmental/operational effects on dynamic properties. *Eng. Struct.* **2015**, *89*, 273–290. [CrossRef]
23. Dai, K.; Wang, Y.; Huang, Y.; Zhu, W.; Xu, Y. Development of a modified stochastic subspace identification method for rapid structural assessment of in-service utility-scale wind turbine towers. *Wind Energy* **2017**, *20*, 1687–1710. [CrossRef]
24. Weijtjens, W.; Verbelen, T.; Sitter, G.D.; Devriendt, C. Foundation structural health monitoring of an offshore wind turbine—A full-scale case study. *Struct. Health Monit.* **2016**, *15*, 389–402. [CrossRef]
25. Loh, C.H.; Loh, K.J.; Yang, Y.S.; Hsiung, W.Y.; Huang, Y.T. Vibration based system identification of wind turbine system. *Struct. Control Health Monit.* **2017**, *24*, e1876. [CrossRef]
26. Dai, K.; Huang, Y.; Gong, C.; Huang, Z.; Ren, X. Rapid seismic analysis methodology for in-service wind turbine towers. *Earthq. Eng. Eng. Vib.* **2015**, *14*, 539–548. [CrossRef]
27. Dai, K.; Sheng, C.; Zhao, Z.; Yi, Z.; Camara, A.; Bitsuamlak, G. Nonlinear response history analysis and collapse mode study of a wind turbine tower subjected to cyclonic winds. *Wind Struct.* **2017**, *25*, 79–100.
28. Qu, L. Analysis on Structural Interaction between Highrise Building and Roof-Mounted Wind Turbine Generator. Master's Thesis, Tongji University, Shanghai, China, 2015. (In Chinese)
29. Bingham, C.; Godfrey, M.; Tukey, J.W. Modern techniques of power spectrum estimation. *IEEE Trans. Audio Electroacoust.* **1967**, *15*, 56–66. [CrossRef]
30. Bendat, J.S.; Piersol, A.G. *Random Data: Analysis and Measurement Procedures*; Willey and Sons, Inc.: Hoboken, NJ, USA, 1971.
31. Van Overschee, P.; De Moor, B. Subspace algorithms for the stochastic identification problem. *Automatica* **1993**, *29*, 649–660. [CrossRef]
32. Peeters, B.; Roeck, G.D. Reference-based stochastic subspace identification for output-only modal analysis. *Mech. Syst. Signal Process.* **1999**, *13*, 855–878. [CrossRef]
33. El-Kafafy, M.; Devriendt, C.; Guillaume, P.; Helsen, J. Automatic tracking of the Modal parameters of an offshore wind turbine drivetrain system. *Energies* **2017**, *10*, 574. [CrossRef]
34. Zhang, Y.; Kurata, M.; Lynch, J.P. Long-term modal analysis of wireless structural monitoring data from a suspension bridge under varying environmental and operational conditions: System design and automated modal analysis. *J. Eng. Mech. ASCE* **2016**, *143*, 04016124. [CrossRef]
35. Cabboi, A.; Magalhães, F.; Gentile, C.; Álvaro, C. Automated modal identification and tracking: Application to an iron arch bridge. *Struct. Control Health Monit.* **2017**, *24*, e1854. [CrossRef]
36. Rodriguez, A.; Laio, A. Clustering by fast search and find of density peaks. *Science* **2014**, *344*, 1492–1496. [CrossRef] [PubMed]

energies

MDPI

Article

Study on Vibration Transmission among Units in Underground Powerhouse of a Hydropower Station

Jijian Lian [1,2,*]**, Hongzhen Wang** [1,2] **and Haijun Wang** [1,2]

[1]   State Key Laboratory of Hydraulic Engineering Simulation and Safety, Tianjin University,
     Tianjin 300072, China; wanghongzhen@tju.edu.cn (H.W.); bookwhj@tju.edu.cn (H.W.)
[2]   School of Civil Engineering, Tianjin University, Tianjin 300072, China
*   Correspondence: jjlian@tju.edu.cn; Tel.: +86-22-2740-1127

Received: 16 October 2018; Accepted: 29 October 2018; Published: 2 November 2018

Abstract: Research on the safety of powerhouse in a hydropower station is mostly concentrated on the vibration of machinery structure and concrete structure within a single unit. However, few studies have been focused on the vibration transmission among units. Due to the integrity of the powerhouse and the interaction, it is necessary to study the vibration transmission mechanism of powerhouse structure among units. In this paper, field structural vibration tests are conducted in an underground powerhouse of a hydropower station on Yalong River. Additionally, the simplified mechanical models are established to explain the transmission mechanism theoretically. Moreover, a complementary finite element (FE) model is built to replicate the testing conditions for comprehensive analysis. The field tests results show that: (1) the transmission of lateral-river vibration is greater than those of longitude-river vibration and vertical vibration; (2) the vibration transmission of the vibrations that is caused by the low frequency tail fluctuation is basically equal to that of the vibrations caused by rotation of hydraulic generator. The transmission mechanism is demonstrated by the simplified mechanical models and is verified by the FE results. This study can provide guidance for further research on the vibration of underground powerhouse structure.

Keywords: vibration transmission mechanism; underground powerhouse; lateral-river vibration; low frequency tail fluctuation; rotation of hydraulic generator

---

1. Introduction

Vibration is a common phenomenon in energy infrastructure structures. Severe vibration can lead to safety problems in rotating machineries and support structures [1–3], such as electrical machines, towers of the wind turbine generators, and parabolic reflective surfaces in the concentrated solar power systems. As a combination of rotating mechanical structures and concrete structures, the powerhouses in hydropower stations usually work under complex hydraulic, electromagnetic, and mechanical loads. Therefore, the safety problems are prone to occur. In recent years, hydropower industry has developed rapidly in China. According to the National Energy Administer (NEA), the installed capacity of hydropower has reached 341 million kW in 2017, accounting for 19.2% of the total installed capacity of electricity in China. The annual hydropower generation has reached 1.19×1012 kWh, accounting for 18.5% of the total electricity generation in China. Hydropower has made great contributions to economic development and reduction of carbon emissions. With the development of hydropower, a group of high-head, large-capacity hydroelectric generators has been commonly used in large-scale hydropower stations. Various powerhouse safety problems that are caused by vibration of units happened in hydropower stations correspondingly. For example, hydropower stations, such as XiaoLangDi, ErTan, and YanTan in China have experienced powerhouse safety problems some extent [4,5]. The most serious safety problem of powerhouse occurred in Russia, the unit #2 of the Sayano-Shushenskaya hydropower station in Yenisei River experienced severe vibration after overload operation, leading to fatigue damage to the cap fixed bolts, and resulting in great casualties and property loss [6,7].

At present, the research on vibration safety of the powerhouse is mostly focused on the vibration of machinery structure [8,9]. For the coupling vibration of unit shaft system, Ma, Song and Zhi, et al. built the FE models of bearing support to analyze the coupling relationship of the foundation and the shafting system [10–13]. Zhou et al. investigate the vibration of the stator frame under the action of electromagnetic forces based on field tests and FE models [14]. Zhang and Wang et al. built the FE models of powerhouse and pumping station to study the vibration under pressure pulsations [15,16]. On the other hand, Lian and He et al. studied the influence of unit on the vibration of the powerhouse structure. The complicated linear and nonlinear coupling vibration rules between the unit and powerhouse structure have been summarized, based on field tests of unit and powerhouse structure of multiple hydropower stations and the FE method [17–21]. Zhang and Mao analyzed the correlation between the vibration response of unit and powerhouse structure [22,23]. The coupling relationship between the unit and the powerhouse structure were discussed based on relevant theories and field tests data.

For the influence and transmission of vibration of powerhouse, Wang and Bai et al. investigated the transmission rules of adjacent units by field tests of a hydropower station [24]. Wei et al. studied the vibration transmission ways between main powerhouse and auxiliary powerhouse by FE method [25]. Ameen et al. studied the effect on dams caused by vibration of powerhouse by ANSYS-CFX model [26]. As for the vibration in underground structure, Gupta et al. investigated the influence of tunnel and soil parameters on vibrations from underground railways [27]. Chen and Xia et al. studied the vibration transmissions that are caused by blasting in underground powerhouse and excavation [28,29]. Kuo et al. studied the effect of a twin tunnel on the propagation of ground-borne vibration from an underground railway theoretically [30]. However, due to the difficulty of field test and complexity of structure, it is hard to explain the mechanism of vibration transmission in underground powerhouse, so the theoretical research is rare.

According to the complexity of powerhouse structure and vibration source mechanism, this paper focuses on two basic problems for vibration transmission among units. The first one is the effect of vibration directions on the vibration transmission ratios, and the second one is the effect of frequency of the vibration source on the vibration transmission ratios. To solve these problems intuitively and accurately, field test, theoretical research, and numerical simulation are employed with appropriate and reasonable simplification.

In this paper, field structural vibration tests of an underground powerhouse in a hydropower station on Yalong River were conducted to investigate the vibration rules. Then, the simplified mechanical vibration models are established to explain the mechanism of the vibration theoretically. Finally, the testing powerhouse structure is simulated and calculated by FE method, the corresponding vibration transmission ratios among units are extracted and compared with the field test results to verify the theoretical analysis. The technology route of this paper is shown in Figure 1. This paper studied the mechanism of the vibration transmission among units systematically. It can provide guidance for further research on the safety of underground powerhouse structure.

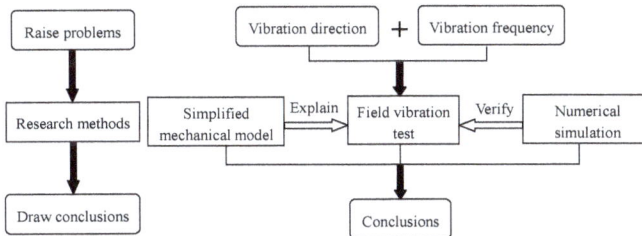

Figure 1. Schematic diagram of technology route.

2. Field Structural Vibration Test

In order to investigate the vibration transmission among units, a series of field structure vibration tests were conducted based on an underground powerhouse in a large hydropower station on Yalong River.

2.1. Field Tests Overview

The hydropower station is located in the main stream of Yalong River at the junction of Yanyuan County and Muli County in Sichuan Province of China. It is the first stage of the five-order hydropower development project in the middle and lower reaches of Yalong River, where the hydropower resources are most concentrated. The hydropower station mainly aims at power generation, and it also has functions of flood control and sand interception. The normal water storage level of the reservoir is 1880 m, the total storage capacity is 7.76 billion m^3, and the adjusted storage capacity is 4.91 billion m^3. The installed capacity of the power station is 3600 MW, the annual utilization hour is 4616 h, and the annual power generation is 166.20 billion kWh. All the units are lined up in the main powerhouse from #1 to #6, and the rated capacity of single unit is 600 MW. Total length of the main powerhouse is 204.52 m, the excavation height is 68.80 m, and the width of the main powerhouse along river is 25.90 m. The main powerhouse is shown as Figure 2.

Figure 2. Main powerhouse of the hydropower station in Yalong River.

To study the vibration transmission rules of powerhouse structure among units, vibration displacement sensors were installed in unit #1 of powerhouse, considering the actual condition. Ds-Net acquisition system and DP type seismic low frequency vibration displacement sensor were used in the field tests.

Ds-Net acquisition system was used as data acquisition instrument. This system includes multi-channel signal acquisition module and instrument fault signal identification module. Data can be acquired and stored in this system simultaneously. The system can eliminate structural background noise and Characteristic parameters of signals, such as maximum, minimum, variance, deviation coefficient, and kurtosis coefficient can be calculated immediately. It also has an intelligent multi-channel display interface during test. This system performs well in low frequency signal, and it is suitable for large-scale structural vibration tests, such as the powerhouse, in this paper. The sampling frequency of field tests is 400 Hz, and each data length is 1 min. In order to minimize the influence of end effect in the data processing, the middle 50 s data was intercepted in the analysis. The data acquisition instrument is shown as Figure 3.

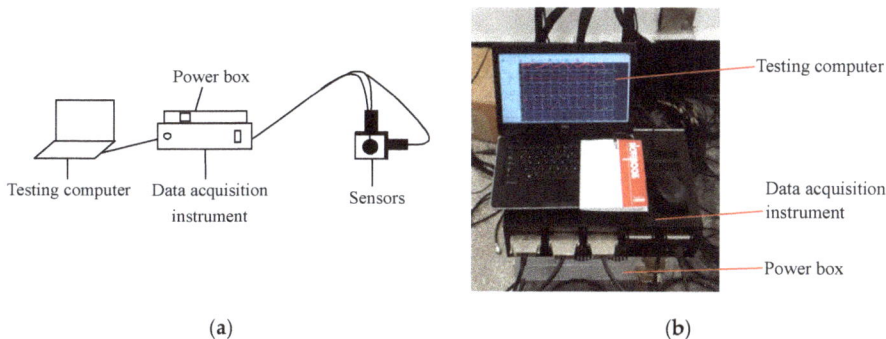

Figure 3. Data acquisition instrument: (**a**) Schematic; and, (**b**) Field tests photo.

DP type seismic low frequency vibration displacement sensor was used in the field tests. A set of low frequency expansion circuits is employed to the seismic detector, in order to reduce the natural frequency of the output characteristics to 1/20–1/100 of the original detector. The sensor has properties of shock resistance, high stability, and good characteristics of low frequency output. The sensitivity of the sensor is high to measure micrometer vibration displacements. Therefore, the sensor is suitable for vibration measurement of large structures, such as powerhouse structure. Frequency response of the sensor that was used in this paper is in the range of 0.35–200 Hz with a sensitivity of 8 mV/μm. The vibration displacement sensors are shown as Figure 4.

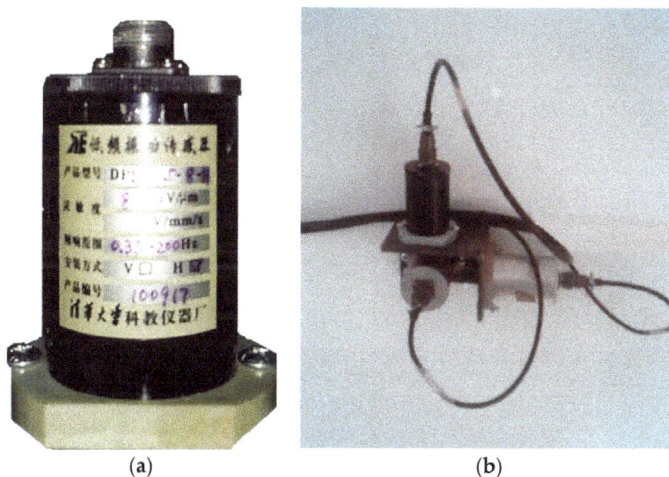

Figure 4. Vibration displacement sensors used in field tests: (**a**) Close-up; and (**b**) Field installation photo.

The vibration displacement sensors were installed in the middle of the main beam on the left side of hydraulic turbine floor of unit #1, as shown in Figure 5. Three sensors were fixed by bolts after drilling to test vibrations in the lateral-river direction, the longitude-river direction and the vertical direction. For the convenience of the following description, the lateral-river direction, as well as the direction of axis of main powerhouse is defined as the X direction. The longitude-river direction is defined as the Y direction, and the vertical direction is defined as the Z direction. As shown in Figure 6.

The investigations were concentrated on unit #2 and unit #3, since they were closed to the sensors in unit #1. In order to study the effects of unit#2 and unit #3 on the vibration of sensors in unit #1, it is necessary to adopt the variables control method. Therefore, when testing the effect of unit #2 on the sensors, unit #2 was in operation, both unit #1 and unit #3 should be in shutdown; the same settings were applied when testing the effects of unit #2 and unit #3. Because the units were connected to the power grid during tests, their operating conditions must meet the needs of power grid, and cannot be controlled as the tests required. Therefore, in the actual tests, a large number of data was acquired. Then, the data in the time period when the unit #4, unit #5, and unit #6 were in shutdown was selected after the tests, as shown in Figure 6. Test results for unit #1, unit #2, and unit #3 in various operating conditions were obtained to investigate the vibration transmission.

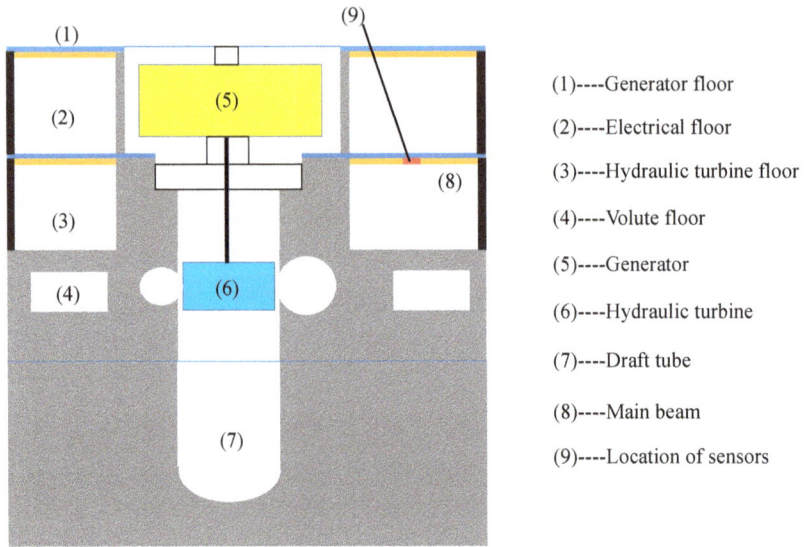

(1)----Generator floor	
(2)----Electrical floor	
(3)----Hydraulic turbine floor	
(4)----Volute floor	
(5)----Generator	
(6)----Hydraulic turbine	
(7)----Draft tube	
(8)----Main beam	
(9)----Location of sensors	

Figure 5. Powerhouse structure and location of sensors.

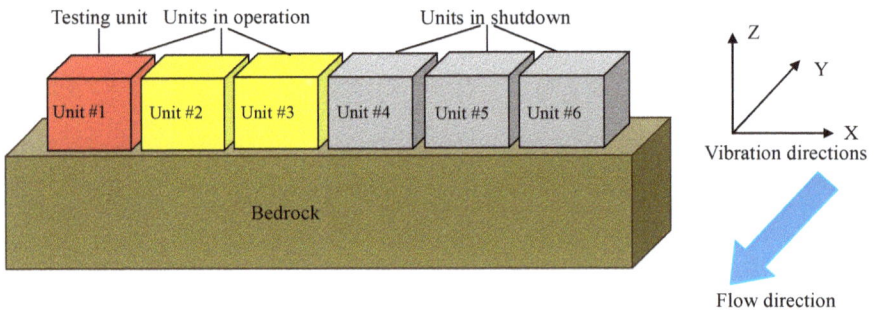

Figure 6. Distribution of units in the main powerhouse and definition of vibration directions.

2.2. Preliminary Tests Results

With the change of operating conditions of unit #1, unit #2, and unit #3, the root mean square (RMS) of vibration displacement of the sensors in unit #1 varies, as shown in Figure 7.

It can be seen from the Figure 7 that the vibration in Z direction is most severe. When the vibration source (unit #1, unit #2, or unit #3) is at 100 MW operating condition, vibrations of structure in X, Y, and Z direction achieve the maximum simultaneously. Therefore, the 100 MW operating condition is the most unfavorable condition in the field tests of powerhouse. This is consistent with the previous tests and research results [17,21]. When the Francis Type Water Turbine-Generator Unit is fewer than 40% of rated load conditions, the tail water vortex belt would be decomposed and split. A large number of irregular small vortexes will replace the spiral vortex belt, and the signal exhibits a noise-like broadband characteristic, which can be seen in the spectrum analysis of tested signal in the following section [31]. Samanta and Vinuesa et al. also studied the characterizations of the flow field through numerical simulations and experiments [32–34]. In the other hand, the preliminary tests results also prove the consistency of each unit as vibration source. Next, all the following data analysis is based on 100 MW operating conditions.

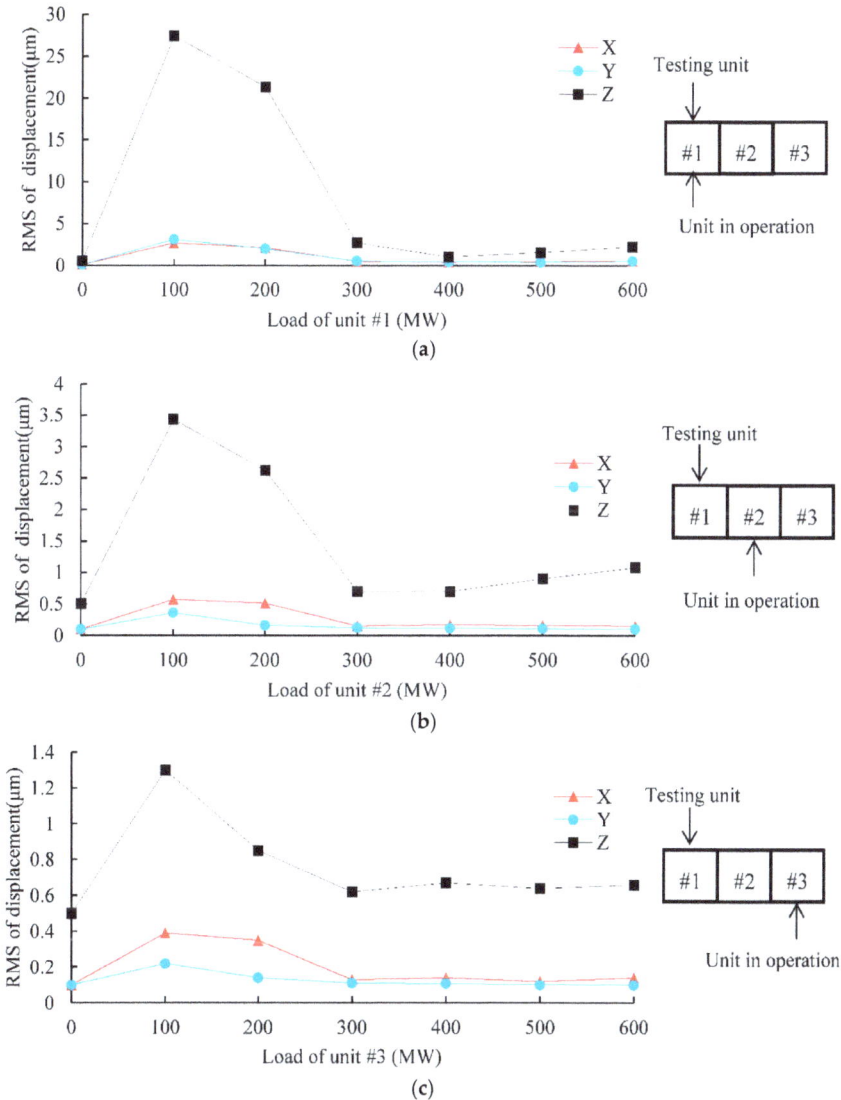

Figure 7. Variation of root mean square (RMS) of displacement with unit #1, unit #2 and unit #3 in operation respectively: (**a**) Unit #1 in operation; (**b**) Unit #2 in operation; and, (**c**) Unit #3 in operation.

2.3. Vibration Transmission Rules of Tests

According to the the two problems raised in the introduction, the variation of vibration intensity in three directions was calculated to study the effect of vibration directions; the signal component was analyzed to study the effect of vibration frequency.

2.3.1. Vibration Intensity

In order to study the rules of vibrations in X, Y, and Z direction with different units, typical time histories of vibration displacements in three directions are shown in Figure 8, when the unit #1, unit #2, or unit #3 is operated as vibration source, respectively.

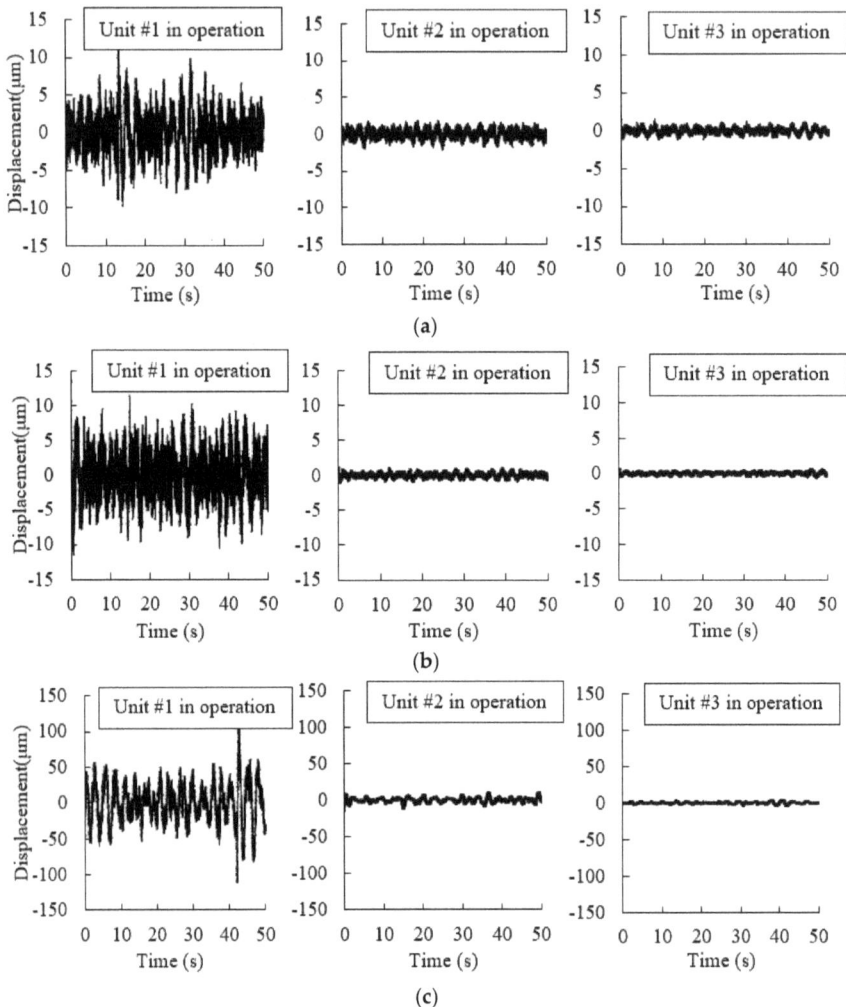

Figure 8. Time histories of vibration displacements in three directions: (**a**) X direction; (**b**) Y direction; and, (**c**) Z direction.

Due to the uncertainty in field vibration tests, it is necessary to minimize the influence of random factors. Therefore, multiple groups of samples were selected in 100 MW operating conditions.

Ten groups of typical data were extracted for analysis, and the RMS values of vibration displacement were calculated. A scatter plot of the RMS of vibrations in X, Y, and Z directions is shown in Figure 9.

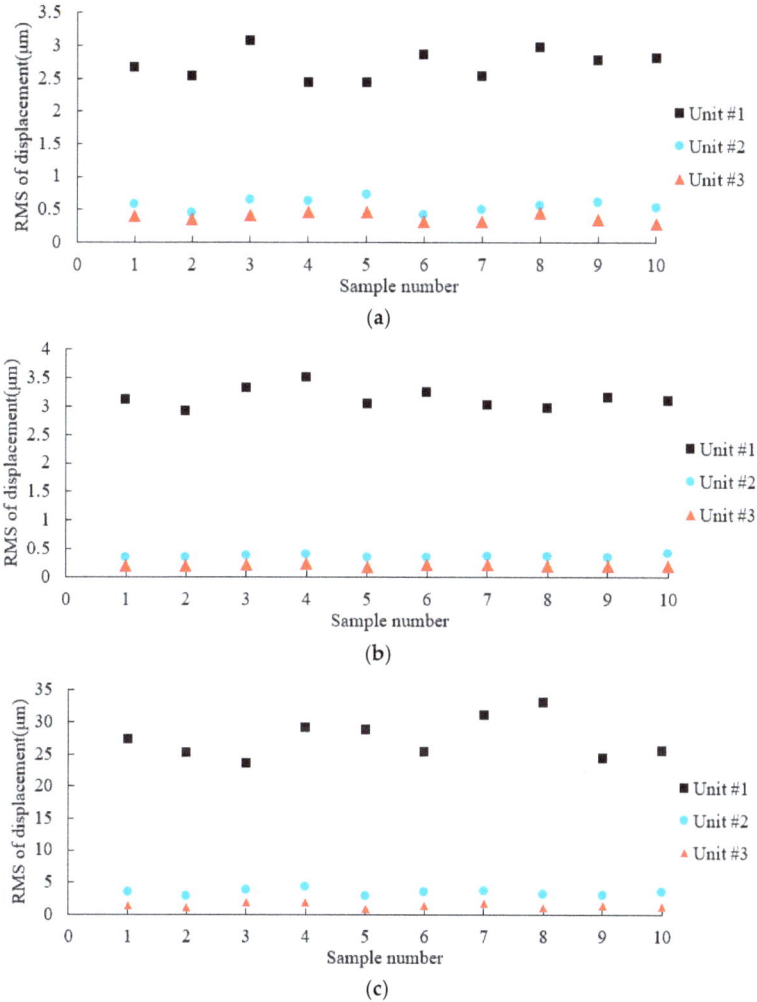

Figure 9. RMS of vibration displacements with different units in three directions: (**a**) X direction; (**b**) Y direction; (**c**) Z direction.

Figure 9 show that the vibrations in three directions of X, Y and Z all present a tendency of decrease with the increase of the distance from the vibration source. The mean values of the ten samples in three directions of X, Y, and Z are calculated respectively. Taking the RMS of vibration displacement caused by unit #1 as reference, the ratios of RMS of vibration displacement caused by different units are attained respectively, as shown in Equation (1).

$$\begin{cases} x_1 : x_2 : x_3 = 1 : 0.1769 : 0.1069 \\ y_1 : y_2 : y_3 = 1 : 0.0953 : 0.0331 \\ z_1 : z_2 : z_3 = 1 : 0.1074 : 0.0320 \end{cases} \quad (1)$$

The ratios are defined as vibration effect ratios by each unit, as shown in Figure 10. For vibration in X direction, the effect ratio of unit #2 is approximately 17.69%, while the ratio of unit #3 is about 10.69%, compared with the vibrations caused by unit #1 as 100%. For vibration in Y direction, the effect ratio of unit #2 is approximately 9.53%, while the ratio of unit #3 is only 3.31%. For vibration in Z direction, the effect ratio of unit #2 is approximately 10.74%, while the ratio of unit #3 is about 3.2%. When considering of the consistency of each unit as vibration source, the vibration effect ratios can also be regarded as vibration transmission ratio.

It can be conducted that the vibrations in X direction caused by adjacent units are greater, and the vibration transmission ratios are bigger, as compared with the vibration in Y and Z directions.

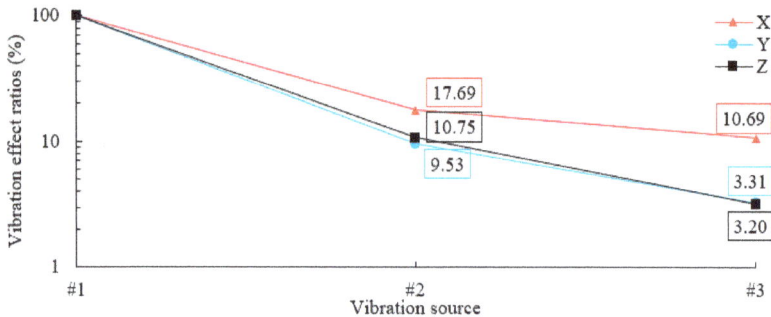

Figure 10. Vibration effect ratios by each unit.

2.3.2. Signal Component

In order to study the variation of signal components in the process of vibration transmission, spectral analysis is performed based on the vibration signals in Figure 8. Power spectral density (PSD) is obtained and shown in Figure 11.

According to the PSD shown in Figure 11, all the signals can be divided into three components: (1) Component A, with frequency between 0.2 and 1.5 Hz; (2) Component B, with frequency of 2.4 Hz; and, (3) Component C, with frequency higher than 5 Hz. For vibration in X direction, the Component A is the main part. Meanwhile, the energy of Component B is less than Component A. For vibration in Y direction, Component A and Component B are two main parts. Energies of the two are basically equal, while the Component A is wider and the peak value of the Component B is larger. For vibration in Z direction, the Component A is the only main part. The Component B is unobvious. For vibrations in all three directions, the Component C is not obvious. The proportions of the three components do not show apparent difference among vibrations caused by unit #1, unit #2 and unit #3.

The components of signals are related to the corresponding vibration sources. Combining previous studies [13,14,16,35], vibration sources of powerhouse structure mainly consist of the following parts: (1) Rotation of hydraulic generator, according to the unit parameters, the rotational frequency of hydraulic generator is 2.4 Hz; (2) Low frequency tail fluctuation, when the unit is in the medium and low load state, severe low frequency tail fluctuation occurs in the draft tube. It is often the main source of vibration for units and powerhouse structure. Its frequency is less than 0.6 times that of rotation frequency; (3) The other medium and high frequency vibration sources, such as volute uneven flow. Combining above, the Component A can be considered to be caused by low frequency tail fluctuation; the Component B can be considered to be caused by the rotation of hydraulic generator; the Component C can be considered to be caused by the other medium and high frequency vibration sources.

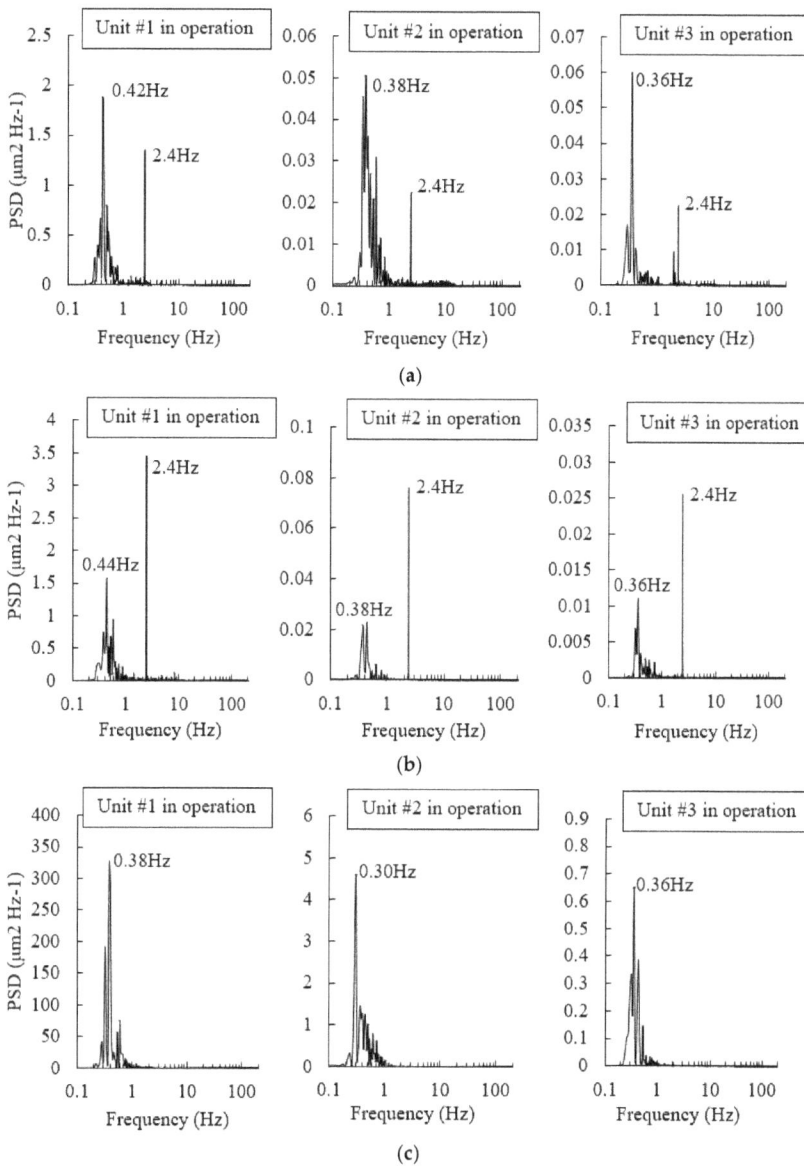

Figure 11. Spectrums of vibration displacements in three directions: (a) X direction; (b) Y direction; and, (c) Z direction.

In order to further quantify the components, the wavelet analysis method was used to component analysis of vibration signals. Wavelet analysis is a local transformation method based on time, space and frequency. It uses horizontal movement and expansion to perform function multi-scale operation, which can realize frequency domain decomposition of data signals. Multi-Resolution Analysis (MRA) was applied in this article. The db3 wavelet was used as the mother wavelet. Firstly, the vibration signals were decomposed by seven-level wavelet transform. Then signals corresponding to different vibration sources were reconstructed from different frequency bands. Finally energies of vibration signals corresponding to different vibration sources were calculated. The frequency range of each frequency band after wavelet decomposition of vibration signals are shown in Table 1.

Table 1. Wavelet decomposition of vibration signals.

Signal Decomposed	a7	d7	d6	d5	d4	d3	d2	d1
Frequency Range (Hz)	0–1.56	1.56–3.13	3.13–6.25	6.25–12.5	12.5–25	25–50	50–100	100–200

Signal corresponding to a7 was reconstructed as the Component A. Signal corresponding to d7 was reconstructed as the Component B. Signals corresponding to d6, d5, d4, d3, d2, and d1 were constructed as the Component C. Variances of the reconstructed signals are calculated to obtain the energy proportions of different vibration components, as shown in Figure 12. It can be shown that energy proportions of Component A and Component B barely change in the progress of vibration transmission among units.

According to the above, it can be concluded: (1) Low frequency tail fluctuation and rotation of hydraulic generator are the two main vibration sources of the vibration of the powerhouse structure; and, (2) They have almost the same transmission ratios among units.

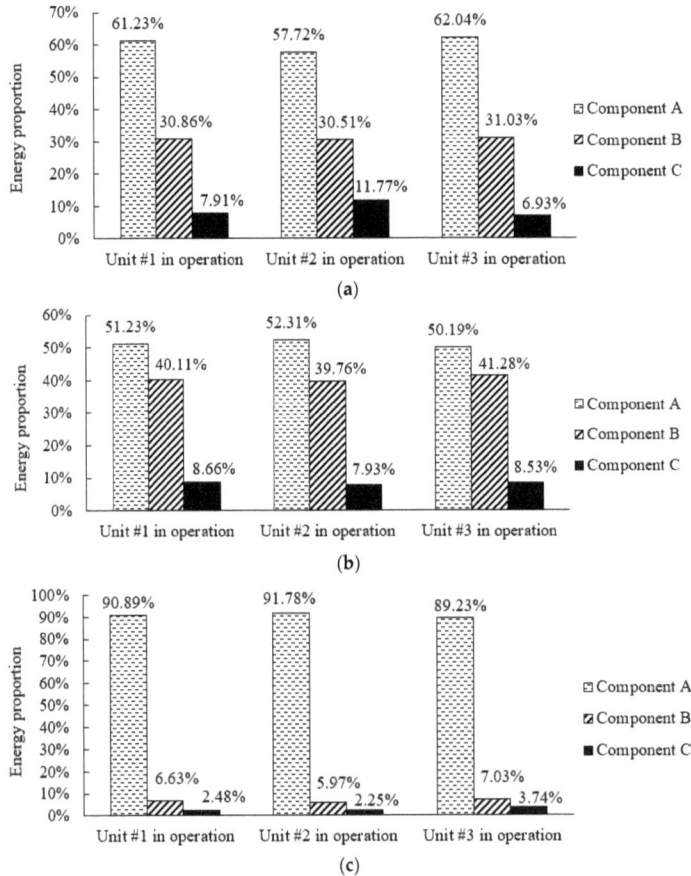

Figure 12. Energy proportions of vibrations in three directions: (**a**) X direction; (**b**) Y direction; and (**c**) Z direction.

3. Study of Vibration Transmission Mechanism

The simplified mechanical vibration models are established to investigate the vibration transmission mechanism of powerhouse structure among units.

3.1. Simplication of Powerhouse Structure

The vibration transmission of underground powerhouse structure among units is essentially a kind of mechanical wave. Its transmission direction is the direction of axis of the main powerhouse (from unit #1 to unit #n), that is the X direction according to the previous definition. When the vibration is in X direction, its direction is consistent with the transmission direction. So, this transmission can be regarded as compression vibration. When the vibration is in Y or Z direction (perpendicular to the direction of transmission), the transmission can be regarded as the shear vibration, as shown in Figure 13. These two different vibration transmissions in powerhouse will be studied in the following section.

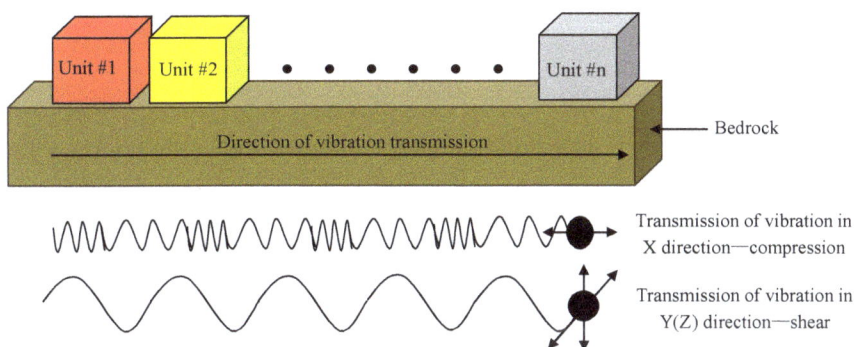

Figure 13. Transmission of vibration in different directions.

According to the structural characteristics of underground powerhouse of the hydropower station, it is known that the main structure of powerhouse is mainly the mass concrete. Therefore the powerhouse structure of each power unit can be regarded as a homogeneous block and fixed on the bedrock. Units are separated by the split seam. Assuming that the effect of split seams on vibration transmission among units is negligible. Only the vibration transmission through the bedrock is considered. When considering the condition of two adjacent units, the main powerhouse structure can be simplified, as shown in Figure 14.

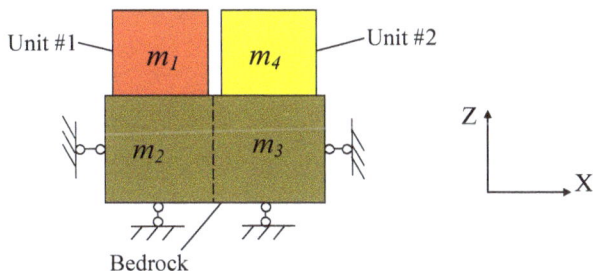

Figure 14. Simplified model of vibration transmission.

As shown in the Figure 14, two units and the bedrocks are plotted in the two-dimensional (2-D) plane. Two adjacent units are represented by two lumped masses m_1 and m_4. The bedrocks below m_1 and m_4 are represented by two homogeneous elastic blocks m_2 and m_3, respectively. The bottom and the lateral sides of bedrocks m_2 and m_3 are restrained by normal constraints. It is easy to know $m_1 = m_4$, let $m_1 = m_4 = m$; similarly, $m_2 = m_3 = M$.

3.2. Establishment of Vibration Models

According to different modes of transmission, two vibration models are established.

Firstly, a horizontal vibration model has been established to study the compression vibration, as shown in Figure 15.

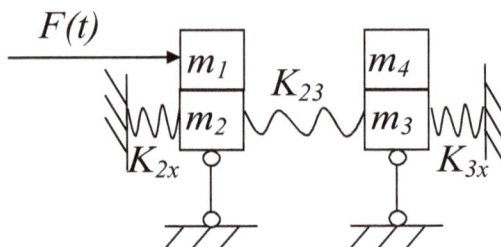

Figure 15. Horizontal vibration model.

By the concentrated mass method, the masses of bedrocks are concentrated at their centroids, and regarded as two lumped mass, m_2 and m_3. They are connected to the left and right boundary by springs; Deformation of the bedrock between m_2 and m_3 is represented by the stretching and compression of a spring to simulate the interaction of axial force. Load $F(t)$ is applied on the lumped mass m_1. x_1, x_2, x_3 and x_4 denote displacements of m_1, m_2, m_3, and m_4, respectively. Considering dynamic load only, equations of motion for the four lumped masses are listed in Equation (2).

$$\begin{cases} F(t) - Q_{12} = m_1 \ddot{x}_1 \\ Q_{12} - K_{2x}x_2 - K_{23}(x_2 - x_3) = m_2 \ddot{x}_2 \\ K_{23}(x_2 - x_3) - Q_{34} - K_{3x}x_3 = m_3 \ddot{x}_3 \\ Q_{34} = m_4 \ddot{x}_4 \end{cases} \tag{2}$$

Q_{12} refers to the shear force between m_1 and m_2; Q_{34} refers to the shear force between m_3 and m_4. K_{2x} refers to the compression stiffness between m_2 and left boundary; K_{3x} refers to the compression stiffness between m_3 and right boundary; K_{23} refers to the compression stiffness between m_2 and m_3. For homogeneous elastic structure, the compression stiffness can be calculated according to $K = EA/l$, then the compression stiffness of horizontal vibration model is obtained as $K_{2x} = K_{3x} = 2Eh/l$, $K_{23} = Eh/l$. l refers to the length of a single unit, h refers to the depth of bedrock considered, and E refers to the elastic modulus of bedrock. Let $K_x = Eh/l$, then $K_{2x} = K_{3x} = 2K_x$, $K_{23} = K_x$. According to the kinematic relationship between units and surrounding rocks, $x_1 = x_2$ and $x_3 = x_4$ can be drawn. After simplification of Equation (2) according to the above formula, Equation (3) is derived.

$$\begin{cases} F(t) - 2K_x x_1 - K_x(x_1 - x_4) = (m + M)\ddot{x}_1 \\ K_x(x_1 - x_4) - 2K_x x_4 = (m + M)\ddot{x}_4 \end{cases} \tag{3}$$

Assuming a simple harmonic load $F(t) = A\sin(\omega t)$, then the expressions of x_1 and x_4 should also be in the simple harmonics form. Let $x_4 = P\sin(\omega t)$, substitute it into the Equation (3). Equation (4) can be derived.

$$\frac{x_4}{x_1} = \frac{K_x}{3K_x - (m + M)\omega^2} \tag{4}$$

The ratio is defined as the vibration transmission ratio of horizontal vibration from x_1 to x_4, to describe the influence on m_4 caused by vibration of m_1 in X direction.

Next, the vertical vibration model is established to study the shear vibration, as shown in Figure 16.

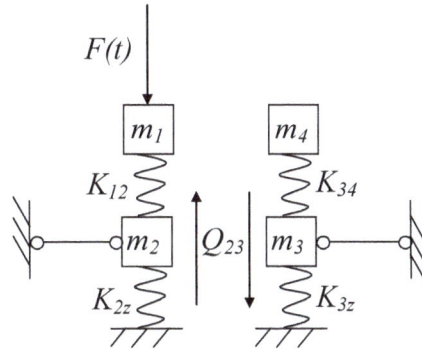

Figure 16. Vertical vibration model.

Similar simplified method is used for the vertical vibration model. The masses of bedrocks are concentrated at their centroids, and regarded as lumped masses m_2 and m_3. They are connected to the bottom boundary and above masses m_1 and m_4 by springs. Interaction between bedrocks and units are represented by shear force Q_{23}. Load $F(t)$ is applied on m_1. Considering dynamic loads only, equations of motion for the four lumped masses are listed in Equation (5).

$$\begin{cases} F(t) - K_{12}(z_1 - z_2) = m_1 \ddot{z}_1 \\ K_{12}(z_1 - z_2) - K_{2z}z_2 - Q_{23} = m_2 \ddot{z}_2 \\ Q_{23} - K_{34}(z_3 - z_4) - K_{3z}z_3 = m_3 \ddot{z}_3 \\ K_{34}(z_3 - z_4) = m_4 \ddot{z}_4 \end{cases} \tag{5}$$

K_{12} refers to the compression stiffness between m_1 and m_2; K_{2z} refers to the compression stiffness between m_2 and the bottom boundary; K_{34} refers to the compression stiffness between m_3 and m_4; K_{3z} refers to the compression stiffness between m_3 and the bottom boundary. According to $K = EA/l$, then the compression stiffness of vertical vibration model is obtained as $K_{12} = K_{2z} = K_{34} = K_{3z} = 2EI/h$. Let $K_z = EI/h$, then $K_{12} = K_{2z} = K_{34} = K_{3z} = 2K_z$. In addition, shear force should be calculated as $Q = K'GA(\partial z/\partial x)$ based on mechanics of materials. For this model, $Q_{23} = K'Gh(z_2 - z_3)/l = G_z(z_2 - z_3)$. Let $G_z = K'Gh/l$, then $Q_{23} = G_z(z_2 - z_3)$. K' refers to the section shape coefficient and G refers to the shear modulus of bedrock. According to actual condition, the units and the bedrocks are always in contact. The relationship between z_1 and z_2 can be derived as $z_1 = 2z_2$, as well as $z_4 = 2z_3$. After simplification of Equation (5), Equation (6) is derived.

$$\begin{cases} F(t) - K_z z_1 - \frac{G_z}{2}(z_1 - z_4) = (m + \frac{1}{2}M)\ddot{z}_1 \\ \frac{G_z}{2}(z_1 - z_4) - K_z z_4 = (m + \frac{1}{2}M)\ddot{z}_2 \end{cases} \tag{6}$$

Then, the ratio is obtained as Equation (7).

$$\frac{z_4}{z_1} = \frac{G_z}{(2K_z + G_z) - (2m + M)\omega^2} \tag{7}$$

The ratio is defined as the vibration transmission ratio of vertical vibration from z_1 to z_4, to describe the influence on m_4 that is caused by the vibration of m_1 in Z direction.

3.3. Rules of Vibration Transmission Ratios

Based on the vibration transmission models above, the transmission ratios of two units in horizontal vibration and vertical vibration are obtained, respectively, as shown in Equation (3) and Equation (6). In the equations, parameters included unit mass m, bedrock mass M, horizontal compression stiffness of bedrock K_x, vertical compression stiffness of bedrock K_z, shear stiffness G_z, and frequency of vibration source load ω. Vibration transmission ratios among units are determined by these parameters in the simplified model.

In order to quantitatively analyze the vibration transmission of horizontal and vertical vibration and study the influence of the vibration frequency on transmission ratio, the parameters are further simplified and calculated. According to previous research experience [25], the depth of the bedrock h is taken as unit length l, then $K_x = K_z = E$ is derived. For the rectangular section, the section shape coefficient K' is 1.2, then $G_z = 1.2G = 1.2E/[2(1 + \mu)]$. After the simplification, Equation (4) and Equation (7) are simplified as two expressions of elastic modulus E, Poisson ratio μ, mass m, and M, and frequency ω. As shown in Equations (8) and (9).

$$\frac{x_4}{x_1} = \frac{E}{3E - (m + M)\omega^2} \tag{8}$$

$$\frac{z_4}{z_1} = \frac{3E}{5(1 + \mu)[(2E + \frac{3E}{5(1+\mu)}) - (2m + M)\omega^2]} \tag{9}$$

For material of bedrock, the Poisson ratio μ is mostly between 0.23–0.27; and, the elastic modulus changes in the range of 20–30 GPa. The masses of unit and bedrock depend on the size of the powerhouse unit and are calculated to be on the order of 10^6 to 10^7 kg for large hydroelectric unit. According to the previous research and load characteristics of the powerhouse, low frequency tail fluctuation and rotation of hydraulic generator are the main vibration sources of powerhouse structural vibration. Their frequencies are within 0–5 Hz, especially in the case of severe vibration. Substituting above data into Equations (8) and (9), it can be found that both $(m + M)\omega^2$ and $(2m + M)\omega^2$ are 1 to 2 orders of magnitude smaller than $3K_x$ and $(2K_z + G_z)$ for low frequency loads. Consequently, the transmission ratio of horizontal vibration in Equation (8) can be approximated, as Equation (10).

$$\frac{x_4}{x_1} \approx \frac{E}{3E} = 0.33 \tag{10}$$

The transmission ratio of vertical vibration in Equation (9) can be approximated as Equation (11) ($\mu = 0.25$).

$$\frac{z_4}{z_1} = \frac{3E}{5(1 + \mu)[(2E + \frac{3E}{5(1+\mu)})]} = 0.19 \tag{11}$$

When compared with field tests, vibration energy of all six units comes from unit #1, as shown in Figure 17. Assuming the same vibration transmission ratio between adjacent units, all the ratios between adjacent units are q. It is calculated that $q_x = 0.25$ for horizontal vibration, and $q_z = 0.16$ for vertical vibration. It can be conclude that, for the vibration transmission among units, the vibration transmission ratio of lateral-river vibration is significantly larger than that of longitude-river vibration and vertical vibration. This is in coincidence with the results obtained from the field tests in Figure 10.

In the theoretical analysis, it is found that the influence of frequency of vibration source ω is negligible as compared with other parameters. This explains why the vibration transmission ratios of the vibration caused by low frequency tail fluctuation and rotation of hydraulic generator in the field tests are basically equal.

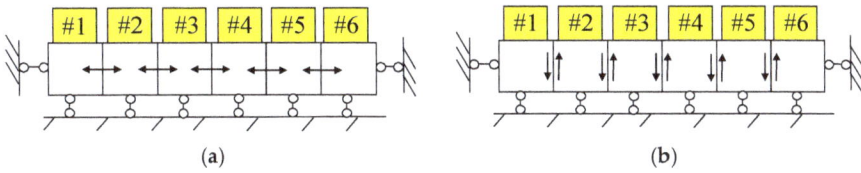

Figure 17. Diagram of simplified models of six units: (**a**) Horizontal vibration; and, (**b**) Vertical vibration.

4. Numerical Simulation

4.1. Establishment of Finite Element Model

The field structure vibration tests are limited by the number and location of sensors, the tests data is rare. In order to further demonstrate the vibration transmission among units, finite element simulations are conducted by commercial software ABAQUS.

A model of single unit is established based on the specific dimensions of the powerhouse structure, as shown in Figure 18. Structures in powerhouse, such as generator pier, floors, beams, and pillars are simulated exactly. The mechanical part is simulated as lumped masses. The bedrock is simulated based on previous research experience and trial calculation. The depth of bedrock is equal to unit length [25]. Material properties are assigned actual values. The units are arranged in an array on the bedrock, and the adjacent units are separated by split seams of 0.2 m width. Normal constraints are applied to the bedrock as boundary conditions to simulate the interactions of rocks.

According to the previous study, the numerical studies are focused on the two parts: (1) The transmission rules among units of vibration in three directions; and, (2) The transmission rules among units caused by low frequency tail fluctuation and rotation of hydraulic generator, respectively.

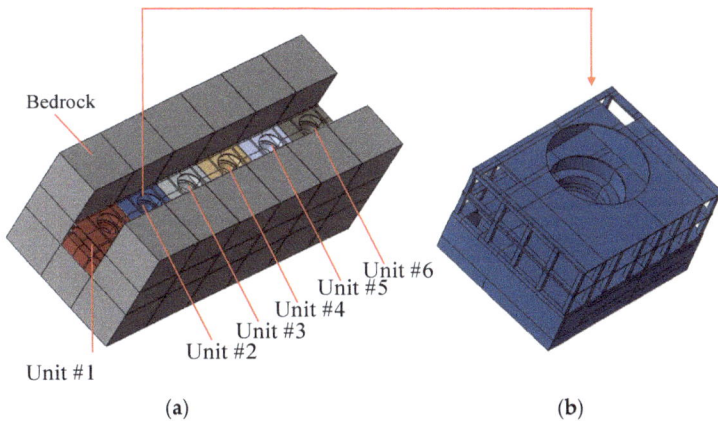

Figure 18. Finite element model: (**a**) Model of whole powerhouse; (**b**) Model of single unit.

4.2. Results of Numerical Simulation

4.2.1. Transmission Rules among Units of Vibration in Three Directions

A harmonic body force is applied to unit #1 of the model as vibration source. According to the characteristics of the vibration signal of field tests, the expression of body force is constructed as Equation (12).

$$F = A \cdot (a \cdot \sin \omega_1 t + b \cdot \sin \omega_2 t) \tag{12}$$

A refers to amplitude of load. ω_1 refers to the frequency of low frequency tail fluctuation with a value between 0.167 and 0.6 times rotational frequency, according to the previous research results and experience [9,31]; it is set as 1 Hz based on the frequency spectrum analysis of field tests in this paper. ω_2 refers to the rotational frequency, set as 2.4 Hz. a and b represent the proportion of two vibration sources, and set as 0.8 and 0.2, respectively, according to the analysis of field tests. Time history of load is shown in Figure 19.

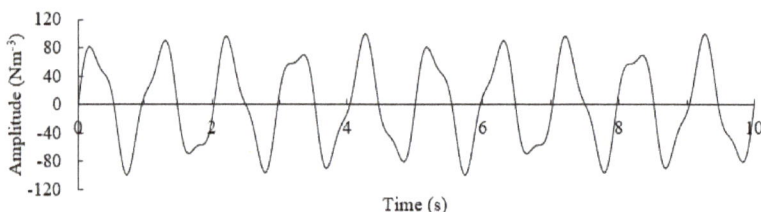

Figure 19. Time history of load.

Load shown as Figure 19 is applied to unit #1 in three directions, respectively. Vibration displacements of the nodes corresponding to the location of sensors are extracted after analysis, RMS values are calculated. For intuitive expression, the RMS of vibration displacements of unit #1 is taken as a reference value to normalize the vibration displacements of different units. These ratios are considered as the vibration transmission ratios, as shown in the Table 2.

It can be seen in Table 2 that the transmission ratios of vibration in X direction are the most significant. The RMS of adjacent unit #2 in X direction reaches 22.55% of that of unit #1. While the RMS of unit #2 in Y and Z directions are only 12.63% and 10.11% of that of unit #1. The comparison is depicted in Figure 20.

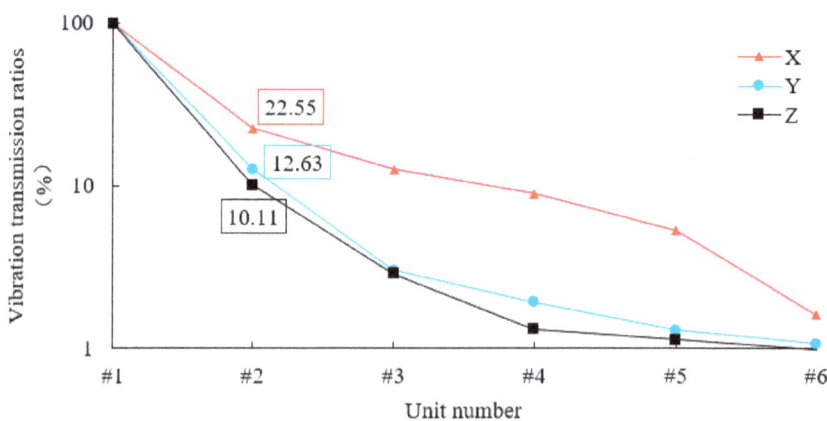

Figure 20. Transmission ratios of vibration in three directions.

Table 2. Ratios of vibration displacements of different units in three directions (%).

Direction	#1	#2	#3	#4	#5	#6
X	100	22.55	12.69	8.89	5.33	1.62
Y	100	12.63	3.02	1.93	1.31	1.08
Z	100	10.11	2.89	1.32	1.13	0.99

\# is used for describing the unit number.

4.2.2. Transmission Rules under Different Vibration Sources

According to the results of field test, two harmonic body forces in the X direction with different frequencies are applied to unit #1 in the model as vibration sources, respectively. The force frequencies ω_1 and ω_2 are set as 1 Hz and 2.4 Hz, which are the typical frequency of low frequency tail fluctuation and rotation of hydraulic generator. After calculation, the vibration displacements of the nodes corresponding to the location of sensors are extracted. The RMS of vibration displacements of unit #1 are taken as reference values. Results of normalized displacements are shown in Table 3.

It can be obtained obviously that the vibration transmission ratios under loads with two low frequencies among units are basically identical. Therefore, the vibration transmission ratios of the vibration caused by low frequency tail fluctuation and rotation of hydraulic generator, respectively, are basically equal. The mechanism that is obtained from simplified model is verified.

Table 3. Ratios of vibration displacements under different frequencies loads (%).

Frequency (Hz)	#1	#2	#3	#4	#5	#6
1	100	22.55	12.69	8.89	5.33	1.62
2.4	100	22.57	12.69	8.90	5.33	1.61

is used for describing the unit number.

5. Discussion

As is well known, the vibration of the unit will influence other adjacent units in the powerhouse. Some researchers have studied the degree of influence, and draw some conclusions [24,25]. But, there are very few studies that are more in-depth. Two influencing factors were raised in this paper: (1) Vibration direction; and, (2) Vibration frequency. The related studies were conducted, which have never been published.

Based on the first factor, all the transmission ratios between two adjacent units derived from simplified model, field test and numerical simulation are shown and compared in Table 4. The results from the numerical simulation basically match with the field test results, indicating that the numerical simulation is capable of simulating this problem.

Table 4. Vibration transmission ratios between two adjacent units derived from three methods (%).

Direction	Simplified Model	Field Test	Numerical Simulation
X	25	17.69	22.55
Y	16	10.69	12.63
Z	16	10.74	10.11

It is obvious that the transmission ratio of vibration in X direction is larger than that of vibration in Y and Z directions. All of the results derived from three methods proved this conclusion. The transmission ratios obtained from simplified model are slightly larger than those by field test and numerical simulation. This is mainly due to that the features such as damping and rock integrity are not taken into consideration in the simplified model.

As for the second factor, the formula of vibration transmission ratio derived from the simplified model directly explains that the effect of frequency is very small, especially for the load with low frequency. In the other hand, according to previous studies [17,18,24], the first order of the modal frequency of most powerhouse is 20–25 Hz. So, both the frequencies of low frequency tail fluctuation and rotation of hydraulic generator are far from the dangerous frequency. Formula for the power amplification factor is shown, as Equation (13).

$$\beta = \frac{1}{1 - \frac{\omega^2}{\theta^2}} \tag{13}$$

The power amplification factors β are 1.002 and 1.009, respectively, which are basically equivalent. Therefore, the vibrations transmission ratios of the two main loads with low frequencies are basically equal, the difference can be ignored.

6. Conclusions

This paper studies the vibration transmission among units in underground powerhouse of a hydropower station through field tests, theoretical analysis, and finite element simulation.

Firstly, the field structural vibration tests in the underground powerhouse of a large-scale hydropower station were designed and conducted, and two preliminary conclusions were raised.

Secondly, based on structural dynamics, the simplified mechanical vibration models were established for the vibration transmission problem among units. The vibration transmission mechanism is elaborated to explain and prove the preliminary conclusions from the theoretical perspective. The previous conclusions can be derived and explained in the model, indicating that the model and assumptions are reasonable.

At last, a complementary FE model for the tested underground powerhouse is established to replicate the tested underground powerhouse. The numerical simulation results verify the previous results.

Through the above work, the following two conclusions of the vibration transmission among units are obtained.

(a) Vibration transmission ratio of lateral-river vibration is significantly larger than those of longitude-river vibration and vertical vibration. The transmission ratio between adjacent units of lateral-river vibration is about 15–25%, while those of longitude-river vibration and vertical vibration are about 10–15%.

(b) Low frequency tail fluctuation and the rotation of hydraulic generator are the main vibration sources of powerhouse structural vibration. Vibration transmission ratios of the vibration caused by the two sources are basically equal.

In general, the vibration transmission among units is widespread exists in underground powerhouse of the hydropower station. It is difficult to completely limit the transmission. However, the research results put forward higher requirements for the monitoring of the structural safety of powerhouse. More attention should be paid to the mutual influence of vibration between units while vibration monitoring. This study has guiding significance for the safe operation of underground powerhouse.

Author Contributions: J.L. and H.W. (Hongzhen Wang) put forward research ideas; H.W. (Haijun Wang) and H.W. (Hongzhen Wang) designed and performed the field tests; H.W. (Hongzhen Wang) built and wrote the majority of the manuscript text; all authors reviewed the manuscript.

Funding: This research was funded by the project of National Key Research and Development Program of China (Grant No. 2016YFC0401905).

Acknowledgments: This work was supported by the project of National Key Research and Development Program of China (Grant No. 2016YFC0401905). All workers from the State Key Laboratory of Hydraulic Engineering Simulation and Safety of Tianjin University are acknowledged. The authors are also grateful for the assistance of the anonymous reviewers.

Conflicts of Interest: The authors declare no conflict of interest.

Abbreviations

A	Amplitude of load
E	Elastic modulus of bedrock
$F(t)$	Load applied on the unit
G	Shear modulus of bedrock
h	Depth of bedrock considered
K'	Section shape coefficient
K_{2x}	Compression stiffness between m_2 and left boundary
K_{3x}	Compression stiffness between m_3 and right boundary
K_{23}	Compression stiffness between m_2 and m_3
K_{12}	Compression stiffness between m_1 and m_2
K_{2z}	Compression stiffness between m_2 and the bottom boundary
K_{34}	Compression stiffness between m_3 and m_4
K_{3z}	Compression stiffness between m_3 and the bottom boundary
l	Length of a single unit
m	Mass of the unit
m_1	Lumped Mass of the unit #1
m_2	Lumped Mass of the bedrock under the unit #1
m_3	Lumped Mass of the bedrock under the unit #2
m_4	Lumped Mass of the unit #2
M	Mass of the bedrock
Q_{12}	Shear force between m_1 and m_2
Q_{34}	Shear force between m_3 and m_4
x_1, x_2, x_3, x_4	Vibration displacement of m_1, m_2, m_3, m_4 in X direction
$\ddot{x}_1, \ddot{x}_2, \ddot{x}_3, \ddot{x}_4$	Vibration acceleration of m_1, m_2, m_3, m_4 in X direction
z_1, z_2, z_3, z_4	Vibration displacement of m_1, m_2, m_3, m_4 in Z direction
$\ddot{z}_1, \ddot{z}_2, \ddot{z}_3, \ddot{z}_4$	Vibration acceleration of m_1, m_2, m_3, m_4 in Z direction
β	Power amplification factor
θ	Frequency of the modal frequency
μ	Poisson ratio
ω	Frequency of vibration source load
FE	Finite element
MRA	Multi-Resolution Analysis
PSD	Power spectral density
RMS	Root mean square

References

1. Xu, X.P.; Han, Q.K.; Chu, F.L. Review of electromagnetic vibration in electrical machines. *Energies* **2018**, *11*, 1779. [CrossRef]
2. Mollasalehi, E.; Wood, D.; Sun, Q. Indicative fault diagnosis of wind turbine generator bearings using tower sound and vibration. *Energies* **2017**, *10*, 1853. [CrossRef]
3. Cachafeiroa, H.; Arevaloa, L.F.; Vinuesaa, R.; Goikoetxeab, J.; Barrigab, J. Impact of solar selective coating ageing on energy cost. *Energy Procedia* **2015**, *69*, 299–309. [CrossRef]
4. Liu, X.; Luo, Y.Y.; Wang, Z.W. A review on fatigue damage mechanism in hydro turbines. *Renew. Sustain. Energy Rev.* **2016**, *54*, 1–14. [CrossRef]
5. Shen, K.; Zhang, Z.Q.; Liang, Z. Hydraulic Vibration Calculation of Yantan Hydropower House. *Water Resour. Power* **2003**, *1*, 73–75. [CrossRef]
6. Kurzin, V.B.; Seleznev, V.S. Mechanism of emergence of intense vibrations of turbines on the Sayano-Shushensk hydro power plant. *J. Appl. Mech. Tech. Phys.* **2010**, *4*, 590–597. [CrossRef]
7. Yang, J.D.; Zhao, K.; Li, L.; Wu, P. Analysis on the causes of units 7 and 9 accidents at Sayano-Shushenskaya hydropower station. *J. Hydroelectr. Eng.* **2011**, *4*, 226–234. (In Chinese)

8.   Dorji, U.; Ghomashchi, R. Hydro turbine failure mechanisms: An overview. *Eng. Fail. Anal.* **2014**, *44*, 136–147. [CrossRef]

9.   Mohanta, R.K.; Chelliah, T.R.; Allamsetty, S.; Akula, A.; Ghosh, R. Sources of vibration and their treatment in hydro power stations—A review. *Eng. Sci. Technol. Int. J.* **2017**, *20*, 637–648. [CrossRef]

10.   Zhi, B.P.; Ma, Z.Y. Disturbance analysis of hydropower station vertical vibration dynamic characteristics: The effect of dual disturbances. *Struct. Eng. Mech.* **2015**, *2*, 297–309. [CrossRef]

11.   Ma, Z.Y.; Dong, Y.X. Dynamic response of hydroelectric set by hydraulic lateral force on turbine runner. *J. Hydroelectr. Eng.* **1990**, *2*, 31–39. (In Chinese)

12.   Song, Z.Q. Research on Coupling Vibration Characteristics of Generator Set and Hydropower House. Ph.D. Thesis, Dalian University of Technology, Dalian, China, 2009. (In Chinese)

13.   Zhi, B.P. Study of Vibration Transmission Path about Hydropower Station Units and Powerhouse with Complex Disturbance. Ph.D. Thesis, Dalian University of Technology, Dalian, China, 2014. (In Chinese)

14.   Zhou, J.Z.; Peng, X.L.; Li, R.H.; Xu, Y.H.; Liu, H.; Chen, D.Y. Experimental and finite element analysis to investigate the vibration of Oblique-Stud stator frame in a large hydropower generator unit. *Energies* **2017**, *10*, 2175. [CrossRef]

15.   Zhang, C.H.; Zhang, Y.L. Nonlinear dynamic analysis of the Three Gorge Project powerhouse excited by pressure fluctuation. *J. Zhejiang Univ. Sci. A* **2009**, *9*, 1231–1240. [CrossRef]

16.   Wang, X.; Li, T.C.; Zhao, L.H. Vibration analysis of large bulb tubular pump house under pressure pulsations. *Water Sci. Eng.* **2009**, *1*, 86–94. [CrossRef]

17.   Lian, J.J.; Qin, L.; He, C.L. Structure vibration of hydropower house based on prototype observation. *J. Tianjin Univ.* **2006**, *2*, 176–180. (In Chinese)

18.   Lian, J.J.; Qin, L.; Wang, R.X.; Hu, Z.G.; Wang, H.J. Study on the dynamic characteristics of the power house structure of two-row placed units. *J. Hydroelectr. Eng.* **2004**, *2*, 55–60. (In Chinese)

19.   Lian, J.J.; Zhang, Y.; Liu, F.; Yu, X.H. Vibration source characteristics of a roof overflow hydropower station. *J. Vib. Shock* **2013**, *18*, 8–14. [CrossRef]

20.   He, L.J.; Lian, J.J.; Ma, B. Intelligent damage identification method for large structures based on strain modal parameters. *J. Vib. Control* **2014**, *12*, 1783–1795. [CrossRef]

21.   He, L.J. Study on Coupled Vibration Characteristics and Response Prediction of Underground Powerhouse. Ph.D. Thesis, Tianjin University, Tianjin, China, 2010. (In Chinese)

22.   Zhang, Y. Vibration Characteristics of the Overflow Powerhouse with Bulb Tubular Unit. Ph.D. Thesis, Tianjin University, Tianjin, China, 2012. (In Chinese)

23.   Mao, L.D.; Wang, H.J. Research on load feedback of structure vibration of underground house of hydropower station. *J. Water Resour. Water Eng.* **2014**, *3*, 79–82. [CrossRef]

24.   Wang, H.J.; Bai, B.; Li, K. Research on vibration propagation regular of adjacent unit-blocks for hydropower house. *J. Water Resour. Water Eng.* **2016**, *1*, 141–146. [CrossRef]

25.   Wei, Y.B.; Chen, J.; Ma, Z.Y. Vibrational travel and behavior of pumped storage power station underground powerhouse. *J. Water Resour. Arch. Eng.* **2017**, *4*, 101–106. [CrossRef]

26.   Ameen, M.S.A.; Ibrahim, Z.; Othman, F.; Al-Ansari, N.; Yaseen, Z.M. Minimizing the principle stresses of powerhoused Rock-Fill dams using control turbine running units: Application of Finite Element Method. *Water-SUI* **2018**, *10*, 1138. [CrossRef]

27.   Gupta, S.; Stanus, Y.; Lombaert, G.; Degrande, G. Influence of tunnel and soil parameters on vibrations from underground railways. *J. Sound Vib.* **2009**, *327*, 70–91. [CrossRef]

28.   Chen, M.; Lu, W.B.; Yi, C.P. Blasting vibration criterion for a rock-anchored beam in an underground powerhouse. *Tunn. Undergr. Space Technol.* **2007**, *22*, 69–79. [CrossRef]

29.   Xia, X.; Li, H.B.; Li, J.C.; Liu, B.; Yu, C. A case study on rock damage prediction and control method for underground tunnels subjected to adjacent excavation blasting. *Tunn. Undergr. Space Technol.* **2013**, *35*, 1–7. [CrossRef]

30.   Kuo, K.A.; Hunt, H.E.M.; Hussein, M.F.M. The effect of a twin tunnel on the propagation of ground-borne vibration from an underground railway. *J. Sound Vib.* **2011**, *330*, 6203–6222. [CrossRef]

31.   Dörfler, P.; Sick, M.; Coutu, A. *Flow-Induced Pulsation and Vibration in Hydroelectric Machinery*, 1st ed.; Springer: London, UK, 2013; pp. 31–60. ISBN 978-7-5684-0070-1.

32.   Samanta, A.; Vinuesa, R.; Lashgari, I.; Schlatter, P.; Brandt, L. Enhanced secondary motion of the turbulent flow through a porous square duct. *J. Fluid Mech.* **2015**, *784*, 681–693. [CrossRef]

33. Vinuesa, R.; Bartrons, E.; Chiu, D.; Dressler, K.M.; Rüedi, J.D.; Suzuki, Y.; Nagib, H.M. New insight into flow development and two dimensionality of turbulent channel flows. *Exp. Fluids* **2017**, *55*, 1759. [CrossRef]
34. Vinuesa, R.; Schlatter, P.; Nagib, H.M. Role of data uncertainties in identifying the logarithmic region of turbulent boundary layers. *Exp. Fluids* **2017**, *55*, 1751. [CrossRef]
35. Wang, H.J. Research on Composite Structural Analysis and Dynamic Identification of Hydropower House. Ph.D. Thesis, Tianjin University, Tianjin, China, 2005. (In Chinese)

Article

energies

MDPI

A Non-Probabilistic Solution for Uncertainty and Sensitivity Analysis on Techno-Economic Assessments of Biodiesel Production with Interval Uncertainties

Zhang-Chun Tang *, Yanjun Xia, Qi Xue and Jie Liu

School of Mechatronics Engineering, University of Electronic Science and Technology of China,
611731 Chengdu, China; yzxia.bbdd.cc@163.com (Y.X.); xueqi_qi@sina.com (Q.X.); liujie_cult@163.com (J.L.)
* Correspondence: tangzhangchun@uestc.edu.cn; Tel.: +86-028-6183-1750

Received: 19 January 2018; Accepted: 1 March 2018; Published: 8 March 2018

Abstract: Techno-economic assessments (TEA) of biodiesel production may comply with various economic and technical uncertainties during the lifespan of the project, resulting in the variation of many parameters associated with biodiesel production, including price of biodiesel, feedstock price, and rate of interest. Engineers may only collect very limited information on these uncertain parameters such as their variation intervals with lower and upper bound. This paper proposes a novel non-probabilistic strategy for uncertainty analysis (UA) in the TEA of biodiesel production with interval parameters, and non-probabilistic reliability index (NPRI) is employed to measure the economically feasible extent of biodiesel production. A sensitivity analysis (SA) indicator is proposed to assess the sensitivity of NPRI with regard to an individual uncertain interval parameter. The optimization method is utilized to solve NPRI and SA. Results show that NPRI in the focused biodiesel production of interest is 0.1211, and price of biodiesel, price of feedstock, and cost of operating can considerably affect TEA of biodiesel production.

Keywords: reliability; non-probabilistic reliability index; sensitivity analysis; techno-economic assessments; life cycle cost

1. Introduction

The global climate, ecological environment, and air quality have been considerably affected by various deleterious emissions and harmful substances including NO_x, SO_x, CO_2, hydrocarbons, carbon monoxide, and particulate matter, resulting in various environmental pollution problems and danger on human health [1–9]. A great number of scientists are investigating other harmless, economic, and clean energy sources for the sake of the reduction of these adverse and negative effects. Being a valuable renewable energy resource, biodiesel is friendly to the natural environment and human health, compared to the traditional fossil fuels [10–15]. Various feedstocks-derived biodiesel production have been reported, for example, palm oil [16], waste cooking oil [17–19], vegetable oils [20,21], soybean oil [22–25], *Jatropha curcas* L. [26], algae [27,28], microalgae [28–30], Oleaginous yeast [31,32], lignocellulosic biomass [33], used frying oil [34], waste cottonseed oil with heterogeneous catalyst [35,36], *Annona squamosa* L. seed oil with heterogeneous catalyst [36,37], Butanol and pentanol [38], etc., and recent advances in biofeedstocks and biofuels have also been reviewed in [39].

Various uncertain factors existing in the biodiesel industry, such as fluctuation in interest rate, may cause instability in biodiesel production, and then may decrease the economical feasibility relevant to biodiesel production [40]. Numerous research works have investigated the techno-economic assessments (TEA) of biodiesel production to ensure the economical feasibility of biodiesel production [41–51], such as TEA for vegetable oil biodiesel production [41], TEA for palm biodiesel production [42], TEA for algal biofuel production [27,43–45], TEA for microalgae biofuel production [46,47], TEA for waste-to-biofuel production [48], TEA for sugarcane biorefineries [49], TEA for lignocellulosic biomass production [51], etc., and recent advances in TEA for biofuel production have also been summarized in [27,39,51]. These works have extensively improved the development in the TEA of biodiesel production on condition that all of the parameters relevant to the TEA are regarded as constant during the project's lifespan. However, real engineering inevitably confronts various uncertain parameters resulting from numerous economic and technical uncertainties when performing TEA of biodiesel production [40], such as variation in the feedstock price [52], fluctuation of biodiesel price [52], and change in the rate of interest [53], and thus it may be more rational to treat these parameters as uncertain parameters. Recently, several works have studied the TEA of biodiesel production subject to many economic and technical uncertainties defined by random variables with probability density functions (PDFs) [54–65], including TEA for algae-derived biodiesel with uncertainties [55,56], TEA for biodiesel production with uncertainties using structural reliability principles [57], TEA for palm biodiesel production with uncertainties [58,59], TEA for inedible Jatropha oil biodiesel production with uncertainties [60], TEA for waste oil biodiesel production with uncertainties [61], probabilistic TEA for microalgae biofuel production [62], TEA for bioethanol production with uncertainties [63], stochastic TEA for alcohol-to-jet fuel production [64], TEA for high-value propylene glycol production with uncertainties [65], etc. These research works have discovered that the uncertainties related to the random parameters have distinct effects on the TEA. The authors also studied the TEA for palm biodiesel production with uncertainties, which were assumed as random variables with uniform distributions, indicating that uncertain parameters are uniformly distributed within variation intervals [58,59].

The previous studies [54–65] consider the effect of the uncertainties, and they consider the uncertainties as random variables following PDFs. Treating uncertainties as random variables may not be reasonable due to the fact that determining the precise PDFs requires a large number of data, but the data in practical engineering is usually limited due to lack of sufficient samples for TEA. This paper will propose a more rational solution for the TEA of palm biodiesel production based on the previous research works of the authors of [58,59]. In previous studies [58,59], the authors only collected very limited data on these uncertain parameters, and only the variation ranges or the lower bounds and upper bounds for uncertain parameters were determined. All of these uncertain parameters were assumed as random variables defined by uniform distributions, distributed uniformly within their variation ranges, and the TEA was done based on this assumption. The estimated results of the TEA may depend on the selected distributions for these uncertain parameters within their variation intervals, and different selection of distributions may lead to completely different estimated results. In order to overcome this difficulty, we will propose a more rational strategy for the TEA of a palm biodiesel production with interval parameters, in which only the lower limit and the upper limit of the parameters are available, being free from the selection of the distributions for uncertain parameters.

The rest of this paper is organized as follows. In Section 2, we propose a novel strategy for the evaluation of the TEA and sensitivity analysis (SA) for palm biodiesel production subject to interval uncertainties, specifically, non-probabilistic reliability index (NPRI) that measures the economically feasible extent of the biodiesel production and the effect of an interval parameter on NPRI. In Section 3, we evaluate the NPRI and SA associated with the TEA of the palm biodiesel production. In Section 4, we summarize our results and make some conclusions.

2. Materials and Methods

In this section, we first introduce several important indicators in the TEA of palm biodiesel production including net present value (NPV), payback period (PP), and total profit for this project, and then some important interval parameters related to the TEA are provided, which are determined by some collected data from some available references. Secondly, we introduce a novel indicator named as non-probabilistic reliability index to rationally measure the economically feasible degree of biodiesel production with interval uncertainties. Then, nonlinear optimization algorithm is employed to solve the NPRI in the TEA of palm biodiesel production. Finally, we develop a new sensitivity analysis (SA) indicator of NPRI with regard to an uncertain parameter, which can measure the effect of a parameter on NPRI and identify important parameters on the TEA.

2.1. Several Important Concepts in the TEA for a Biodiesel Production

We focus on the TEA for palm biodiesel production originally proposed in [42], in which economic and technical uncertainties are not considered. The mathematical formulations of total profit, payback period, and net present value for this problem are defined as [42,58,59]:

$$
\begin{aligned}
\text{TotalProfit} &= -\text{LCC} + (\text{TBS} - \text{TAX}) \times n = -\text{LCC} + (\text{TBS} - \text{TAX}) \times 20 \\
&= (\text{TBS}_i - \text{TAX}_i) \times 20 - \text{LCC}
\end{aligned}
\tag{1}
$$

$$
\text{PP} = \frac{\text{CC}}{(\text{TotalProfit}/n)} = \frac{n \times \text{CC}}{\text{TotalProfit}}
\tag{2}
$$

$$
\begin{aligned}
\text{NPV} &= \sum_{i=1}^{n} \frac{(\text{TBS}_i - \text{TAX}_i)}{(1+r)^i} - \text{LCC} \\
&= -\text{LCC} + \sum_{i=1}^{n} \frac{(\text{TBS} - \text{TAX})}{(1+r)^i} = -\text{LCC} + \sum_{i=1}^{n} \frac{\text{TotalProfit} + \text{LCC}}{n(1+r)^i} \\
&= -\text{LCC} + \frac{\text{TotalProfit} + \text{LCC}}{n} \sum_{i=1}^{n} \frac{1}{(1+r)^i}.
\end{aligned}
\tag{3}
$$

with

$$
\begin{aligned}
\text{LCC} &= \text{CC} + \text{MC} + \text{FC} + \text{OC} - \text{BPC} - \text{SV} \\
&= \text{CC} + \sum_{i=1}^{n} \frac{\text{FC}_i + \text{OC}_i + \text{MC}_i}{(1+r)^i} - \frac{\text{SV}}{(1+r)^n} - \sum_{i=1}^{n} \frac{\text{BPC}_i}{(1+r)^i}
\end{aligned}
\tag{4}
$$

where TotalProfit is total profit of the project, PP is payback period, and NPV represents net present value; MC, CC, FC, LCC, OC, BPC, and SV represent maintenance cost, capital cost, feedstock cost, life cycle cost, operating cost, byproduct credit, and salvage value indicating the remaining value of the components and the assets of the plant at the end of the project's lifetime, respectively; TBS is annual total biodiesel sale, TAX is annual total taxation, $n = 22$ years is project's lifetime, and r represents rate of interest which takes values from 4.44% to 13.53% [66], i.e., $r \in [4.44\%, 13.53\%]$; MC_i, FC_i, OC_i, BPC_i, TAX_i and TBS_i are maintenance cost, feedstock cost, operating cost, byproduct credit, total taxation, and total biodiesel sale for the ith year, respectively.

The annual production capacity for this plant is 50 kt, that is, PC = 50 kt, and its capital cost should take values between \$9 million and \$15 million, that is, $\text{CC} \in [\$9 \text{ million}, \$15 \text{ million}]$ [42]. The corresponding FC, OC, MC, SV, BPC, TBS, and TAX are defined by:

$$
\text{FC} = \sum_{i=1}^{n} \text{FC}_i = \sum_{i=1}^{n} \frac{\text{FP} \times \text{FU}}{(1+r)^i} = \sum_{i=1}^{n} \frac{\text{FP} \times \frac{\text{PC} \times 1000}{\text{CE}}}{(1+r)^i}
\tag{5}
$$

$$
\text{OC} = \sum_{i=1}^{n} \text{OC}_i = \sum_{i=1}^{n} \frac{\text{OR} \times \text{PC} \times 1000}{(1+r)^i}
\tag{6}
$$

$$
\text{MC} = \sum_{i=1}^{n} \text{MC}_i = \sum_{i=1}^{n} \frac{\text{MR} \times \text{CC}}{(1+r)^i}
\tag{7}
$$

$$
\text{SV} = \text{RC} \times (1-d)^{n-1} \times \text{PWF}_n = \frac{\text{RC} \times (1-d)^{n-1}}{(1+r)^n}
\tag{8}
$$

$$\text{BPC} = \sum_{i=1}^{n} \text{BPC}_i = \sum_{i=1}^{n} \frac{\text{GP} \times \text{GCF} \times \text{PC} \times 10^6}{(1+r)^i} \tag{9}$$

$$\text{TBS} = \text{PC} \times 10^6 / \rho \times \text{BP} \tag{10}$$

$$\text{TAX} = \text{TBS} \times \text{TR} \tag{11}$$

where FC commonly makes up about 80–90% of life cycle cost [67], and OC generally accounts for not more than 15% of life cycle cost [68]; FP is feedstock price or crude palm oil price, which takes values between $200/t and $1200/t in the past years [42], that is, FP ∈ [$200/t, $1200/t]; FU is annual total feedstock consumption; CE is conversion efficiency from palm oil to biodiesel which commonly takes values between 96% and 99% [69], that is, CE ∈ [96%, 99%]; OR is the operating rate, indicating operating cost of per-ton biodiesel production, which varies from $37.5/t to $225/t evaluated by feedstock price FP ∈ [$200/t, $1200/t] [42] when FC makes up 80% of life cycle cost [67] and OC accounts for 15% of life cycle cost [68], that is, OR ∈ [$37.5/t, $225/t]; MR is maintenance rate, varying from 1% to 2%, i.e., MR ∈ [1%, 2%] [41,42]; d and RC represent depreciation rate and replacement cost respectively, that is, RC = $10 million and d = 5% [42]; GP and GCF represent glycerol price and glycerol conversion factor, that is, GP ∈ [$0.08/kg, $0.2/kg] [70] and GCF = 0.0985 [42]; BP is biodiesel price, that is, BP ∈ [$0.66/L, $1.58/L] [71]; ρ is biodiesel density, i.e., ρ = 0.95 kg/L; and TR = 15% is tax rate for biodiesel sale.

The important quantities involved in the TEA, such as life cycle cost, net present value, payback period, and total profit, unavoidably meet with various economic and technical uncertainties within the project lifespan. Table 1 gives the variation intervals for these uncertain parameters, which are obtained by the collected data from many available research works.

Table 1. Variation intervals of uncertain parameters for biodiesel production.

Uncertain Parameters	Variation Intervals $[\underline{x}_i, \bar{x}_i]$
Capital cost (CC: x_1) [42]	[$9 million, $15 million]
Interest rate (r: x_2) [66]	[4.44%, 13.53%]
Operating rate (OR: x_3) [42,67,68]	[$37.5/t, $225/t]
Feedstock price (FP: x_4) [42]	[$200/t, $1200/t]
Glycerol price (GP: x_5) [70]	[$0.08/kg, $0.2/kg]
Maintenance rate (MR: x_6) [41,42]	[1%, 2%]
Biodiesel conversion efficiency (CE: x_7) [69]	[96%, 99%]
Biodiesel price (BP: x_8) [71]	[$0.66/L, $1.58/L]

2.2. NPRI for Measuring Economically Feasible Extent of Biodiesel Production

In this section, we will first introduce a NPRI, which is commonly employed to measure the reliable level of practical engineering problems subject to interval uncertainties. Then, NPRI is further extended to measure the economically feasible degree in the TEA of biodiesel production.

2.2.1. NPRI for Problems with Interval Parameters

For a system with interval input parameters $x = (x_1, x_2, \dots, x_n)$, the corresponding output y is defined by:

$$y = g(x) \tag{12}$$

where x represents the input parameters with interval uncertainties, and y commonly is the continuous function of the inputs $x = (x_1, x_2, \dots, x_n)$. Obviously, y varies within an interval with a lower bound $\underline{y}$ and an upper bound $\bar{y}$. In general, $\Omega_s = \{x|y = g(x) \geq 0; x = (x_1, x_2, \dots, x_n)\}$ indicates the safe region, and $\Omega_f = \{x|y = g(x) < 0; x = (x_1, x_2, \dots, x_n)\}$ represents the failure region. In addition, $y = g(x) = 0$ is named as limit state function (LSF) or limit state curve (LSC), separating the whole space into two regions, that is, the safe region and failure region. Non-probabilistic reliability index has been employed for measuring the reliable level related to the system with interval parameters as [72–75]:

$$\eta = y^c / y^r \tag{13}$$

with

$$y^c = \left(\overline{y} + \underline{y}\right)/2$$

and

$$y^r = \left(\overline{y} - \underline{y}\right)/2$$

where η is NPRI; $\underline{y}$ and $\overline{y}$ are lower limit and upper limit of the output y. When $\eta \geq 1$ holds, one can have $\underline{y} \geq 0$ holds, indicating that the output is always larger than or equals to zero and thus the system is absolutely safe. Condition $\eta \leq -1$ will lead to $\overline{y} \leq 0$, implying that the system is completely a failure. Accordingly, $-1 < \eta < 1$ corresponds to $\underline{y} < 0 < \overline{y}$, which indicates that a part of the output will lie in the failure space and the system is not reliable. Thus, η can be employed to measure the reliable degree associated with a system with interval uncertainties, and a larger value of η corresponds to a more reliable system and vice versa [72–75]. In general, engineers focus on the situation with $\eta \geq 0$. The following will further discuss the physical significance in the NPRI.

For x_i, we first do the following standard transformation [72–75]:

$$x_i = x_i^c + x_i^r q_i = \frac{(\overline{x}_i - \underline{x}_i)}{2} q_i + \frac{(\overline{x}_i + \underline{x}_i)}{2} \tag{14}$$

where $q_i \in [-1, 1]$ is the normalized interval for x_i. Substituting Equation (14) into Equation (12) can lead to normalized formulation for y as

$$y = g(\mathbf{q}) = g(q_1, q_2, \ldots, q_n) \tag{15}$$

Obviously, the normalized intervals $\mathbf{q}$ of Equation (15) vary in the domain $\Omega_{\mathbf{q}} = \{\mathbf{q} \mid |q_i| \leq 1; i = 1, 2, \ldots, n\}$, which is a hyperbox. Figure 1 illustrates the representative figure of $\Omega_{\mathbf{q}}$ in a two-dimension situation, in which $\Omega_{\mathbf{q}}$ is a square centered at coordinate origin and its side-length is 2, representing the set consisting of all the possible values of the two normalized intervals. When the square box enlarges proportionally in two directions, all the possible values of the two interval variables will locate in the reliable domain until the square box is tangential to normalized LSC $y = g(\mathbf{q}) = 0$. The maximum allowable variability can be defined by the shortest distance between LSC $y = g(\mathbf{q}) = 0$ and the coordinate origin in the normalized space in the form of infinite norm [72–75], which can be employed to measure the reliable extent of the system, i.e., non-probabilistic reliability index. More discussions on non-probabilistic reliability can be found in [76–81].

According to the discussion in Figure 1, another mathematical definition of NPRI η can be provided by [72–75]:

$$\eta = \min(\|\mathbf{q}\|_\infty) \\ \text{S.t. } g(\mathbf{q}) = g(q_1, q_2, \ldots, q_n) = 0 \tag{16}$$

with

$$\|\mathbf{q}\|_\infty = \max(|q_1|, |q_2|, \ldots, |q_n|)$$

where $\min(\bullet)$ is the operation of taking the minimum of the set, $\|\bullet\|_\infty$ represents the operation of infinite norm, $\max(\bullet)$ is the operation of taking the maximum of the set, and $|\bullet|$ denotes the operation of taking the absolute value. If a system has m outputs $y_j = g_j(x) (j = 1, 2, \ldots, m)$ which corresponds to m failure modes, then failure associated with anyone of them will lead to the failure of the whole system. Thus, NPRI η_s for system is provided as:

$$\eta_s = \min\{\eta_1, \eta_2, \ldots, \eta_m\} \tag{17}$$

where $\eta_j (j = 1, 2, \ldots, m)$ is NPRI associated with $y_j = g_j(x)(j = 1, 2, \ldots, m)$.

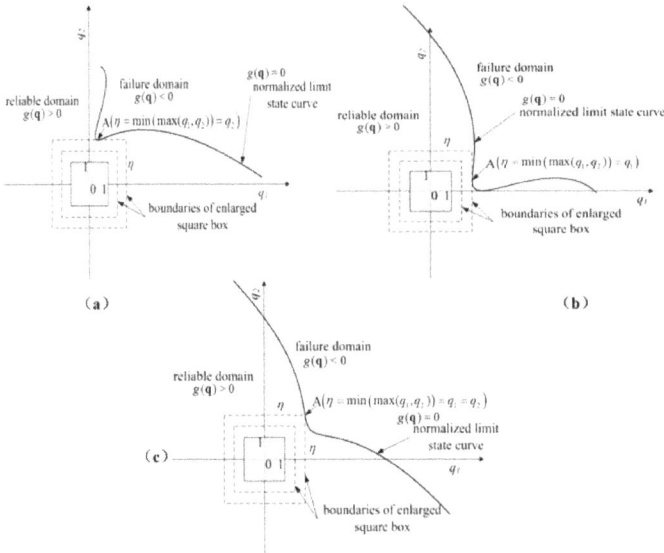

Figure 1. Diagrammatic presentation for non-probabilistic reliability index (NPRI) of a system with two intervals, (**a**) normalized limit state function (LSF) intersects with enlarged square box at one side; (**b**) normalized limit state curve (LSC) intersects with enlarged square box at another side; and (**c**) normalized limit state curve intersects with enlarged square box at cater-corner point.

2.2.2. NPRI for Economically Feasible Degree in the TEA of Biodiesel Production

Total profit defined in Equation (1) is expected to be larger than zero, specifically,

$$\text{TotalProfit} \geq 0. \tag{18}$$

Meanwhile, payback period given in Equation (2) is expected to be less than the allowable upper bound, that is,

$$PP = \frac{CC}{(\text{TotalProfit}/n)} = \frac{n \times CC}{\text{TotalProfit}} \leq PP^u. \tag{19}$$

where PP^u is the permitted upper limit, and here PP^u is one third of project's lifespan, that is, $PP^u = n/3 = 20/3$ years. Then, Equation (19) is transformed into Equation (20):

$$\frac{n \times CC}{\text{TotalProfit}} \leq PP^u \Rightarrow \text{TotalProfit} \geq \frac{n \times CC}{PP^u}. \tag{20}$$

Finally, NPV given by Equation (3) must be larger than zero, specifically,

$$NPV = -LCC + \frac{\text{TotalProfit} + LCC}{n} \sum_{i=1}^{n} \frac{1}{(1+r)^i} \geq 0. \tag{21}$$

Then, Equation (21) can be transformed into Equation (22):

$$\text{TotalProfit} \geq \left(\frac{n}{\sum\limits_{i=1}^{n} \frac{1}{(1+r)^i}} - 1 \right) \times LCC. \tag{22}$$

Thus, Equations (18), (20), and (22) should simultaneously hold to ensure that biodiesel production is economically feasible. For the sake of convenience, Equations (18), (20) and (22) can be written into the following forms:

$$y_1 = g_1(x) = \text{TotalProfit} \tag{23}$$

$$y_2 = g_2(x) = \text{TotalProfit} - \frac{n \times \text{CC}}{\text{PP}^u} \tag{24}$$

$$y_3 = g_3(x) = \text{TotalProfit} - \left(\frac{n}{\sum_{i=1}^{n} \frac{1}{(1+r)^i}} - 1 \right) \times \text{LCC} \tag{25}$$

where $y_1 = g_1(x)$, $y_2 = g_2(x)$, and $y_3 = g_3(x)$ are LSFs, and x represents the vector consisting of interval parameters, as shown in Table 1.

Uncertainties involved in the interval parameters in Table 1 will lead to the variability of the TotalProfit, payback period, and NPV defined in Equations (1)–(3), and then one, two, or all of Equations (23)–(25) may not hold. Any one of the three LSFs in Equations (23)–(25) not being feasible will lead to the result that biodiesel production will not be economically feasible. In other words, biodiesel production is economically feasible if and only if the three LSFs in Equations (23)–(25) simultaneously apply. Thus, according to Equation (17), the following indicator can be employed to measure the economically feasible degree of biodiesel production with interval parameters:

$$\eta_s = \min\{\eta_1, \eta_2, \eta_3\} \tag{26}$$

where η_s represents NPRI for measuring economical feasibility of biodiesel production; $\eta_j (j = 1, 2, 3)$ is NPRI for $y_j = g_j(x)$ given in Equations (23)–(25). The significance relevant to η_s will be discussed in the following.

When $\eta_s \geq 1$ holds, the minimum of η_1, η_2, and η_3 will be larger than or equal to one, then $\underline{y}_1 \geq 0, \underline{y}_2 \geq 0$, and $\underline{y}_3 \geq 0$ in Equations (23)–(25) hold, indicating biodiesel production with interval uncertainties is absolutely feasible in terms of economical feasibility. When $0 < \eta_s < 1$ holds, the minimum of η_1, η_2, and η_3 will be less than one, then $\bar{y}_1 \geq 0, \bar{y}_2 \geq 0$, and $\bar{y}_3 \geq 0$ hold, while at least one of $\underline{y}_1 < 0, \underline{y}_2 < 0$, and $\underline{y}_3 < 0$ holds, implying that biodiesel production with interval uncertainties is partially feasible. When $\eta_s < 0$ holds, the minimum of η_1, η_2, and η_3 will be less than 0, and at least one of $\bar{y}_1 < 0, \bar{y}_2 < 0$, and $\bar{y}_3 < 0$ holds, indicating that biodiesel production with interval uncertainties is completely infeasible. Thus, η_s can be employed to measure the economical feasibility relevant to biodiesel production with interval uncertainties, and a larger value of η_s corresponds to a better economical feasibility of biodiesel production with interval uncertainties and vice versa.

2.3. Evaluation Procedure of the NPRI

According to the definition of NPRI in Equation (13), we need to first evaluate $\underline{y}_j$ and $\bar{y}_j$ for the evaluation of $\eta_j (j = 1, 2, 3)$ for Equations (23)–(25). The following two equations can be utilized to calculate $\underline{y}_j$ and $\bar{y}_j$ as:

$$
\begin{aligned}
\underline{y}_j &= \min_x g_j(x) \\
\text{S.t.} \quad & \underline{x}_i \leq x_i \leq \bar{x}_i \\
& x = (x_1, x_2, \dots, x_8) \\
& j = 1, 2, 3 \\
& i = 1, 2, \dots, 8
\end{aligned}
\tag{27}
$$

and

$$\bar{y}_j = \max_x g_j(x)$$
$$\text{S.t.} \quad \underline{x}_i \le x_i \le \bar{x}_i$$
$$x = (x_1, x_2, \dots, x_8) \tag{28}$$
$$j = 1, 2, 3$$
$$i = 1, 2, \dots, 8$$

In this paper, an available optimization function of Matlab, i.e., fmincon, is employed to evaluate $\underline{y}_j$ and $\bar{y}_j$ defined in Equations (27) and (28), then NPRI $\eta_j (j = 1, 2, 3)$ for Equations (23)–(25) can be estimated, and NPRI η_s for measuring economical feasibility of biodiesel production with interval uncertainties can be calculated by Equation (26).

2.4. SA of NPRI for Economical Feasibility of Biodiesel Production with Regards to Uncertain Interval Parameter

When an interval parameter $x_i (i = 1, 2, \dots, 8)$ is fixed at $x_{ij} \in [\underline{x}_i, \bar{x}_i] (j = 1, 2, \dots, p)$, i.e., $x_i = x_{ij}$, indicating that x_i takes a value within the lower bound $\underline{x}_i$ and the upper bound $\bar{x}_i$, the uncertainty associated with x_i is eliminated, and original NPRI η_s will become $\eta_{s|x_i=x_{ij}}$. The absolute difference $\Delta\eta_{s|x_i=x_{ij}}$ between original NPRI η_s and $\eta_{s|x_i=x_{ij}}$ can reflect the effect of the elimination of uncertainty related to x_i, which can be defined by:

$$\Delta\eta_{s|x_i=x_{ij}} = \left| \eta_s - \eta_{s|x_i=x_{ij}} \right| (j = 1, 2, \dots, p), \tag{29}$$

where $\eta_{s|x_i=x_{ij}}$ can be evaluated by the method given in Section 2.3, similar to the evaluation procedure for η_s. When $x_i (i = 1, 2, \dots, 8)$ takes different values, i.e., $x_{i1}, x_{i2}, \dots, x_{ip}$, the original NPRI η_s will become $\eta_{s|x_i=x_{i1}}, \eta_{s|x_i=x_{i2}}, \dots, \eta_{s|x_i=x_{ip}}$, and then p absolute differences can be obtained by Equation (29), i.e., $\Delta\eta_{s|x_i=x_{i1}}, \Delta\eta_{s|x_i=x_{i2}}, \dots, \Delta\eta_{s|x_i=x_{ip}}$. The average of the p absolute differences, i.e., $\Delta\eta_{s|x_i=x_{i1}}, \Delta\eta_{s|x_i=x_{i2}}, \dots, \Delta\eta_{s|x_i=x_{ip}}$, can be employed to define the sensitivity of NPRI with regards to x_i, which can measure the effect of x_i on NPRI:

$$IM_i = \frac{1}{p} \sum_{j=1}^{p} \Delta\eta_{s|x_i=x_{ij}} (j = 1, 2, \dots, p) \tag{30}$$

where IM_i represents the average shift in the NPRI due to the elimination of uncertainty in x_i.

Similar to IM_i, the average difference rate in the NPRI because of eliminating uncertainty associated with x_i can be defined as:

$$IMR_i = \frac{1}{p} \sum_{j=1}^{p} \Delta\eta R_{s|x_i=x_{ij}} (j = 1, 2, \dots, p) \tag{31}$$

with

$$\Delta\eta R_{s|x_i=x_{ij}} = \frac{\left| \eta_s - \eta_{s|x_i=x_{ij}} \right|}{\eta_s} (j = 1, 2, \dots, p) \tag{32}$$

where $\Delta\eta R_{s|x_i=x_{ij}}$ measures the absolute difference rate between η_s and $\eta_{s|x_i=x_{ij}}$ with regard to η_s when $x_i = x_{ij} (x_{ij} \in [\underline{x}_i, \bar{x}_i])$.

The important interval parameters and non-important ones can be identified by the values of IM_i and IMR_i. An interval parameter with large values of IM_i and IMR_i belongs to the important interval parameters, while one with small values of IM_i and IMR_i is considered as the non-important parameters. If x_i has small values for IM_i and IMR_i, x_i can be fixed to any value within its variation interval, which will not considerably affect NPRI η_s.

3. Results and Discussion

In this section, we first evaluate NPRI η_s for biodiesel production with the eight interval parameters shown in Table 1. Then, the corresponding sensitivity analysis of NPRI η_s with regard to interval parameter x_i, i.e., IM_i and IMR_i, is estimated. Finally, the interval parameters are classified into the important ones and non-important ones by the size of the values of IM_i and IMR_i.

3.1. Evaluation of NPRI for Biodiesel Production

Biodiesel production has eight interval parameters because of economic and technical uncertainties when performing techno-economic assessments, and all interval parameters have been summarized in Table 1. The uncertainty in these interval parameters will result in the variation of the total profit, net present value, and payback period of biodiesel production. Figure 2 has shown the variation intervals for total profit (USD) expressed in Equation (1), net present value (USD) formulated in Equation (3), and $y_j = g_j(x)$ given in Equations (23)–(25). Two important observations have been revealed in Figure 2. The first observation is that total profit, net present value, and $y_j = g_j(x)(j = 1, 2, 3)$ have exhibited variability owing to the effect of the uncertainties related to the interval parameters, i.e., TotalProfit $\in [-3.8935 \times 10^8, 1.3296 \times 10^9]$, NPV $\in [-9.2579 \times 10^8, 1.1808 \times 10^9]$, $y_1 = g_1(x) \in [-3.8935 \times 10^8, 1.3296 \times 10^9]$, $y_2 = g_2(x) \in [-4.3435 \times 10^8, 1.3026 \times 10^9]$, and $y_3 = g_3(x) \in [-9.2579 \times 10^8, 1.1808 \times 10^9]$. Secondly, we can find that a part of total profit, net present value, and $y_j = g_j(x)$ have been less than zero because of the effect of the uncertainty in these interval parameters, implying that biodiesel production is partially economically feasible, and has the possibility of being infeasible in the presence of the economic and technical uncertainties.

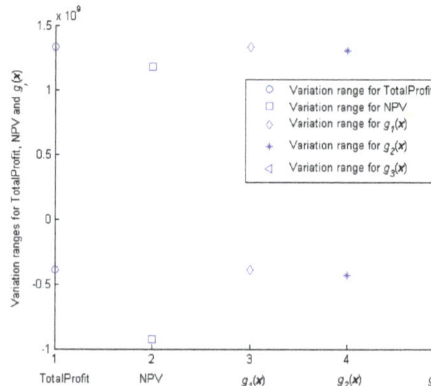

Figure 2. Variation ranges of total profit (TotalProfit: USD), net present value (NPV: USD), and $y_j = g_j(x)$ due to economic and technical uncertainties.

Figure 2 depicts the variation intervals of $y_j = g_j(x)(j = 1, 2, 3)$, including lower limit $\underline{y}_j$ and upper limit $\overline{y}_j$ for three LSFs defined in Equations (23)–(25). Substituting $\underline{y}_j$ and $\overline{y}_j$ into Equation (15) leads to NPRI $\eta_j(j = 1, 2, 3)$ of Equations (23)–(25). Finally, the estimated value of NPRI can be obtained as 1.2104×10^{-1} by using Equation (26). A value of 1.2104×10^{-1} for η_s implies that the project will not be profitable to a great extent, in other words, a considerable part of the outcomes may be economically infeasible under the uncertain interval parameters shown in Table 1.

In our previous work [59], all the uncertain parameters are assumed as random variables following uniform distributions within their ranges, and we propose economical infeasibility probability (EIP) to measure economical feasibility for biodiesel production. For the same problem, the estimated value for EIP is 0.3676, implying that the project is partially economically feasible and the plant may be profitable with the probability of 0.6324, and in other words, 63.24 out of 100 outcomes will be economically feasible under the assumed probabilistic distribution [59]. Here, we perform the TEA in terms of the non-probabilistic perspective being free from the probabilistic distribution assumption, and the estimated result for NPRI is 1.2104×10^{-1}, also indicating that the project is partially economically feasible, according to the discussion in Section 2.2. Thus, the two methods have the same decisions. It is noted that the introduced method in this work is more rational than that in the previous work [59], which is subjected to the assumption on probabilistic distribution and different assumptions can lead to different results for EIP.

The previous results reveal that interval parameters resulting from uncertainties can remarkably affect the TEA of biodiesel production. We will further quantify the effect of an interval uncertain parameter on the economical feasibility by the sensitivity analysis proposed in Section 2.4.

3.2. Evaluation of Sensitivity Analysis for Biodiesel Production with Respect to Interval Parameter

In Table 2, we have provided the results of IM_i and IMR_i relevant to $x_i (i = 1, \ldots, 8)$. The results show that x_3 (operating rate), x_4 (price of feedstock), x_8 (price of biodiesel), and x_7 (biodiesel conversion efficiency) can produce remarkable influences on the economic feasibility of biodiesel production, while the rest of the parameters may generate very lower effects. The importance ranking of the interval parameters can be further gained by the results in Table 2 as: $x_4 > x_8 > x_3 > x_7 > x_2 > x_1 > x_5 > x_6$. Compared with the previous results, in which all of the uncertain parameters have been assumed as random variables uniformly distributed within their variation ranges [59], the same importance ranking of sensitivity parameters has been obtained.

Table 2. Results of the proposed sensitivity analysis $IM_i (i = 1, 2, \ldots, 8)$ and $IMR_i (i = 1, 2, \ldots, 8)$.

Parameters	$IM_i(i = 1,2,\ldots,8)$	$IMR_i(i = 1,2,\ldots,8)$
Capital cost (CC: x_1)	4.454×10^{-3}	3.680×10^{-2}
Interest rate (r: x_2)	5.231×10^{-3}	4.322×10^{-2}
Operating rate (OR: x_3)	4.961×10^{-2}	4.099×10^{-1}
Feedstock price (FP: x_4)	4.858×10^{-1}	4.013×10^{0}
Glycerol price (GP: x_5)	2.865×10^{-3}	2.367×10^{-2}
Maintenance rate (MR: x_6)	6.643×10^{-4}	5.488×10^{-3}
Biodiesel conversion efficiency (CE: x_7)	9.302×10^{-3}	7.685×10^{-2}
Biodiesel price (BP: x_8)	3.257×10^{-1}	2.691×10^{0}

The important interval parameters and the non-important ones have been identified by the results in Table 2, specifically: x_4, x_8, x_3, and x_7 belong to the important group while x_2, x_1, x_5 and x_6 belong to the non-important group. Figure 3 shows the comparison between original NPRI η_s and conditional NPRI $\eta_{s|x_i=x_{ij}}$ with $x_i = x_{ij}$, in which x_i is fixed to a value x_{ij} within its variation interval $[\underline{x}_i, \overline{x}_i]$. Figure 4 shows the change rate between $\eta_{s|x_i=x_{ij}}$ and η_s with respect to η_s, i.e., $\left(\eta_{s|x_i=x_{ij}} - \eta_s \right) / \eta_s (j = 1, 2, \ldots, 10)$, in which x_i is fixed to a value $x_{ij} \in [\underline{x}_i, \overline{x}_i]$, i.e., $x_i = x_{ij}$. Here, x_{ij} takes the following values, i.e., $x_{ij} = \underline{x}_i + (\overline{x}_i - \underline{x}_i)/(10 - 1) \times (j - 1)(j = 1, 2, \ldots, 10)$. Figures 3 and 4 show that removing the uncertainty related to a non-important parameter $x_i (i = 1, 2, 5, 6)$ and fixing it to any value $x_{ij} (i = 1, 2, 5, 6)$ within its interval $[\underline{x}_i, \overline{x}_i]$ will not exert distinct influence on NPRI η_s, while eliminating the uncertainty associated with an important parameter $x_i (i = 3, 4, 7, 8)$ can cause considerable variation of NPRI η_s.

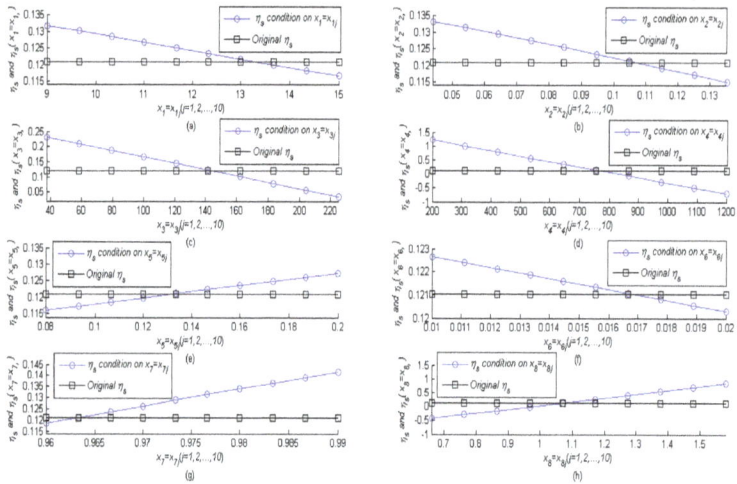

Figure 3. Original NPRI η_s and conditional NPRI $\eta_{s|x_i=x_{ij}}(j=1,2,\ldots,10)$, in which x_i is fixed to a value $x_{ij} \in [\underline{x}_i, \overline{x}_i]$, i.e., $x_i = x_{ij}(j=1,2,\ldots,10)$.

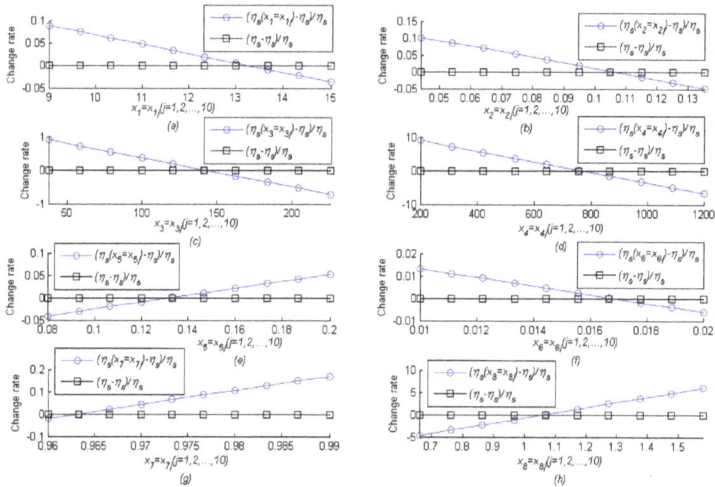

Figure 4. Change rate between $\eta_{s|x_i=x_{ij}}$ and η_s with respect to η_s, i.e., $\left(\eta_{s|x_i=x_{ij}} - \eta_s\right)/\eta_s(j=1,2,\ldots,10)$, in which x_i is fixed to a value $x_{ij} \in [\underline{x}_i, \overline{x}_i]$, i.e., $x_i = x_{ij}$.

Figures 5 and 6 have further shown the point figures of $\eta_{s|x_i=x_{ij}}$ and $\left(\eta_{s|x_i=x_{ij}} - \eta_s\right)/\eta_s$ with $x_{ij} = \underline{x}_i + (\overline{x}_i - \underline{x}_i)/(10-1) \times (j-1)(j=1,2,\ldots,10)$ for all interval parameters $x_i(i=1,2,\ldots,8)$. The results shown in Figures 5 and 6 have drawn the same conclusions as Figures 3 and 4.

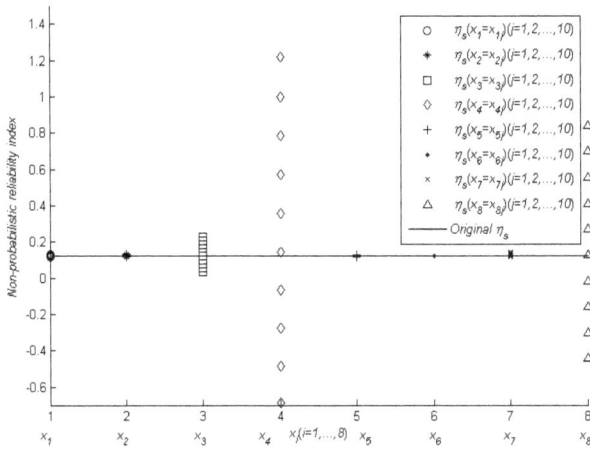

Figure 5. Point figure for conditional NPRI $\eta_{s|x_i=x_{ij}}$ with $x_i = x_{ij}(i = 1, 2, \ldots, 8; j = 1, 2, \ldots, 10)$.

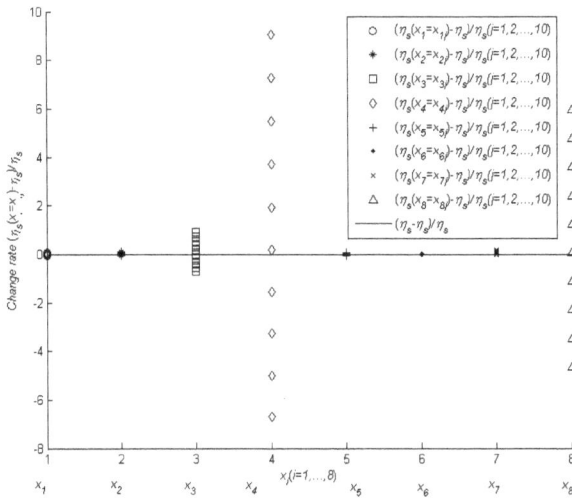

Figure 6. Point figure for change rate between $\eta_{s|x_i=x_{ij}}$ and η_s with respect to η_s, $\left(\eta_{s|x_i=x_{ij}} - \eta_0\right)/\eta_0$ with $x_i = x_{ij}(i = 1, 2, \ldots, 8; j = 1, 2, \ldots, 10)$.

The previous results show that engineers should focus more concern on these important interval parameters within the project's lifespan to ensure that biodiesel production is economically feasible. For these non-important interval parameters, taking any value within their ranges will not create remarkable effect on the TEA.

4. Conclusions

This paper employs NPRI to measure the economically feasible extent in the TEA of biodiesel production with uncertainties. Sensitivity analysis of NPRI with regard to uncertain parameters is developed. The final results show that NPRI for biodiesel production is 1.2104×10^{-1} with the interval parameters summarized in Table 1. Price of biodiesel, price of feedstock, and operating cost can cause distinct influence on the economical feasibility of biodiesel production. Compared with our previous study [59], this work has the same decision on TEA and the same importance ranking for uncertain parameters. This method is free of the assumption on distribution, but the previous method is subjected to this assumption in which different assumptions on distribution can result in different results for EIP.

Acknowledgments: Authors gratefully thank the support of the National Natural Science Foundation of China under Grant Nos. NSFC 51405064. Comments and suggestions from all reviewers and the Editor are very much appreciated.

Author Contributions: Zhang-Chun Tang proposed the new method, analyzed the data and wrote the paper; Yanjun Xia wrote computing programs and plotted the figures, Qi Xue and Jie Liu did literature survey.

Conflicts of Interest: The authors declare no conflict of interest.

Nomenclature

BP	biodiesel price
BPC	byproduct credit
BPC_i	byproduct credit of the ith year
CC	capital cost
CE	conversion efficiency from feedstock to biodiesel
d	depreciation rate
FC	feedstock cost
FC_i	feedstock cost of the ith year
FP	feedstock price
FU	annual total feedstock consumption
GCF	glycerol conversion factor
GP	glycerol price
LCC	life cycle cost
MC	maintenance cost
MC_i	maintenance cost of the ith year
MR	maintenance rate
NPRI	non-probabilistic reliability index
OC	operating cost
OC_i	operating cost of the ith year
OR	operating rate or operating cost of per-ton crude-palm-oil-derived biodiesel production
PC	production capacity
PP	payback period of the biodiesel production
PP^u	allowable upper limit of payback period
PWF_n	worth factor in the year n
RC	replacement cost
r	interest rate
SA	sensitivity analysis
SV	salvage value
TAX	annual total taxation
TBS	annual total biodiesel sales
TEA	techno-economic assessments
TotalProfit	total profit
TR	tax rate
UA	uncertainty analysis
ρ	density of the biodiesel

References

1. Höök, M.; Tang, X. Depletion of fossil fuels and anthropogenic climate change—A review. *Energy Policy* **2013**, *52*, 797–809. [CrossRef]
2. Höök, M. Energy, Climate and Society. 2013. Available online: http://cemusstudent.se/wp-content/uploads/2012/02/Week-49-Energy-Climate-and-Society-Mikael-Höök.pdf (accessed on 3 January 2018).
3. Yan, Y.J.; Li, X.; Wang, G.L.; Gui, X.H.; Li, G.L.; Su, F.; Wang, X.F.; Liu, T. Biotechnological preparation of biodiesel and its high-valued derivatives: A review. *Appl. Energy* **2014**, *113*, 1614–1631. [CrossRef]
4. Capellán-Pérez, I.; Mediavilla, M.; Castro, C.; Carpintero, O.; Miguel, L.J. Fossil fuel depletion and socio-economic scenarios: An integrated approach. *Energy* **2014**, *77*, 641–666. [CrossRef]
5. Ellabban, O.; Abu-Rub, H.; Blaabjerg, F. Renewable energy resources: Current status, future prospects and their enabling technology. *Renew. Sustain. Energy Rev.* **2014**, *39*, 748–764. [CrossRef]
6. Mohr, S.H.; Wang, J.; Ellem, G.; Ward, J.; Giurco, D. Projection of world fossil fuels by country. *Fuel* **2015**, *141*, 120–135. [CrossRef]
7. Abas, N.; Kalair, A.; Khan, N. Review of fossil fuels and future energy technologies. *Futures* **2015**, *69*, 31–49. [CrossRef]
8. Nicoletti, G.; Arcuri, N.; Nicoletti, G.; Bruno, R. A technical and environmental comparison between hydrogen and some fossil fuels. *Energy Convers. Manag.* **2015**, *89*, 205–213. [CrossRef]
9. Day, C.; Day, G. Climate change, fossil fuel prices and depletion: The rationale for a falling export tax. *Econ. Model.* **2017**, *63*, 153–160. [CrossRef]
10. Mahmudul, H.M.; Hagos, F.Y.; Mamat, R.; Adam, A.A.; Ishak, W.F.W.; Alenezi, R. Production, characterization and performance of biodiesel as an alternative fuel in diesel engines—A review. *Renew. Sustain. Energy Rev.* **2017**, *72*, 497–509. [CrossRef]
11. No, S.Y. Application of straight vegetable oil from triglyceride based biomass to IC engines—A review. *Renew. Sustain. Energy Rev.* **2017**, *69*, 80–97. [CrossRef]
12. Othman, M.F.; Adam, A.; Najafi, G.; Mamat, R. Green fuel as alternative fuel for diesel engine: A review. *Renew. Sustain. Energy Rev.* **2017**, *80*, 694–709. [CrossRef]
13. Efe, S.; Ceviz, M.A.; Temur, H. Comparative engine characteristics of biodiesels from hazelnut, corn, soybean, canola and sunflower oils on DI diesel engine. *Renew. Energy* **2018**, *119*, 142–151. [CrossRef]
14. Ruhul, A.M.; Kalam, M.A.; Masjuki, H.H.; Shahir, S.A.; Alabdulkarem, A.; Teoh, Y.H.; How, H.G.; Reham, S.S. Evaluating combustion, performance and emission characteristics of Millettia pinnata and Croton megalocarpus biodiesel blends in a diesel engine. *Energy* **2017**, *141*, 2362–2376. [CrossRef]
15. Silva, M.A.V.D.; Ferreira, B.L.G.; Marques, L.G.D.C.; Murta, A.L.S.; Freitas, M.A.V.D. Comparative study of NOx emissions of biodiesel-diesel blends from soybean, palm and waste frying oils using methyl and ethyl transesterification routes. *Fuel* **2017**, *194*, 144–156. [CrossRef]
16. Ding, H.; Ye, W.; Wang, Y.Q.; Wang, X.Q.; Li, L.J.; Liu, D.; Gui, J.Z.; Song, C.F.; Ji, N. Process intensification of transesterification for biodiesel production from palm oil: Microwave irradiation on transesterification reaction catalyzed by acidic imidazolium ionic liquids. *Energy* **2018**, *144*, 957–967. [CrossRef]
17. Zou, C.J.; Zhao, P.W.; Shi, L.H.; Huang, S.B.; Luo, P.Y. Biodiesel fuel production from waste cooking oil by the inclusion complex of heteropoly acid with bridged bis-cyclodextrin. *Bioresour. Technol.* **2013**, *146*, 785–788. [CrossRef] [PubMed]
18. Ali, C.H.; Qureshi, A.S.; Mbadinga, S.M.; Liu, J.F.; Yang, S.Z.; Mu, B.Z. Biodiesel production from waste cooking oil using onsite produced purified lipase from Pseudomonas aeruginosa FW_SH-1: Central composite design approach. *Renew. Energy* **2017**, *109*, 93–100. [CrossRef]
19. Tangy, A.; Pulidindi, I.N.; Perkas, N.; Gedanken, A. Continuous flow through a microwave oven for the large-scale production of biodiesel from waste cooking oil. *Bioresour. Technol.* **2017**, *224*, 333–341. [CrossRef] [PubMed]
20. Allesina, G.; Pedrazzi, S.; Tebianian, S.; Tartarini, P. Biodiesel and electrical power production through vegetable oil extraction and byproducts gasification: Modeling of the system. *Bioresour. Technol.* **2014**, *170*, 278–285. [CrossRef] [PubMed]
21. Wang, S.X.; Yuan, H.R.; Wang, Y.Z.; Shan, R. Transesterification of vegetable oil on low cost and efficient meat and bone meal biochar catalysts. *Energy Convers. Manag.* **2017**, *150*, 214–221. [CrossRef]

22. Tang, S.K.; Zhao, H.; Song, Z.Y.; Olubajo, O. Glymes as benign co-solvents for CaO-catalyzed transesterification of soybean oil to biodiesel. *Bioresour. Technol.* **2013**, *139*, 107–112. [CrossRef] [PubMed]

23. Celante, D.; Schenkel, J.V.D.; Castilhos, F. Biodiesel production from soybean oil and dimethyl carbonate catalyzed by potassium methoxide. *Fuel* **2018**, *212*, 101–107. [CrossRef]

24. Luna, M.D.G.D.; Cuasay, J.L.; Tolosa, N.C.; Chung, T.W. Transesterification of soybean oil using a novel heterogeneous base catalyst: Synthesis and characterization of Na-pumice catalyst, optimization of transesterification conditions, studies on reaction kinetics and catalyst reusability. *Fuel* **2017**, *209*, 246–253. [CrossRef]

25. Li, K.; Fan, Y.L.; Zeng, L.P.; Han, X.T.; Yan, Y.J. Burkholderia cepacia lipase immobilized on heterofunctional magnetic nanoparticles and its application in biodiesel synthesis. *Sci. Rep.* **2017**, *7*, 16473. [CrossRef] [PubMed]

26. Nisar, J.; Razaq, R.; Farooq, M.; Iqbal, M.; Khan, R.A.; Sayed, M.; Shah, A.; Rahman, I. Enhanced biodiesel production from Jatropha oil using calcined waste animal bones as catalyst. *Renew. Energy* **2017**, *101*, 111–119. [CrossRef]

27. Quinn, J.C.; Davis, R. The potentials and challenges of algae based biofuels: A review of the techno-economic, life cycle, and resource assessment modeling. *Bioresour. Technol.* **2015**, *184*, 444–452. [CrossRef] [PubMed]

28. Rodríguez, R.P.; Borroto, Y.S.; Espinosa, E.A.M.; Verhelst, S. Assessment of diesel engine performance when fueled with biodiesel from algae and microalgae: An overview. *Renew. Sustain. Energy Rev.* **2017**, *69*, 833–842. [CrossRef]

29. Taparia, T.; Mvss, M.; Mehrotra, R.; Shukla, P.; Mehrotra, S. Developments and challenges in biodiesel production from microalgae: A review. *Biotechnol. Appl. Biochem.* **2015**, *63*, 715–726. [CrossRef] [PubMed]

30. Piligaev, A.V.; Sorokina, K.N.; Samoylova, Y.V.; Parmon, V.N. Lipid production by microalga *Micractinium* sp. IC-76 in a flat panel photobioreactor and its transesterification with cross-linked enzyme aggregates of Burkholderia cepacia lipase. *Energy Convers. Manag.* **2018**, *156*, 1–9. [CrossRef]

31. Probst, K.V.; Schulte, L.R.; Durrent, T.P.; Rezac, M.E.; Vadlani, P.V. Oleaginous yeast: A value-added platform for renewable oils. *Crit. Rev. Biotechnol.* **2016**, *36*, 942–955. [CrossRef] [PubMed]

32. Sitepu, I.R.; Garay, L.A.; Sestric, R.; Levin, D.; Block, D.E.; German, J.B.; Mills, K.L.B. Oleaginous yeasts for biodiesel: Current and future trends in biology and production. *Biotechnol. Adv.* **2014**, *32*, 1336–1360. [CrossRef] [PubMed]

33. Zhang, K.; Pei, Z.J.; Wang, D.H. Organic solvent pretreatment of lignocellulosic biomass for biofuels and biochemicals: A review. *Bioresour. Technol.* **2016**, *199*, 21–33. [CrossRef] [PubMed]

34. Lam, S.S.; Mahari, W.A.W.; Jusoh, A.; Chong, C.T.; Lee, C.L.; Chase, H.A. Pyrolysis using microwave absorbents as reaction bed: An improved approach to transform used frying oil into biofuel product with desirable properties. *J. Clean. Prod.* **2017**, *147*, 263–272. [CrossRef]

35. Malhotra, R.; Ali, A. Lithium-doped ceria supported SBA−15 as mesoporous solid reusable and heterogeneous catalyst for biodiesel production via simultaneous esterification and transesterification of waste cottonseed oil. *Renew. Energy* **2018**, *119*, 32–44. [CrossRef]

36. Abdullah, S.H.Y.S.; Hanapi, N.H.M.; Azid, A.; Umar, R.; Juahir, H.; Khatoon, H.; Endut, A. A review of biomass-derived heterogeneous catalyst for a sustainable biodiesel production. *Renew. Sustain. Energy Rev.* **2017**, *70*, 1040–1051. [CrossRef]

37. Singh, V.; Sharma, Y.C. Low cost guinea fowl bone derived recyclable heterogeneous catalyst for microwave assisted transesterification of *Annona squamosa* L. seed oil. *Energy Convers. Manag.* **2017**, *138*, 627–637. [CrossRef]

38. Vinod, B.M.; Madhu, M.K.; Amba, P.R.G. Butanol and pentanol: The promising biofuels for CI engines—A review. *Renew. Sustain. Energy Rev.* **2017**, *78*, 1068–1088.

39. Singh, L.K.; Chaudhary, G. *Advances in Biofeedstocks and Biofuels*; John Wiley & Sons: Hoboken, NJ, USA, 2017.

40. Sotoft, L.F.; Rong, B.G.; Christensen, K.V.; Norddahl, B. Process simulation and economical evaluation of enzymatic biodiesel production plant. *Bioresour. Technol.* **2010**, *101*, 5266–5274. [CrossRef] [PubMed]

41. Haas, M.J.; McAloon, A.J.; Yee, W.C.; Foglia, T.A. A process model to estimate biodiesel production costs. *Bioresour. Technol.* **2006**, *97*, 671–678. [CrossRef] [PubMed]

42. Ong, H.C.; Mahlia, T.M.I.; Masjuki, H.H.; Honnery, D. Life cycle cost and sensitivity analysis of palm biodiesel production. *Fuel* **2012**, *98*, 131–139. [CrossRef]

43. Xin, C.H.; Addy, M.M.; Zhao, J.Y.; Cheng, Y.L.; Cheng, S.B.; Mu, D.Y.; Liu, Y.H. Comprehensive techno-economic analysis of wastewater-based algal biofuel production: A case study. *Bioresour. Technol.* **2016**, *211*, 584–593. [CrossRef] [PubMed]

44. Kern, J.D.; Hise, A.M.; Characklis, G.W.; Gerlach, R.; Viamajala, S.; Gardner, R.D. Using life cycle assessment and techno-economic analysis in a real options framework to inform the design of algal biofuel production facilities. *Bioresour. Technol.* **2017**, *225*, 418–428. [CrossRef] [PubMed]

45. Thomassen, G.; Dael, M.V.; Lemmens, B.; Passel, S.V. A review of the sustainability of algal-based biorefineries: Towards an integrated assessment framework. *Renew. Sustain. Energy Rev.* **2017**, *68*, 876–887. [CrossRef]

46. Dutta, S.; Neto, F.; Coelho, M.C. Microalgae biofuels: A comparative study on techno-economic analysis & life-cycle assessment. *Algal Res.* **2016**, *20*, 44–52.

47. Hpffman, J.; Pate, R.C.; Drennen, T.; Quinn, J.C. Techno-economic assessment of open microalgae production systems. *Algal Res.* **2017**, *23*, 51–57. [CrossRef]

48. Xin, C.H.; Addy, M.M.; Zhao, J.Y.; Cheng, Y.L.; Ma, Y.W.; Liu, S.Y.; Mu, D.Y.; Liu, Y.H.; Chen, P.; Ruan, R. Waste-to-biofuel integrated system and its comprehensive techno-economic assessment in wastewater treatment plants. *Bioresour. Technol.* **2018**, *250*, 523–531. [CrossRef] [PubMed]

49. Mandegari, M.A.; Farzad, S.; Görgens, J.F. Recent trends on techno-economic assessment (TEA) of sugarcane biorefineries. *Biofuel Res. J.* **2017**, *4*, 704–712. [CrossRef]

50. Tao, L.; Milbrandt, A.; Zhang, Y.N.; Wang, W.C. Techno-economic and resource analysis of hydroprocessed renewable jet fuel. *Biotechnol. Biofuels* **2017**, *10*, 261. [CrossRef] [PubMed]

51. Patel, M.; Zhang, X.L.; Kumar, A. Techno-economic and life cycle assessment on lignocellulosic biomass thermochemical conversion technologies: A review. *Renew. Sustain. Energy Rev.* **2015**, *53*, 1486–1499. [CrossRef]

52. Busse, S.; Brümmer, B.; Ihle, R. Price formation in the German biodiesel supply chain: A Markov-switching vector error-correction modeling approach. *Agric. Econ.* **2012**, *43*, 545–560. [CrossRef]

53. Mankiw, N.G. *Principles of Economics*, 6th ed.; South-Western Cengage Learning: Mason, OH, USA, 2011.

54. Borgonovo, E.; Peccati, L. Uncertainty and Global Sensitivity Analysis in the Evaluation of Investment Projects. *Int. J. Prod.* **2006**, *104*, 62–73. [CrossRef]

55. Brownbridge, G.; Azadi, P.; Smallbone, A.; Bhave, A.; Taylor, B.; Kraft, M. The future viability of algae-derived biodiesel under economic and technical uncertainties. *Bioresour. Technol.* **2014**, *151*, 166–173. [CrossRef] [PubMed]

56. López, P.P.; Montazeri, M.; Feijoo, G.; Moreira, M.T.; Eckelman, M.J. Integrating uncertainties to the combined environmental and economic assessment of algal biorefineries: A Monte Carlo approach. *Sci. Total Environ.* **2018**, *626*, 762–775. [CrossRef] [PubMed]

57. Abubakar, U.; Sriramula, S.; Renton, N.C. Stochastic techno-economic considertations in biodiesel production. *Sustain. Energy Technol. Assess.* **2015**, *9*, 1–11.

58. Tang, Z.C.; Lu, Z.Z.; Liu, Z.W.; Xiao, N.C. Uncertainty analysis and global sensitivity analysis of techno-economic assessments for biodiesel production. *Bioresour. Technol.* **2015**, *175*, 502–508. [CrossRef] [PubMed]

59. Xia, Y.J.; Tang, Z.C. A novel perspective for techno-economic assessments and effects of parameters on techno-economic assessments for biodiesel production under economic and technical uncertainties. *RSC Adv.* **2017**, *7*, 9402–9411. [CrossRef]

60. Sajid, Z.; Zhang, Y.; Khan, F. Process design and probabilistic economic risk analysis of biodiesel production. *Sustain. Prod. Consum.* **2016**, *5*, 1–15. [CrossRef]

61. Liew, W.H.; Hassim, M.H.; Ng, D.K.S. Sustainability assessment framework for chemical production pathway: Uncertainty analysis. *J. Environ. Chem. Eng.* **2016**, *4*, 4878–4889. [CrossRef]

62. Batan, L.Y.; Graff, G.D.; Bradley, T.H. Techno-economic and Monte Carlo probabilistic analysis of microalgae biofuel production system. *Bioresour. Technol.* **2016**, *219*, 45–52. [CrossRef] [PubMed]

63. Ochoa, M.P.; Estrada, V.; Maggio, J.D.; Hoch, P.M. Dynamic global sensitivity analysis in bioreactor networks for bioethanol production. *Bioresour. Technol.* **2016**, *200*, 666–679. [CrossRef] [PubMed]

64. Yao, G.L.; Staples, M.D.; Malina, R.; Tyner, W.E. Stochastic techno-economic analysis of alcohol-to-jet fuel production. *Biotechnol. Biofuels* **2017**, *10*, 18. [CrossRef] [PubMed]

65. Garay, A.G.; Miquel, M.G.; Gosalbez, G.G. High-Value Propylene Glycol from Low-Value Biodiesel Glycerol: A Techno-Economic and Environmental Assessment under Uncertainty. *ACS Sustain. Chem. Eng.* **2017**, *5*, 5723–5732. [CrossRef]

66. Trading Economics. Malaysia Interest Rate 1996–2014. 2014. Available online: http://zh.tradingeconomics.com/malaysia/bank-lending-rate (accessed on 3 January 2018).

67. Hitchcock, G. The Economics of Biofuel Production and Use. 2014. Available online: http://www.sts-technology.com/docs/Economics-of-biofuel-production-and-use.ppt (accessed on 3 January 2018).

68. Duncan, J. Costs of Biodiesel Production. 2003. Available online: http://www.globalbioenergy.org/uploads/media/0305_Duncan_-_Cost-of-biodiesel-production.pdf (accessed on 3 January 2018).

69. Nagi, J.; Ahmed, S.K.; Nagi, F. Palm Biodiesel an Alternative Green Renewable Energy for the Energy Demands of the Future. In Proceedings of the International Conference on Construction and Building Technology, Kuala Lumpur, Malaysia, 16–20 June 2008; pp. 79–94.

70. Nanda, M.R.; Yuan, Z.; Qin, W.; Poirier, M.A.; Chunbao, X. Purification of Crude Glycerol using Acidification: Effects of Acid Types and Product Characterization. *Austin J. Chem. Eng.* **2014**, *1*, 1–7.

71. FarmdocDAILY. Recent Trends in Biodiesel Prices and Production Profits. 2013. Available online: http://farmdocdaily.illinois.edu/2013/09/recent-trends-in-biodiesel.html (accessed on 3 January 2018).

72. Cremona, C.; Gao, Y. The possibilistic reliability theory: Theoretical aspects and applications. *Struct. Saf.* **1997**, *19*, 173–201. [CrossRef]

73. Guo, S.X.; Lu, Z.Z.; Feng, Y.S. A non-probabilistic model of structural reliability based on interval analysis. *Chin. J. Comput. Mech.* **2001**, *18*, 56–60.

74. Guo, S.X.; Li, Y. Non-probabilistic reliability method and reliability-based optimal LQR design for vibration control of structures with uncertain-but-bounded parameters. *Acta Mech. Sin.* **2013**, *29*, 864–874. [CrossRef]

75. Kang, Z.; Luo, Y. Non-probabilistic reliability-based topology optimization ofgeometrically nonlinear structures using convex models. *Comput. Methods Appl. Mech. Eng.* **2009**, *198*, 3228–3238. [CrossRef]

76. Ben-Haim, Y.; Elishakoff, I. *Convex Models of Uncertainty in Applied Mechanics*; Elsevier Press: Amsterdam, The Netherlands, 1990.

77. Ben-Haim, Y. A non-probabilistic measure of reliability of linear systems based on expansion of convex models. *Struct. Saf.* **1995**, *17*, 91–109. [CrossRef]

78. Ben-Haim, Y. *Robust Reliability in the Mechanics Sciences*; Springer: Berlin, Germany, 1996.

79. Ben-Haim, Y. Robust reliability of structures. *Adv. Appl. Mech.* **1997**, *33*, 1–41.

80. Elishakoff, I. Essay on uncertainties in elastic and viscoelastic structures: From A. M. Freudenthal's criticisms to modern convex modeling. *Comput. Struct.* **1995**, *56*, 871–895. [CrossRef]

81. Elishakoff, I. *Are Probabilistic and Anti-Optimization Approaches Compatible? Whys and Hows in Uncertainty Modelling: Probability, Fuzziness and Antioptimization*; Springer: New York, NY, USA, 1999.

energies

MDPI

Article

Remaining Useful Life Estimation of Aircraft Engines Using a Modified Similarity and Supporting Vector Machine (SVM) Approach

Zhongzhe Chen [1],*, Shuchen Cao [2] and Zijian Mao [1]

[1]   School of Mechanical and Electrical Engineering, University of Electronic and Science Technology of China, Chengdu 611731, China; 2016210801442@std.uestc.edu.cn
[2]   Department of Mathematics, University of Illinois at Urbana-Champaign, Urbana, IL 61801, USA; shuchen8@illinois.edu
*   Correspondence: zhzhchen@uestc.edu.cn; Tel.: +86-139-8179-1418

Received: 25 November 2017; Accepted: 18 December 2017; Published: 23 December 2017

Abstract: As the main power source for aircrafts, the reliability of an aero engine is critical for ensuring the safety of aircrafts. Prognostics and health management (PHM) on an aero engine can not only improve its safety, maintenance strategy and availability, but also reduce its operation and maintenance costs. Residual useful life (RUL) estimation is a key technology in the research of PHM. According to monitored performance data from the engine's different positions, how to estimate RUL of an aircraft engine by utilizing these data is a challenge for ensuring the engine integrity and safety. In this paper, a framework for RUL estimation of an aircraft engine is proposed by using the whole lifecycle data and performance-deteriorated parameter data without failures based on the theory of similarity and supporting vector machine (SVM). Moreover, a new state of health indicator is introduced for the aircraft engine based on the preprocessing of raw data. Finally, the proposed method is validated by using 2008 PHM data challenge competition data, which shows its effectiveness and practicality.

Keywords: prognostics; residual useful life; similarity-based approach; supporting vector machine (SVM)

1. Introduction

Recent developments of complex systems, such as aircraft engines, engineering machines, high-speed vehicles and computer numerical control (CNC) systems have been emphasized by the increasing requirements of on-line health monitoring for the purpose of maximizing its operational reliability and safety [1–3]. As the core part and power source of aircrafts, the reliable operation of an aero engine is critical for ensuring the reliability and safety of the aircraft, and to maintain its availability, and reduce its maintenance costs [4–6]. Among them, prognostics and health management (PHM) is an effective approach and one of the most commonly-used [7,8]. In particular, residual useful life (RUL) estimation is a key technology for PHM. In general, RUL estimation is to indicate the system/component lifetime before it can no longer perform its function, which is also an important way to reduce production loss, save maintenance costs and avoid fatal machine breakdowns of the equipment before its failure [9–12].

Since the aircraft engine is a complex system, there are various monitored performance data from different positions during its operation. How to estimate RUL of an aircraft engine by utilizing these data has become the focus of most engine industries. Until now, approaches to predict system lifetime can be broadly categorized into three types: physics-based models, data-driven approaches and hybrid approaches [12–14]. Generally, a physics-based model utilizes the failure physical model of the system/component to estimate its RUL, which is usually based on the system/component's physics of failure or physics of dynamics deeply [15–19]. It can usually obtain reasonable and accurate predictions of RUL based on physical models with limited historical data [20]. However, it is usually different or too expensive to apply a physics-based model to a complex system. Besides, this approach has shown significant limitations due to the assumptions and simplifications of the adopted models [21]. The data-driven approach utilizes the monitored operational data relating to system health for RUL estimation [22,23], which is preferred when the system's failure physics is complicated or unavailable but systems' degradation procedure and degradation data are available. Note from [3] that the data-driven approach provides accurate RUL predictions for a complex system, which can be applied quickly and cheaply compared to the physics-based model. Furthermore, recent development of sensor technology and simulation capabilities enables us to continuously monitor the healthy situation of a complex system and obtain the related large amount of performance index data. In addition, data-driven approaches can be divided into three categories: statistical techniques and artificial intelligence (AI) techniques. The former includes regression methods such as the auto-regressive and moving average (ARMA) models, and the later includes neural networks and supporting vector machine (SVM), fuzzy logic, etc. The third approach, the so-called hybrid approach proposed by Hansen et al. [24], is the combination of physics-based and data-driven models, in which prognostics results are claimed to be more reliable and accurate, but few studies have been reported [20].

Data-driven RUL prediction models, which are most widely applied in the field of prognostics or PHM, mainly include extrapolation models and statistical models. The extrapolation model is usually used to fit a curve of a system degradation evolution by regression, extrapolate the curve to the failure threshold and obtain the RUL between the current moment and the predicted failure time [25]. The statistical model establishes the relationship between a system's failure likelihood and its degradation indicator from collected CM (condition maintenance) and failure data [26]. The statistical model approach is classified into the models based on the direct CM data and indirect CM data. The models based on the direct CM data include the proportional hazards model [27,28], proportional covariate model (PCM) [29], Wiener processes, Gamma processes and Markovian-based models. The models based on the indirect CM data include stochastic filtering-based models, covariate-based hazard models and hidden Markov model (HMM) [30], hidden semi-Markov models (HSMM), etc. Statistical models are the most effective ones for RUL estimation when system failure procedure is invisible. Most research has been conducted in RUL estimation based on data-driven models. Stetter and Witczak [31] explored various degradation modeling techniques and how to select the degradation indicator to estimate the RUL. Lee et al. [32] reviewed various methodologies and techniques in PHM research and proposed the systematic PHM design methodology, namely 5S methodology. Moreover, current methodologies of RUL estimation can be summarized as three classes as shown in Figure 1.

Referring to the previous literature and existing methods, a structured form of methodology for RUL prediction is expressed as shown in Figure 1.

When utilizing the data-driven approach for RUL estimation, the whole run-to-failure data of systems are normally needed, but it is difficult to obtain enough run-to-failure data for the long-life systems with high reliability. Thus, it might lead to a large error if the available system history data are lacking. The same problem will arise when the ARMA model is employed. However, if there are some similar systems to the researched system, the failure and performance-deteriorated information of these similar systems are useful for RUL estimation of the researched system. In general, the principle of similarity-based RUL prediction approach is given as follows: if an operating system has similar performance to the reference system during a time range, then assume that they have a similar RUL. Because this reference system is an identical system with the operating system physically, moreover, they operate under the same working conditions and reference systems that have already failed. In addition, if there are more reference systems similar to the researched one, the similarity-based approach can be introduced through a weighted average of the reference systems' RUL as the researched one's RUL [33], while the weight is proportional to the similarity between the researched and reference systems. According to this, the similarity-based RUL prediction model gives more reasonable results without modeling the deteriorated process of the researched system. Besides, with the development of PHM, there are abundant historical deteriorated data before failure that could be utilized to perform PHM.

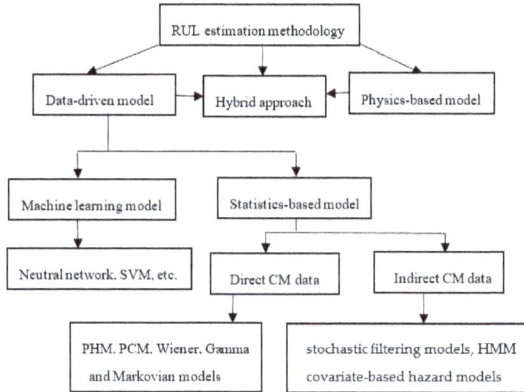

Figure 1. Methodologies for RUL estimation. RUL: residual useful life; SVM: supporting vector machine; CM: condition maintenance; PHM: prognostics and health management; PCM: proportional covariate model; HMM: hidden Markov model.

Zio et al. [21] developed a similarity-based approach to predict the RUL by comparing its evolution data to the trajectory patterns of reference samples through fuzzy similarity analysis, and aggregating their time to failure in a weighted sum, which accounts for their similarity to the developing pattern [21]. Gebraeel et al. [22] presents a stochastic process by combining with a data analysis method and deterioration modeling of the components for RUL prediction.

For the traditional similarity-based RUL prediction method, current and past degradation parameters of reference systems have an equal weight when calculating the similarity measure. However, as we all know, a system's most recent performance to its current health/state is more relative than its earlier performance, and provides more information for its RUL than its earlier performance. Therefore, it is reasonable to assign more weight to a system's most recent sampling point than its earlier sampling point of performance parameters when measuring its similarity with other systems. However, the traditional similarity-based method ignored this situation. Accordingly, this paper adopts a modified similar-based methodology which introduces a weight-adjusted coefficient α to embody the different effect on the calculation of similarity degree from different time ranges while calculating the similarity measure. The more recent sampling point of performance, the bigger weight of the parameter is given. In addition, the earlier value of performance, the parameter is given smaller weight and this paper provides an approach to optimize the weight α.

Until now, most research on the similarity-based model for RUL prediction are based on run-to-failure data, but sometimes there are only deteriorated performance data without run-to-failure data. How to utilize these deteriorated performance data, which do not work to failure, to estimate RUL of equipment by similarity-based method, is lacking and expected. Suspension history condition monitoring data usually contain useful information revealing the degradation situation of the system, including environmental factors and loading variations in actual situations, such as degradations and variations of stress amplitudes [10–12,34,35]. If these data are properly used, it is helpful to estimate RUL more accurately, particularly when the failure data are insufficient and unavailable in some cases [36,37]. Li et al. [38] used the suspension data to promote the prediction precision of a neural network. However, how to utilize these suspension data to predict RUL of the equipment has not been deeply studied.

This paper attempts to develop a modified similarity and SVM-based method to predict the RUL of an aircraft engine, including two schemes with different reference samples. The first scheme adopts a modified similarity-based method for estimating the RUL of the engine with abundant run-to-failure data of referenced samples, which is named as the modified similarity methodology based on run-to-failure data. The second scheme utilizes deteriorated data of samples without running to failure to estimate the RUL of the operating sample based on SVM and similarity methodology, named as the modified similarity and SVM methodology based on deteriorated data. The structure of this paper is as follows. Section 2 provides a detailed description of two approaches aimed for RUL estimation under two situations. Section 3 introduces how to utilize the proposed approaches to estimate the RUL of an aircraft engine. Section 4 concludes the current research.

2. Proposed Methodology for RUL Estimation

This section is devoted to introducing a similarity-based methodology including two schemes for RUL estimation. The first scheme is to estimate the RUL with abundant run-to-failure data of referenced samples. The other scheme is to estimate the RUL of aircraft engines with some deteriorated data of referenced samples which have no run-to-failure data.

2.1. The Scheme of the Modified Similarity Methodology Based on Run-to-Failure Data

The RUL of an operating sample is the weighted average of RUL of referenced samples. The weights are determined by the similarity degree between referenced samples and the operating one. In particular, the similarity degree is calculated by the weighted average of similarity degrees of sampling points between the reference and operating equipment. This subsection tends to introduce the framework of the modified similarity methodology based on run-to-failure data as shown in Figure 2.

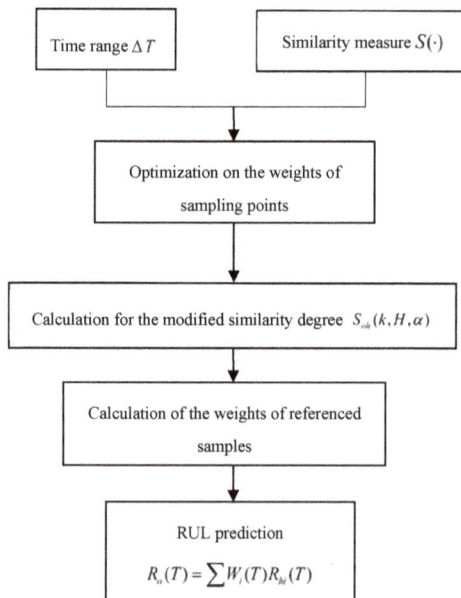

Figure 2. Framework for the modified similarity methodology based on run-to-failure data.

2.1.1. Determination of Time Range for Similarity Measurement

In this analysis, the first step is to set up the time range Δt for similarity measurement, namely, to determine the number of sampling points of the operating system for similarity measurement:

$$X(k, H) = [x(k \cdot \Delta t), \cdots, x((k - H) \cdot \Delta t)] \tag{1}$$

where H is the number of sampling points; Δt represents the time range in which similarity degree between a referenced sample and the operating one; $x(k \cdot \Delta t)$ denotes the degradation indicator of the operating sample at the kth sampling point since its operation.

Generally, most of the recent sampling points of the operating system represent its current state. In the traditional similarity-based method, any consecutive sampling points of the condition monitor a reference system before its failure can be used for similarity measurement [15]. In addition, a reasonably long time range can be determined based on operational experience in the lack of prior knowledge. The sampling points $X(k, H)$ in sampling time range are equally considered to be fully representative of the system's RUL. In this paper, the sampling points of the reference system are confined in the same time range as the operating system, namely, $[k - H, k]$.

2.1.2. Calculation of the Similarity Measure

The second step is to define and calculate the similarity measure, which indicates the similarity degree between the operating and reference systems, and then quantify the degradation duration of the ith reference system that is most similar as the duration of the operating system. The similarity measure S is the function of degradation indicators of the system, which measures the similarity between referenced and operating systems. Note that it may be Euclidean distance, probability function [27] or membership function in fuzzy logic theory [26]. In this paper, the Euclidean distance of degradation indicators between the reference systems and the operating system is introduced as the similarity measure function. The traditional Euclidean distance is expressed as:

$$S_{ohi}(k, H, m) = \sum_{v=0}^{H} [X_0((k - v) \cdot \Delta t) - X_{hi}((m - v) \cdot \Delta t)]^2 / (H + 1) \tag{2}$$

where $S_{ohi}(k, H, m)$ is the similarity measure between the operating system's degradation process in the time range $[(k - H)\Delta t, k\Delta t]$ and reference system's degradation process in the time range $[(m - H)\Delta t, m\Delta t]$; where $M_i \Delta t$ is the failure time of the ith reference system, and $H \leq m \leq M$. $X_0((k - v) \cdot \Delta t)$ denotes the degradation indicator of the operating system at the vth sampling point from the kth sampling point. $X_{hi}((m - v) \cdot \Delta t)$ denotes the degradation indicator of the ith reference system at the vth sampling point from the mth sampling point.

The similarity degree between the operating system and the ith reference system at time $T = k\Delta t$ is defined as:

$$S_{ohi}(k) = \frac{1}{\min_{H \leq m \leq M_i} S(k, H, m)} \tag{3}$$

In this analysis, more weights are assigned to the recent sampling point of degradation indicator than its former sampling points, thus, the Euclidean distance as the similarity measure for illustration is defined as:

$$S_{ohi}(k, H, \alpha) = \sum_{v=0}^{H} \left\{ \alpha^v [X_0((k - v) \cdot \Delta t) - X_{hi}((k - v) \cdot \Delta t)]^2 \right\} / (H + 1) \tag{4}$$

where α is a weight-adjusting coefficient ranging from 0 to 1. A smaller α corresponds to a smaller weight assigned to the former sampling point than recent sampling points of reference systems. α can be obtained by optimization for minimal predicting error of operating system's RUL. An example to obtain α is elaborated in Section 3.1.

2.1.3. Definition of the Weight Function

The third step is to define the weight function based on the similarity measure. As aforementioned, the weight is a function of similarity-degree, which is assigned to the reference systems according similarity degree to calculate the RUL of the operating system. The weight of the *ith* reference system is given by

$$W_i(k) = \frac{S_{ohj}(k)}{\sum_{j=1}^{n} S_{ohj}(k)} \tag{5}$$

2.1.4. RUL Estimation of the Operating System

The last step is to estimate the RUL of the operating system. As aforementioned, the RUL of an operating sample at time $t = k\Delta t$ is the weighted mean value of reference systems at the kth sampling point, and can be obtained by

$$RUL(k) = \sum_{i=1}^{n} W_i(k) RUL_i(k) \tag{6}$$

where n is the number of available reference systems.

The real RUL of the reference system at $t = k \cdot \Delta t$ is $RUL_i(k) = (M_i - k) \cdot \Delta t$, then the operating system's RUL can be calculated by

$$SUL(k) = \sum_{i=1}^{n} W_i(k)(M_i - k) \cdot \Delta t \tag{7}$$

2.1.5. Optimization of the Weight-Adjust Coefficient α

In order to embody different effects of sampling points of reference systems at different time on RUL estimation of the operating system, the weight-adjust coefficient α is introduced in this analysis. The weight-adjust coefficient α leads the recent sampling points of reference systems with more weights for the similarity degree calculation, which tends to provide more accurate prediction, specifically, α can be obtained by optimization under the goal function

$$MAPE_\alpha(k) = \min \left[\sum_{j=k_0}^{j=k-1} MAPE_\alpha(j) / (k - 1 - k_0) \right] \tag{8}$$

where k_0 denotes the first sampling point to deteriorate; $MAPE_\alpha(j)$ is the estimated percentage error at the value of α.

2.2. The Scheme of the Similarity and SVM Methodology Based on Deteriorated Data

Research shows that the similarity-based method gives effective and accurate estimation under abundant run-to-failure data of reference samples. However, most equipment operates with high reliability and long life, especially in aerospace applications; the reference samples with enough run-to-failure data are seldom. For this limited or no run-to-failure reference samples, whether the similarity-based method can be used or not needs to be explored. In practices, there is abundant performance deteriorating data and maintenance data during its operating process. These suspension data include useful degradation information relating to the operating system. However, the degradation indicators after halting operating and lifetime of reference samples are unknown since they did not work until failure. Thus, the trend of the degradation indicators and the lifetime of reference samples need to be collected and analyzed. This methodology for RUL estimation consists two essential preprocessing procedures: performance assessment for reference samples and

RUL estimation based on reference systems, and then the RUL of the operating system can be derived. In particular, the implementation flowchart is given in Figure 3.

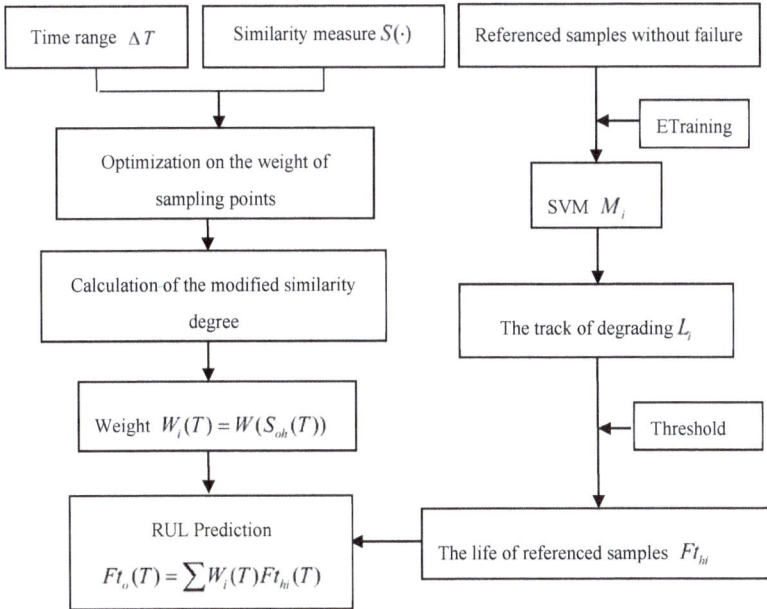

Figure 3. The framework of the similarity and SVM methodology based on deteriorated data.

Particularly, SVM is adopted to perform the degradation trend assessment of reference samples and estimate their lifetimes. As is well known, SVM has been commonly used for handling the data of small samples and multiple dimensions. The monitored degradation indicators of reference samples are used to train SVM and obtain the performance-deteriorated pattern of these samples, and fit their relation curve of degradation indicator with time. Based on the curve, the relation function is estimated by using the maximum likelihood estimation method. Once it reaches the failure threshold, the reference systems are considered as failure, so the lifetime of these reference samples can be estimated in this way. The estimated precision regarding the lifetime of these reference systems is the basis for calculating weights of similarity degree. When the estimated precision is higher, the weight assigned to this reference sample is higher. The rest steps are same as that of the modified similarity methodology based on run-to-failure data in Section 2.1.

3. Model Applications to an Aero Engine

This section provides two cases to illustrate the proposed two approaches for RUL estimation of an aero engine.

3.1. The Estimation of RUL for an Airplane Engine with Run-to-Failure Data Though the Modified Similarity Methodology

In this section, the similarity methodology based on run-to-failure data is applied to estimate the RUL of the aircraft engine with multidimensional degraded parameters. The 2008 PHM Data Challenge Competition is introduced for model validation and comparison. The data sets include 21 monitored parameters under 3 different operating modes at a sequence of time, in which three operating modes are flight height (Alt: 0–42 k feet), Mach number (M: 0–0.84) and throttle resolver angle (TRA: 20–100), which reflect the whole operational state of an aero engine. The 21 monitored parameters are different under different operating modes. The raw data of performance parameters from different parts of an aircraft engine is multiple and fluctuated largely without evident regular patterns. It is difficult to estimate the RUL based on these raw data. This paper puts forward a new indicator to characterize the health of the engine based on these raw data. The following section introduces the procedure to obtain the new health indicator.

Firstly, the 11 performance parameters that have shown evident changing trend with time are selected after inspecting 21 performance parameters. For the 11 performance parameters, a principal component analysis (PCA) is used to extract the main performance parameters that represent healthy state and degradation trend of the engine system from 11 performance parameters. PCA can reduce the data dimensions. Under different operating modes, the PCA result for 11 parameters is listed as shown in Table 1. Through PCA, the main two-dimensional performance parameters, which occupy more than 98% in all 11 parameters, are derived. Then a new status indicator is established based on the residual two-dimensional performance parameters referring to [22]. The new status indicator is built using the Euclidean distance between the projection of two-dimensional performance parameters at a certain cycle on the failure space and the center of the failure space projection dot in an operating mode.

Table 1. The detailed occupancy of the main two components in different modes.

Mode	Mode 1	Mode 2	Mode 3	Mode 4	Mode 5	Mode 6
PC1	0.6082	0.5892	0.7959	0.7185	0.6087	0.5363
PC2	0.3803	0.4005	0.1903	0.2641	0.3034	0.4329

The detailed steps to construct this healthy status index are shown as follows:

(1) Build the failure space (two-dimensional space) and calculate the projection of the failure values in the failure space, as shown as the hollow dots in Figure 3;
(2) Calculate the center of these projection dots, as shown as star dot in Figure 3;
(3) Calculate the projection dot of the performance parameters on the failure space at a certain cycle;
(4) Calculate the Euclidean distance between the projection dot of the performance parameters in the failure space at a certain cycle and the center of the projected dots in the failure space in an operating mode, which is shown in Figure 4.

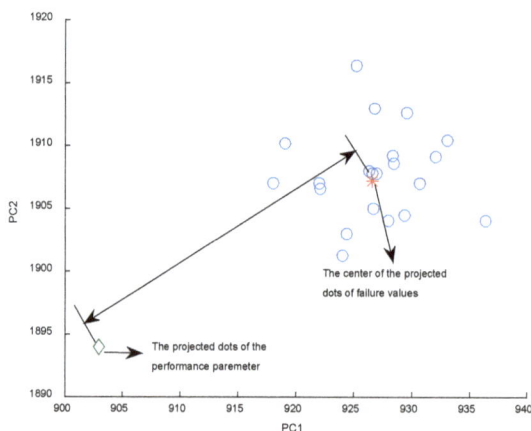

Figure 4. Definition of the new health index.

Further Euclidean distance means a healthier status of the engine, so this distance is defined as a new healthy status indicator, which represents the engine healthy state and degradation level.

Figure 4 plots the curve of the new health status index of the 196th reference sample. Though this healthy status index shows a certain changing trend, it is still fluctuated intensively. Accordingly, Karman filtering is utilized to further handle this healthy status index. Figure 5 reflects the compared curve of the 196th sample after and before Karman filtering. The red and thick curve is the new degradation indicator of the 196th sample after Karman filtering.

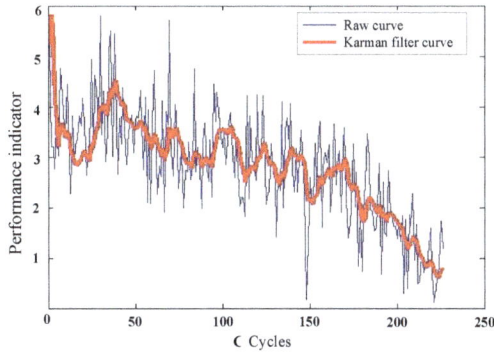

Figure 5. Trend curve of the 196th sample after and before Karman filtering.

This paper predicted the RUL of five samples No. 196–200 at the 50th cycles, 30th cycles and 10th cycles before failures using the first scheme. An example prediction of the196th samples is given as Table 2.

Table 2. The five-sample point of the degradation indicator values of the 196th sample.

Run Time	156 Cycles	161 Cycles	166 Cycles	171 Cycles	176 Cycles
RUL	2.5986	2.4292	2.3260	2.4667	2.3691

Firstly, the five sample values from the degradation indicator curve after Karman filtering at every 5 cycles before the 176th cycle are given in Table 2. Then the 10 samples which are most similar with the operated samples are selected as the reference samples according to the new degradation indicator in Equation (1). Time range is $\Delta T = 5$, sampling interval is $\Delta T = 1$. The weights of these reference samples are calculated by using Equation (4). The results and other information on these 10 reference samples are shown in Table 3.

Table 3. Information of the reference samples.

Ranking	Sample Number	Sampling Interval	Lifetime	RUL	Weight
1	38	224–228	287	59	0.192924
2	82	193–197	223	26	0.164397
3	115	211–215	260	45	0.162683
4	12	120–124	242	118	0.093304
5	29	164–168	228	60	0.082048
6	103	219–223	243	20	0.080489
7	64	122–126	154	28	0.060018
8	53	205–209	259	50	0.058619
9	78	176–180	228	48	0.055197
10	34	244–248	286	38	0.050321

The predicted RUL of the 196th sample with its actual lifetime 226 cycles are based on these 10 reference samples is given in Table 4. Meanwhile, the estimated RUL of the 196th sample by the traditional similarity method and modified similarity method are compared in Table 4. The weight-adjusted coefficient α is preliminarily set as 0.4 in this analysis.

Table 4. RUL estimation of the 196th sample.

Operating Time	Traditional Similarity Method		Modified Similarity Method	
	Predicted RUL	Error (%)	Predicted RUL	Error (%)
176	225.69	0.1358	214.50	5.0871
177	238.11	5.3600	219.63	2.8201
178	213.58	5.4953	221.18	2.1310
179	209.86	7.1432	219.50	2.8742
180	203.53	9.9426	221.56	1.9642
. . .	. . .	. . .	. . .	. . .
196	197.23	12.7305	221.62	1.9388
197	200.43	11.3132	219.35	2.9427
198	205.42	9.1069	220.53	2.4208
199	200.55	11.2619	225.75	0.1121
200	194.23	14.0578	226.77	0.3401
. . .	. . .	. . .	. . .	. . .
216	188.03	16.8012	226.80	0.3558
217	184.37	18.4209	226.17	0.0741
218	182.68	19.1695	225.31	0.3036
219	183.04	19.0067	224.09	0.8448
220	181.41	19.7290	224.15	0.8169

As can be seen from Table 4, the modified method provides better predictions than the traditional one. Moreover, the error by the traditional method increases with time. The prediction precision by the modified method tends to be better when the time is closer to failure. Since the traditional method chooses reference samples that are most similar with the operating sample during a certain time interval in their whole life, the operating sample and this similar reference sample maybe are in different degradation epochs. The modified method constrains the same time range to seek the most similar reference samples. In addition, the modified method assigns larger weight to the more recent sampling point.

Finally, the weight-adjusting coefficient of sampling points is optimized. Through assigning different values to get different predicted precision, the optimized weight-adjusting coefficient value can be obtained at $\alpha = 0.6$, as shown in Figure 6.

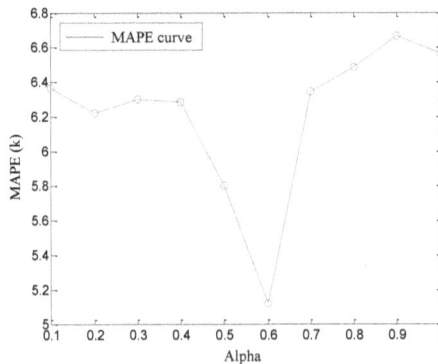

Figure 6. The MAPE corresponding to weight-adjusting coefficient.

3.2. The Estimation of RUL for an Aero Engine with Deteriorated Data Though the Similarity and SVM Methodology

This methodology for RUL estimation includes two essential procedures: assessment for performance of reference samples and RUL estimation of reference samples. The assessment for performance of reference samples is implemented using SVM in this paper. The data are extracted from the same data sets as the previous case, but these whole life data of the original samples are cut off the rear part and only the front part data are applied in this scheme. The degradation indicator pattern of the No. 1 aircraft engine trained by SVM is shown in Figure 7.

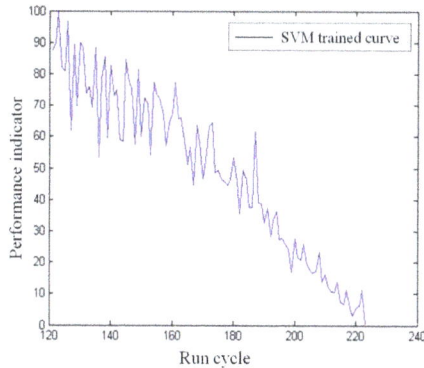

Figure 7. The SVM trained result of the No. 1 sample.

The predicted lifetime of all the 20 reference samples by SVM are shown in Table 5. Chi-square test is used to measure the prediction precision, which are used for calculating the weights of reference samples. The No. 11, 13, 18, 17 and No. 3 samples with higher prediction precision are selected to calculate the operating sample's lifetime as the reference samples. The calculated weights of the reference samples are given in Table 6. The lifetime of No. 196 sample is predicted as shown in Table 7.

Table 5. The predicted lifetime of 20 trained samples.

Sample Number	Lifetime				
$Ft_1 \sim Ft_5$	231.12598	289.93651	214.7592	299.89567	372.86392
$Ft_6 \sim Ft_{10}$	232.23493	174.22486	290.63039	183.51086	239.97266
$Ft_{11} \sim Ft_{15}$	214.47475	262.17492	215.14943	238.10682	297.89302
$Ft_{15} \sim Ft_{20}$	301.69223	236.68347	201.39653	243.82267	255.27005

Table 6. The weights of the reference samples.

Reference Samples	W_{11}	W_{13}	W_{18}	W_{17}	W_3
Weights	0.2526	0.2258	0.1957	0.1678	0.1581

It is worth noting from Table 7 that, the proposed methodology has shown better predictions than the traditional one. In particular, the prediction precision is higher when the operational time is closer to the failure point.

Table 7. Model predicted lifetime and error of the 196th sample.

Work Time	Predicted Failure Time	Error (%)
121–125	201.4377	10.86
126–130	189.9478	15.95
131–135	188.2144	16.71
136–140	187.5556	17.01
141–145	188.0059	16.81
146–150	192.7102	14.73
151–155	203.6575	9.88
156–160	227.415	0.62
161–165	227.8171	0.80
166–170	230.5832	2.02
171–175	219.4822	2.88
176–180	217.1443	3.91
181–185	220.0715	2.62
186–190	224.6852	0.58
191–195	219.9376	2.68
196–200	232.6292	2.93
201–205	231.4903	2.42
206–210	234.244	3.64
211–215	233.7521	3.43
216–220	230.2518	1.88
221–225	228.2501	0.99

4. Conclusions

The RUL prediction of an aircraft engine can not only improve its safety, maintenance, and availability, but also reduce its operation and maintenance costs. This paper presents two schemes to estimate the RUL of an aircraft engine under different situations. The first scheme adopts a modified similarity-based method for estimating the RUL of the aero engine with abundant run-to-failure data of referenced samples. The second scheme utilizes deteriorated data of samples without up-to-failure data to estimate the RUL of the operating sample with less deteriorated performance data than the reference systems. The two schemes are utilized for RUL estimation of an aircraft engine. The model prediction precision shows these two schemes are effective and suitable for RUL estimation of aero engines. More specifically, it is suitable to adopt the modified similarity-based methodology when failed historical samples are abundant and the similarity and SVM methodology is suitable under limited historical samples conditions.

Acknowledgments: The authors acknowledge the support of the National Natural Science Foundation of China (No. 11672070 and 51505067).

Author Contributions: Zhongzhe Chen supervised the projects, and developed the innovative methods and prepared this manuscript. Shuchen Cao and Zijian Mao contributed to the analysis of data, and reviewed and read the final manuscript.

Conflicts of Interest: The authors declare no conflict of interest.

References

1. Wang, D.; Peter, W.T.; Tsui, K.L. An enhanced Kurtogram method for fault diagnosis of rolling element bearings. *Mech. Syst. Signal Process.* **2013**, *35*, 176–199. [CrossRef]
2. Zhu, S.P.; Huang, H.Z.; Li, Y.; Liu, Y.; Yang, Y. Probabilistic modeling of damage accumulation for time-dependent fatigue reliability analysis of railway axle steels. *Proc. Inst. Mech. Eng. Part F* **2015**, *229*, 23–33. [CrossRef]
3. Wang, D.; Tsui, K.L. Brownian motion with adaptive drift for remaining useful life prediction: Revisited. *Mech. Syst. Signal Process.* **2018**, *99*, 691–701. [CrossRef]

4. Zhu, S.P.; Huang, H.Z.; Peng, W.; Wang, H.; Mahadevan, S. Probabilistic physics of failure-based framework for fatigue life prediction of aircraft gas turbine discs under uncertainty. *Reliab. Eng. Syst. Saf.* **2016**, *146*, 1–12. [CrossRef]
5. Zhu, S.P.; Yang, Y.J.; Huang, H.Z.; Lv, Z.; Wang, H. A unified criterion for fatigue-creep life prediction of high temperature components. *Proc. Inst. Mech. Eng. Part G* **2017**, *231*, 677–688. [CrossRef]
6. Yu, Z.Y.; Zhu, S.P.; Liu, Q.; Liu, Y. A new energy-critical plane damage parameter for multiaxial fatigue life prediction of turbine blades. *Materials* **2017**, *10*, 513. [CrossRef] [PubMed]
7. Wang, D.; Sun, S.; Peter, W.T. A general sequential Monte Carlo method-based optimal wavelet filter: A Bayesian approach for extracting bearing fault features. *Mech. Syst. Signal Process.* **2015**, *52*, 293–308. [CrossRef]
8. Wang, D.; Tsui, K.L.; Miao, Q. Prognostics and Health Management: A Review of Vibration-based Bearing and Gear Health Indicators. *IEEE Access* **2017**, in press. [CrossRef]
9. Wang, D.; Peter, W.T.; Guo, W.; Miao, Q. Support vector data description for fusion of multiple health indicators for enhancing gearbox fault diagnosis and prognosis. *Meas. Sci. Technol.* **2010**, *22*, 25102. [CrossRef]
10. Peng, W.; Li, Y.; Yang, Y.J.; Zhu, S.; Huang, H. Bivariate analysis of incomplete degradation observations based on inverse Gaussian processes and copulas. *IEEE Trans. Reliab.* **2016**, *65*, 624–639. [CrossRef]
11. Peng, W.; Li, Y.F.; Yang, Y.J.; Mi, J.; Huang, H. Bayesian degradation analysis with inverse Gaussian process models under time-varying degradation rates. *IEEE Trans. Reliab.* **2017**, *66*, 84–96. [CrossRef]
12. Peng, W.; Shen, L.; Shen, Y.; Sun, Q. Reliability analysis of repairable systems with recurrent misuse-induced failures and normal-operation failures. *Reliab. Eng. Syst. Saf.* **2018**, *171*, 87–98. [CrossRef]
13. Chiang, L.H.; Russel, E.; Braatz, R. *Fault Detection and Diagnosis in Industrial Systems*; Springer: London, UK, 2001.
14. Si, X.S.; Wang, W.B.; Hu, C.H.; Zhou, D.H. Remaining useful life estimation: A review on the statistical data driven approaches. *Eur. J. Oper. Res.* **2011**, *213*, 1–14. [CrossRef]
15. Kacprzynski, G.J.; Sarlashkar, A.; Roemer, M.J.; Hess, A.; Hardman, W. Predicting remaining life by fusing the physics of failure modeling with diagnostics. *J. Miner.* **2004**, *56*, 29–35. [CrossRef]
16. Li, C.J.; Lee, H. Gear fatigue crack prognosis using embedded model, gear dynamic model and fracture mechanics. *Mech. Syst. Signal Process.* **2005**, *19*, 836–846. [CrossRef]
17. Zhu, S.P.; Huang, H.Z.; He, L.; Liu, Y.; Wang, Z. A generalized energy-based fatigue-creep damage parameter for life prediction of turbine disk alloys. *Eng. Fract. Mech.* **2012**, *90*, 89–100. [CrossRef]
18. Wang, R.Z.; Zhang, X.C.; Tu, S.T.; Zhu, S.; Zhang, C. A modified strain energy density exhaustion model for creep-fatigue life prediction. *Int. J. Fatigue* **2016**, *90*, 12–22. [CrossRef]
19. Zhu, S.P.; Foletti, S.; Beretta, S. Probabilistic framework for multiaxial LCF assessment under material variability. *Int. J. Fatigue* **2017**, *103*, 371–385. [CrossRef]
20. Ahmadzade, F.; Lundberg, J. Remaining useful life estimation: Review. *Int. J. Syst. Assur. Eng. Manag.* **2014**, *5*, 461–474. [CrossRef]
21. Zio, E.; Maio, F.D. A data-driven fuzzy approach for predicting the remaining useful life in dynamic failure scenarios of an unclear system. *Reliab. Eng. Syst. Saf.* **2001**, *95*, 49–57. [CrossRef]
22. Gebraeel, N.; Lawley, M.; Liu, R.; Parmeshwaran, V. Residual life predictions from vibration-based degradation signals: A neural network approach. *IEEE Trans. Ind. Electron.* **2004**, *51*, 694–700. [CrossRef]
23. Huang, R.; Xi, L.; Li, X.; Liu, C.R.; Qiu, H.; Lee, J. Residual life predictions for ball bearings based on self-organizing map and back propagation neural network methods. *Mech. Syst. Signal Process.* **2007**, *21*, 193–207. [CrossRef]
24. Hansen, R.J.; Hall, D.L.; Kurtz, S.K. New approach to the challenge of machinery prognostics. *J. Eng. Gas Turbines Power* **1995**, *117*, 320–325. [CrossRef]
25. Son, K.; Fouladirad, M.; Barros, A.; Levrat, E.; Iung, B. Remaining useful life estimation based on stochastic deterioration models: A comparative study. *Reliab. Eng. Syst. Saf.* **2013**, *112*, 165–175. [CrossRef]
26. Wang, T.; Yu, J.; Siegel, D.; Lee, J. A similarity-based prognostics approach for remaining useful life estimation of engineered systems. In Proceedings of the International Conference on Prognostics and Health Management, Denver, CO, USA, 6–9 October 2008; pp. 1–6.
27. Yan, J.; Koc, M.; Lee, J. A prognostic algorithm for machine performance assessment and its application. *Prod. Plan. Control* **2004**, *15*, 796–801. [CrossRef]

28. Roy, A.; Tangirala, S. Stochastic modeling of fatigue crack dynamics for on-line failure prognostics. *IEEE Trans. Control Syst. Technol.* **1996**, *4*, 443–451. [CrossRef]

29. Sun, Y.; Ma, L.; Mathew, J.; Wang, W.; Zhang, S. Mechanical systems hazard estimation using condition monitoring. *Mech. Syst. Signal Process.* **2006**, *20*, 1189–1201. [CrossRef]

30. Gu, J.; Barker, D.; Pecht, M. Prognostics implementation of electronics under vibration loading. *Microelectron. Reliab.* **2007**, *47*, 1849–1856. [CrossRef]

31. Stetter, R.; Witczak, M. Degradation modelling for health monitoring systems. *J. Phys. Conf. Ser.* **2014**, *570*, 62002. [CrossRef]

32. Lee, J.; Wu, F.J.; Zhao, W.Y.; Ghaffari, M.; Liao, L.; Siegel, D. Prognostics and health management design for rotary machinery systems-Reviews, methodology and applications. *Mech. Syst. Signal Process.* **2014**, *42*, 314–334. [CrossRef]

33. You, M.Y.; Meng, G. A generalized similarity measure for similarity-based residual life prediction. *Proc. Inst. Mech. Eng. Part E* **2011**, *225*, 151–160. [CrossRef]

34. Zhu, S.P.; Lei, Q.; Wang, Q.Y. Mean stress and ratcheting corrections in fatigue life prediction of metals. *Fatigue Fract. Eng. Mater. Struct.* **2017**, *40*, 1343–1354. [CrossRef]

35. Zhu, S.P.; Lei, Q.; Huang, H.Z.; Yang, Y.; Peng, W. Mean stress effect correction in strain energy-based fatigue life prediction of metals. *Int. J. Damage Mech.* **2017**, *26*, 1219–1241. [CrossRef]

36. Zhu, S.P.; Liu, Q.; Lei, Q.; Wang, Q.Y. Probabilistic fatigue life prediction and reliability assessment of a high pressure turbine disc considering load variations. *Int. J. Damage Mech.* **2017**, in press. [CrossRef]

37. Yu, Z.Y.; Zhu, S.P.; Liu, Q.; Liu, Y. Multiaxial fatigue damage parameter and life prediction without any additional material constants. *Materials* **2017**, *10*, 923. [CrossRef] [PubMed]

38. Li, Y.; Billington, S.; Zhang, C.; Kurfess, T.; Danyluk, S.; Liang, S. Adaptive prognostics for rolling element bearing condition. *Mech. Syst. Signal Process.* **1999**, *13*, 103–113. [CrossRef]

MDPI

St. Alban-Anlage 66

4052 Basel

Switzerland

Tel. +41 61 683 77 34

Fax +41 61 302 89 18

www.mdpi.com

Energies Editorial Office

E-mail: energies@mdpi.com

www.mdpi.com/journal/energies

www.ingramcontent.com/pod-product-compliance
Lightning Source LLC
Chambersburg PA
CBHW051845210326
41597CB00033B/5786